AF360905

Laser Welding

Laser Welding

Special Issue Editors

João Pedro Oliveira
Zhi Zeng

MDPI • Basel • Beijing • Wuhan • Barcelona • Belgrade • Manchester • Tokyo • Cluj • Tianjin

Special Issue Editors

João Pedro Oliveira
Universidade Nova de Lisboa
Portugal

Zhi Zeng
University of Electronic Science and Technology of China
China

Editorial Office
MDPI
St. Alban-Anlage 66
4052 Basel, Switzerland

This is a reprint of articles from the Special Issue published online in the open access journal *Metals* (ISSN 2075-4701) (available at: https://www.mdpi.com/journal/metals/special_issues/laser_Welding).

For citation purposes, cite each article independently as indicated on the article page online and as indicated below:

LastName, A.A.; LastName, B.B.; LastName, C.C. Article Title. *Journal Name* **Year**, *Article Number*, Page Range.

ISBN 978-3-03928-861-8 (Pbk)
ISBN 978-3-03928-862-5 (PDF)

Contents

About the Special Issue Editors

João Pedro Oliveira was born on the 14th of January 1989 in Sintra, Portugal. He completed his M.Sc. degree in Materials Science and Engineering at Faculdade de Ciências e Tecnologia, Universidade NOVA de Lisboa, Portugal, in 2012 with a classification of 18/20. He then pursued his Ph.D. here on "Laser Welding of Shape Memory Alloys", which was unanimously approved by the Ph.D. jury in June 2016. João joined the Welding Engineering Program at the Department of Materials Science and Engineering, Ohio State University, USA, as Postdoctoral Researcher in September 2016. Prior to this appointment, he was an Invited Researcher for a total of 8 months (split between 2014 and 2016) at the Center for Advanced Materials Joining, University of Waterloo, Canada, under the supervision of Professor Norman Zhou, where he developed new welding strategies for dissimilar joining of shape memory alloys. João is also an experienced user of synchrotron facilities, where he actively participated in a total of 10 approval-funded proposals for beamtime allocation in both DESY (Germany) and LNLS (Brazil). His research activities and participation in seminars and training schools allowed acquiring competences in several characterization techniques, such as SEM, TEM, and X-ray diffraction using both conventional and synchrotron radiation sources, thermal analysis, and mechanical testing, among others. During his Ph.D., João has published a total of 19 papers in international scientific journals on welding of shape memory alloys. Currently, he has published more than 60 peer-reviewed papers. He has also presented several oral communications and posters at international conferences. He is currently serving on the Editorial Board of Materials & Design and is an active reviewer for several journals, including Additive Manufacturing; International Journal of Fatigue, Smart Materials and Structures; Materials Letters; and Optics and Laser Technology. His research interests are on the weldability and correlation between microstructure and mechanical behavior of different classes of materials such as shape memory alloys, Co-based superalloys, high-entropy alloys, and stainless steels. Additionally, João has a special interest in additive manufacturing techniques. João now serves as Assistant Professor at the Department of Mechanical and Industrial Engineering at Faculdade de Ciências e Tecnologia, Universidade NOVA de Lisboa, where he supervises 3 Ph.D. students and numerous M.Sc. students.

Zhi Zeng was born on the 21th of October 1982 in Hebei, China. He completed his Ph.D. in Materials Processing Engineering at the School of Materials Science and Engineering, Tianjin University, China on "Study on Welding Stress and Deformation Behavior of Complicated Aluminum Alloy Structure" in June 2009. Zhi Zeng joined the School of Mechanical and Electrical Engineering at the University of Electronic Science and Technology of China in September 2009, where he now serves as Professor and supervises 3 Ph.D. and 8 M.Sc. students. During February 2014–February 2015, he was sponsored by the China Scholarship Council to be a Visiting Scholar at the Center for Advanced Materials Joining, University of Waterloo, Canada, under the supervision of Professor Norman Zhou, where he developed dissimilar welding of shape memory alloys. Currently, Zhi Zeng is a committee member of the China Welding Association, where his research interests are on the welding and joining of advanced lightweight materials, such as Ti, Al alloys, and shape memory alloys. Another area of interest is additive manufacturing techniques, including selective laser melting and wire arc additive manufacturing. He has published more than 50 peer-reviewed papers and presented more than 10 oral communications at international conferences. He was awarded the Prize for Progress in Science and Technology of Sichuan province in both 2015 and 2019.

Preface to "Laser Welding"

Welding technologies are critical to most relevant engineering applications. Laser welding is a key joining technology characterized by small heat-affected and fusion zones, as well as minimal or non-existent distortions. As a result, laser welding is often used for joining advanced metallic alloys.

The metallurgical alterations that occur as a result of the welding process determine the mechanical properties of the welded part. As such, a fundamental understanding of the process–microstructure–property relationships is necessary. Given the non-equilibrium solidification conditions found in laser welding, a thorough understanding of the associated welding metallurgy is even more important. This knowledge can then be used to optimize the joining process, with the aim of improving the properties of welded joints.

The present Special Issue on "Laser Welding" was a success, with a total of 16 original research works published after peer review. Different topics were discussed within this Special Issue: modeling and simulation of laser welding; porosity control by means of high-speed imaging and microscopy techniques; the effect of processing parameters on the microstructure and mechanical properties of laser-welded joints for different metallic systems such as AZ31 alloy, steels, Ti-based alloys, and Al-based alloys; and, finally, dissimilar laser welding of aluminum to steel.

Laser welding is one of the most important and versatile welding techniques for joining advanced materials. Exciting developments in this field are continuously being presented, pushing the boundaries for the application of this technique. The need to develop modeling and simulation tools and understanding the welding metallurgy associated with the process and process control, among other features, will require continuous efforts by researchers in this field.

João Pedro Oliveira, Zhi Zeng
Special Issue Editors

Editorial

Laser Welding

J. P. Oliveira [1,*] and Zhi Zeng [2,*]

1   UNIDEMI, Departamento de Engenharia Mecânica e Industrial, Faculdade de Ciências e Tecnologia, Universidade Nova de Lisboa, 2829-516 Caparica, Portugal
2   School of Mechanical and Electronic Engineering, University of Electronic Science and Technology of China, Chengdu 611731, Sichuan, China
*   Correspondence: jp.oliveira@fct.unl.pt (J.P.O.); zhizeng@uestc.edu.cn (Z.Z.)

Received: 8 January 2019; Accepted: 11 January 2019; Published: 11 January 2019

1. Introduction and Scope

Welding technologies are critical to most relevant engineering applications. Laser welding is a key joining technology characterized by small heat affected and fusion zones, as well as minimal or non-existent distortions. As a result, laser welding is often used for joining advanced metallic alloys.

The metallurgical alterations that occur as a result of the welding process determine the mechanical properties of the welded part. As such, a fundamental understanding of process-microstructure-properties relationships is necessary. Given the non-equilibrium solidification conditions found in laser welding, a thorough understanding of the associated welding metallurgy is even more important. This knowledge can then be used to optimize the joining process, aiming at improving the properties of the welded joints.

2. Contributions

The present special issue on "Laser Welding" was a success with a total of 16 original research works published after peer- review. Different topics were discussed within this special issue: modelling and simulation of laser welding were presented in [1–4]; porosity control by means of high speed imaging and microscopy techniques was studied and discussed [5]; the effect of processing parameters on the microstructure and mechanical properties of laser-welded joints was evaluated for different metallic systems such as AZ31 alloy [6], steels [7–10], Ti-based alloys [11–13], and Al-based alloys [14]; and finally, dissimilar laser welding of aluminum to steel was presented [15,16].

3. Conclusions and Outlook

Laser welding is one of the most important and versatile welding techniques for joining advanced materials. Exciting developments in this field are continuously being presented, pushing the boundaries of the application of the technique. The need to develop modelling and simulation tools and understanding the welding metallurgy associated with the process and process control, among other features, will require a continuous effort by researchers in this field.

As guest editors, we would like to express our gratitude to all the authors who submitted their contributions, making this special issue a success. Furthermore, we gratefully acknowledge and the peer-reviewers who contributed to improving the quality of the manuscripts. Finally, special thanks to Natalie Sun and the remaining Metals editorial team for their support and assistant during the call period of this special issue.

Conflicts of Interest: The authors declare no conflict of interest.

References

1. Wu, J.; Zhang, H.; Feng, Y.; Luo, B. 3D Multiphysical Modelling of Fluid Dynamics and Mass Transfer in Laser Welding of Dissimilar Materials. *Metals* **2018**, *8*, 443. [CrossRef]
2. Ruan, X.; Zhou, Q.; Shu, L.; Hu, J.; Cao, L. Accurate Prediction of the Weld Bead Characteristic in Laser Keyhole Welding Based on the Stochastic Kriging Model. *Metals* **2018**, *8*, 486. [CrossRef]
3. Dal, M.; Peyre, P. Multiphysics Simulation and Experimental Investigation of Aluminum Wettability on a Titanium Substrate for Laser Welding-Brazing Process. *Metals* **2017**, *7*, 218. [CrossRef]
4. D'Ostuni, S.; Leo, P.; Casalino, G. FEM Simulation of Dissimilar Aluminum Titanium Fiber Laser Welding Using 2D and 3D Gaussian Heat Sources. *Metals* **2017**, *7*, 307. [CrossRef]
5. Popescu, A.; Delval, C.; Leparoux, M. Control of Porosity and Spatter in Laser Welding of Thick AlMg5 Parts Using High-Speed Imaging and Optical Microscopy. *Metals* **2017**, *7*, 452. [CrossRef]
6. Laser, F.; Az, W. Study on the Size Effects of H-Shaped Fusion Zone of Fiber Laser Welded AZ31 Joint. *Metals* **2018**, *8*, 198. [CrossRef]
7. Górka, J.; Stano, S. Microstructure and Properties of Hybrid Laser Arc Welded Joints (Laser Beam-MAG) in Thermo-Mechanical Control Processed S700MC Steel. *Metals* **2018**, *8*, 132. [CrossRef]
8. Ridha Mohammed, G.; Ishak, M.; Ahmad, S.; Abdulhadi, H. Fiber Laser Welding of Dissimilar 2205/304 Stainless Steel Plates. *Metals* **2017**, *7*, 546. [CrossRef]
9. Xue, X.; Pereira, A.; Amorim, J.; Liao, J. Effects of Pulsed Nd:YAG Laser Welding Parameters on Penetration and Microstructure Characterization of a DP1000 Steel Butt Joint. *Metals* **2017**, *7*, 292. [CrossRef]
10. Evin, E.; Tomáš, M. The Influence of Laser Welding on the Mechanical Properties of Dual Phase and Trip Steels. *Metals* **2017**, *7*, 239. [CrossRef]
11. Zeng, Z.; Oliveira, J.P.; Bu, X.; Yang, M.; Li, R.; Wang, Z. Laser Welding of BTi-6431S High Temperature Titanium Alloy. *Metals* **2017**, *7*, 504. [CrossRef]
12. Sánchez-Amaya, J.; Pasang, T.; Amaya-Vazquez, M.; Lopez-Castro, J.; Churiaque, C.; Tao, Y.; Botana Pedemonte, F. Microstructure and Mechanical Properties of Ti5553 Butt Welds Performed by LBW under Conduction Regime. *Metals* **2017**, *7*, 269. [CrossRef]
13. Caiazzo, F.; Alfieri, V.; Corrado, G.; Argenio, P.; Barbieri, G.; Acerra, F.; Innaro, V. Laser Beam Welding of a Ti–6Al–4V Support Flange for Buy-to-Fly Reduction. *Metals* **2017**, *7*, 183. [CrossRef]
14. Zhang, X.; Li, L.; Chen, Y.; Yang, Z.; Zhu, X. Experimental Investigation on Electric Current-Aided Laser Stake Welding of Aluminum Alloy T-Joints. *Metals* **2017**, *7*, 467. [CrossRef]
15. Cui, L.; Chen, B.; Qian, W.; He, D.; Chen, L. Microstructures and Mechanical Properties of Dissimilar Al/Steel Butt Joints Produced by Autogenous Laser Keyhole Welding. *Metals* **2017**, *7*, 492. [CrossRef]
16. Casalino, G.; Leo, P.; Mortello, M.; Perulli, P.; Varone, A. Effects of Laser Offset and Hybrid Welding on Microstructure and IMC in Fe–Al Dissimilar Welding. *Metals* **2017**, *7*, 282. [CrossRef]

 *metals*

Article

Accurate Prediction of the Weld Bead Characteristic in Laser Keyhole Welding Based on the Stochastic Kriging Model

Xiongfeng Ruan [1], Qi Zhou [2], Leshi Shu [1], Jiexiang Hu [1,3] and Longchao Cao [1,3,*]

[1]   The State Key Laboratory of Digital Manufacturing Equipment and Technology, School of Mechanical Science and Engineering, Huazhong University of Science & Technology, Wuhan 430074, China; ruanxf@hust.edu.cn (X.R.); shuleshi@hust.edu.cn (L.S.); jiexianghuhust@gmail.com (J.H.)
[2]   School of Aerospace Engineering, Huazhong University of Science & Technology, Wuhan 430074, China; qizhouhust@gmail.com
[3]   George W. Woodruff School of Mechanical Engineering, Georgia Institute of Technology, Atlanta, GA 30332, USA
*   Correspondence: longchaocao@hust.edu.cn; Tel.: +1-404-736-4876

Received: 25 May 2018; Accepted: 18 June 2018; Published: 25 June 2018

Abstract: As an important index of weld quality, the weld bead geometry is closely related to the welding process parameters (WPP). Therefore, it is crucial to establish the relationships between the WPP and weld bead shape to serve as an indicator of the weld quality. However, it is difficult to predict the weld bead shape accurately due to uncertainty. In this paper, laser keyhole welding (LKW) experiments are conducted on 2205 stainless steel at sample points generated by the optimal Latin hypercube sampling (OLHS). Then the relationships between WPP and weld width (WW) are constructed using stochastic kriging model (SKM), considering the randomness of the welding process. To verify the effectiveness of the SKM, two validation approaches, the additional experiments validation and k-fold cross-validation, are used to compare the prediction performance of SKM and the traditional kriging model. SKM is superior to the kriging model at the whole five additional test points with smaller relative error. As to k-fold cross-validation, SKM provides a smaller root mean square error at four in five groups of the data. In addition, SKM can provide the variations of the entire weld bead shape. Overall, the SKM is very prominent in predicting the weld bead shape, considering fluctuations of WPP.

Keywords: LKW; WW; prediction; accuracy; SKM; WPP

---

1. Introduction

With the obvious advantages of high efficiency, a narrow heat-affected zone, and small distortion, laser keyhole welding (LKW) has been widely applied in aerospace, aviation, automobile, shipbuilding, and other fields. As intelligent manufacturing develops, accurate predicting of weld bead is becoming crucial to accomplishing industrial automation [1,2]. Therefore, it is necessary to establish the relationships between the laser keyhole welding process parameters (WPP) and weld bead geometry to predict the weld bead quality [3–5]. However, the relationship between the WPP and the weld bead is highly nonlinear [6]. The traditional methods of determining process parameters are based on the experience and manual, but the process is difficult to quantify, and the quality of the parameters depends largely on the ability of the engineers [7]. An effective way is to describe the relationships between process parameters and weld bead using the approximate model, which is also called the surrogate model. Gunaraj et al. [8] applied a response surface methodology (RSM) to devise a four-factor five-level central composite rotatable design matrix to fabricate pipes of different specifications in submerged arc welding. Nagesh et al. [9] explored the connection between the shielded metal-arc welding process parameters and the characteristics of the welding bead and penetration utilizing artificial neural networks model. Srivastava et al. [10] employed polynomial response surface (PRS) model to study how the gas metal arc WPP influenced welding quality. Samantaray et al. [11] selected six process parameters and virtual values of two sensor signals to construct six distinguished types of radial basis function network models. Zhou et al. [12] put forward an ensemble of surrogate models to optimize the laser keyhole WPP of stainless steel 316 L.

However, the aforementioned studies assume that the process parameters are stable during the welding process, which is not consistent with practical engineering application scenarios. In the real continuous LKW process, certain parameters fluctuations are inevitable, which will lead to the instability of welding width. Therefore, selecting certain cross sections to represent the entire weld bead's characteristics is unreliable. The groundbreaking article of Ankenman et al. [13] expanded kriging to include stochastic kriging by classifying the uncertainties in stochastic simulations as extrinsic and intrinsic uncertainty. Chen et al. [14] studied the effect of Common Random Numbers (CRN) on the performance of stochastic kriging predictions, indicating that the introduction of CRN was not conducive to prediction, but better gradient parameter estimation was obtained. The current research on stochastic kriging mainly focuses on theoretical development [15–17], while practical engineering applications are rare. In this paper, the SKM is used to predict weld width in LKW by considering the fluctuations of the welding process. Firstly, the experimental sample points are generated using optimal Latin hypercube sampling (OLHS). The image processing techniques are used to extract weld bead features from the specimens. Then two validation methods are used to compare the prediction performance of stochastic kriging and kriging.

The structure of the rest of the paper is as follows. In Section 2, the LKW experimental platform and the experimental process are described in detail. The general framework of the proposed approach and the construction process of SKM are introduced in Section 3. In Section 4, data processing and the evaluation of the predictive performance of the SKM are presented. Section 5 provides the conclusion.

2. Laser Keyhole Welding Procedure

2.1. Problem Definition

According to lots of theoretical research and engineering experience, the weld width largely depends on process parameters such as laser power, welding speed, laser focal position, gap width and shielding gas [18,19]. In this work, three main process parameters (i.e., laser power, welding speed, and focal position) are selected. Low laser power and high welding speed will make the molten pool too small, resulting in incomplete welding and a poor welding formation. On the other hand, excessively high laser power and a low welding speed can cause sag geometry, which will increase the deformation of the weldment and decrease mechanical properties of the weld bead. When the laser focal position is too large, the power density on the workpiece surface is too low to melt required levels of material. According to engineering experience and our previous researches [2,20–22], the WPP are selected as follows:

$$2.0(kW) \leq LP \leq 3.5(kW), \ 2.5(m/min) \leq WS \leq 3.5(m/min), \ and \ -2(mm) \leq FP \leq 0(mm) \quad (1)$$

Figure 1 depicts the LKW process. As mentioned earlier, the weld width may fluctuate along the welding direction because of changes of the process parameters. D_1 and D_2 are the different values of weld width under the same process parameters.

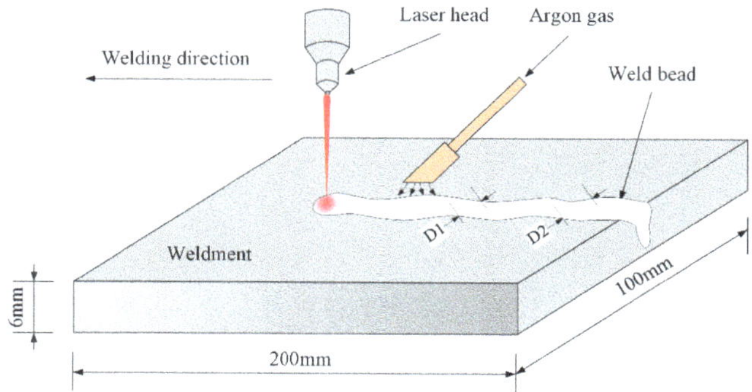

Figure 1. Schematic diagram of laser keyhole welding process.

2.2. Materials

With the characteristics of high strength, good impact toughness and good stress corrosion resistance, the 2205 stainless steel was selected as the experimental material. Its chemical composition is shown in Table 1 [23]. The size of the steel sample used in this experiment was $200 \times 100 \times 6$ mm^3. To avoid the influence of the oxide film and oil stain, the sample surface was pretreated with organic solvent acetone.

Table 1. Chemical composition of 2205 stainless steel.

Chemical Elements	C	Cr	Ni	Mo	Mn	Si	P	S	N
Composition (%)	0.025	22.73	5.53	3.03	1.89	0.34	0.028	0.002	0.170

2.3. Laser Keyhole Welding Process

The experimental setup used for LKW is shown in Figure 2. The laser generation device is IPG YLR-4000 (IPG Photonics Corp., Boston, MA, USA). The laser with a wavelength of 1.07 µm passes through the optical fiber to the laser head mounted on the ABB IRB4400 robot. The beam parameter product is 6.3 mm · mrad. To prevent oxidation during the welding process, the weld bead is protected with argon gas with a flow rate of 30 L/min. Furthermore, the laser power is set through the operation panel of the laser generation device, and the welding path, the welding speed, and the focus point can be controlled by setting the parameters of the welding robot.

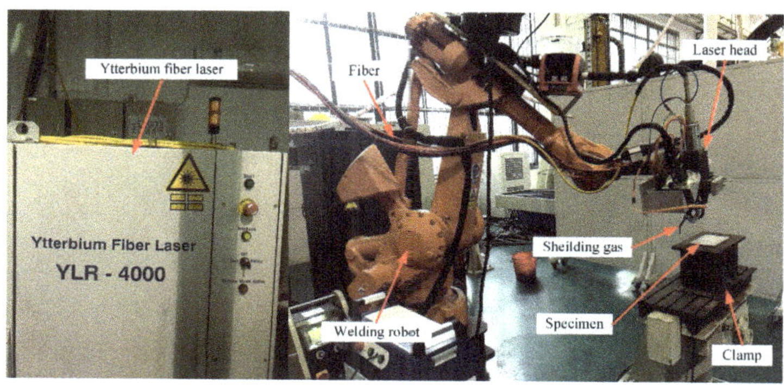

Figure 2. Laser keyhole welding equipment.

3. The Proposed Approach

3.1. General Framework

Figure 3 shows the general framework of the proposed approach. The first step consists of design of experiment (DOE) and LKW experiments. The second step is to obtain the mean and variance of weld width and then to construct SKM through these data. The last step is to validate SKM. In this step, additional welding experiments and cross-validation are adopted to compare the weld width prediction accuracies between kriging and SKM.

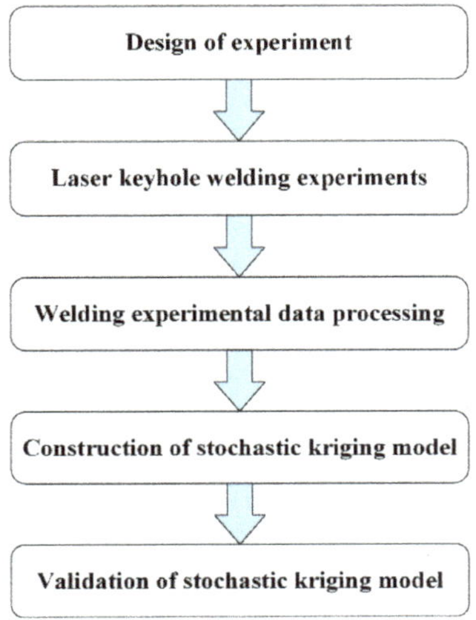

Figure 3. General framework of the proposed approach.

3.2. Theory of Stochastic Kriging

Deterministic computer experiments do not take into account the noise and errors of the simulation process [24]. Stochastic kriging is an extension of kriging metamodel by considering the uncertainty of the simulation. For the n-dimensional input point x, response of stochastic kriging on jth simulation replication is

$$y_j = Y(x) + \varepsilon_j(x) = f(x)^{\mathrm{T}}\beta + M(x) + \varepsilon_j(x) \tag{2}$$

where $Y(x)$ indicates the response of the kriging model without simulated errors at the input point x. $f(x)$ and β represent the corresponding dimensional vectors. $M(x)$ is a second-order stationary Gaussian random field with mean zero. $\varepsilon_j(x)$ denotes the simulation error at replication j. According to classic literature on stochastic kriging [13], we call $M(x)$ and $\varepsilon_j(x)$ the extrinsic and intrinsic uncertainty, respectively.

The experimental design of stochastic kriging includes the position of the sample point x_i and the corresponding independent number of replication n_i, which constitutes the data pair (x_i, n_i), $i = 1, 2, \ldots, k$. The sample mean at design point x_i is

$$\bar{y}(x_i) = \frac{1}{n_i}\sum_{j=1}^{n_i} y_j(x_i) = Y(x_i) + \frac{1}{n_i}\sum_{j=1}^{n_i}\varepsilon_j(x_i) \tag{3}$$

Let $k \times k$ matrix Σ_ε record the intrinsic uncertainty correlation with (i, h)th entry

$$\Sigma_\varepsilon(x_i, x_h) = \text{Cov}(\frac{1}{n_i}\sum_{=1}^{n_i}\varepsilon_j(x_i), \frac{1}{n_h}\sum_{j=1}^{n_h}\varepsilon_j(x_h)) \tag{4}$$

According to the previous assumption about $\varepsilon_j(x_i)$, Σ_ε can be simplified to a k-dimensional diagonal matrix $\text{diag}\{\sigma_1^2, \sigma_2^2, \dots, \sigma_k^2\}$ with

$$\sigma_i^2 = Var(\frac{1}{n_i}\sum_{j=1}^{n_i}\varepsilon_j(x_i)), \, i = 1, 2, \dots, k \tag{5}$$

Assume that extrinsic uncertainties $M(x)$ possess covariance, which is used to express the spatial correlation between any two points x and x'. The covariance is

$$\text{Cov}(M(x), M(x')) = \sigma^2 R(x - x', \theta) = \sigma^2 \prod_{j=1}^{n} R_j(\theta_j, x_j - x'_j) \tag{6}$$

where σ^2 can be explicated as the process variance, and $R(x - x', \theta)$ stands for correlation model that is determined by spatial distance $x - x'$.

Define a $k \times k$ matrix Σ_M to express the extrinsic spatial correlation between every two design points, whose (i, j)th element is $\text{Cov}(M(x_i), M(x_j))$. Similarly, $k \times 1$ vector $\Sigma_M(x_0, \cdot)$ is created to connect the point of interest x_0 with design point x_i. Specifically, they are

$$\Sigma_M = \sigma^2 \begin{pmatrix} 1 & \text{Cov}(M(x_1), M(x_2)) & \cdots & \text{Cov}(M(x_1), M(x_k)) \\ \text{Cov}(M(x_2), M(x_1)) & 1 & \cdots & \text{Cov}(M(x_2), M(x_k)) \\ \vdots & \vdots & \ddots & \vdots \\ \text{Cov}(M(x_k), M(x_1)) & \text{Cov}(M(x_k), M(x_2)) & \cdots & 1 \end{pmatrix} \tag{7}$$

and

$$\Sigma_M(x_0, \cdot) = \sigma^2 \Big(\text{Cov}(M(x_0), M(x_1)) \quad \text{Cov}(M(x_0), M(x_2)) \quad \cdots \quad \text{Cov}(M(x_0), M(x_k)) \Big)^{\text{T}} \tag{8}$$

Stochastic kriging obtains the unbiased predicted value $\hat{Y}(x_0)$ at point x_0 while minimizing the mean squared error (MSE) for predicting. The expression of $\hat{Y}(x_0)$ is

$$\hat{Y}(x_0) = f(x_0)^{\text{T}}\hat{\beta} + \Sigma_M(x_0, \cdot)^{\text{T}}[\Sigma_M + \Sigma_\varepsilon]^{-1}(\bar{y} - F\hat{\beta}) \tag{9}$$

and the corresponding MSE of $\hat{Y}(x_0)$ is

$$\text{MSE} = \sigma^2 - \Sigma_M(x_0, \cdot)^{\text{T}}[\Sigma_M + \Sigma_\varepsilon]^{-1}\Sigma_M(x_0, \cdot) + \eta^{\text{T}}(F^{\text{T}}[\Sigma_M + \Sigma_\varepsilon]^{-1}F)^{-1}\eta \tag{10}$$

where $F = (f(x_1)^{\text{T}}, f(x_2)^{\text{T}}, \dots, f(x_k)^{\text{T}})^{\text{T}}$ and $\eta = f(x_0) - F^{\text{T}}[\Sigma_M + \Sigma_\varepsilon]^{-1}\Sigma_M(x_0, \cdot)$.

In practice, stochastic parameters $(\beta, \theta, \sigma^2)$ need to be estimated based on maximum likelihood estimation. By substituting the parameters into Equations (9) and (10), the predicted value and MSE will be obtained.

4. Result and Discussion

4.1. Design of Experiment

There have been many DOE methods that can fill the experimental design space, such as orthogonal design (UD) [25], center composite design (CCD) [26], and optimal Latin hypercube sampling (OLHS) [27]. OLHS is an improved Latin hypercube sampling (LHS) which evaluates the maximum and minimum distances of points in the search space through a random evolution algorithm to obtain spatially filled sampling points. In this paper, OLHS was used. According to the description in Section 2.1, the laser power, welding speed, and focal position were selected and 25 sample points were generated. The spatial distribution of the generated sample points is shown in Figure 4. The weld width of the corresponding weldments under different process parameters are shown in Table 2.

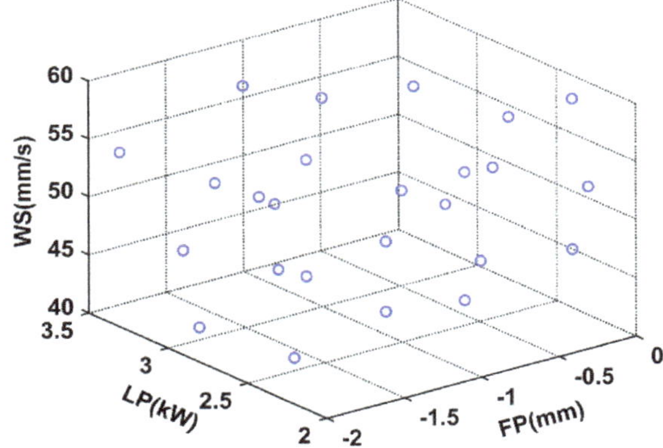

Figure 4. Spatial distribution diagram of the generated sample points.

Table 2. Weld width under different process parameters.

No.	FP (mm)	LP (kW)	WS (mm/s)	Weld Width (mm)	Variance (10^{-6})
1	−2	2.2	44	1.40	1.09
2	−2	2.3	51	1.35	2.1
3	−2	3.3	55	1.37	1.33
4	−2	2.8	43	1.46	1.49
5	−2	2.9	49	1.37	1.50
6	−2	2.7	56	1.29	0.89
7	−1	2.0	50	1.59	1.06
8	−1	2.1	46	1.56	3.54
9	−1	2.1	57	1.56	1.29
10	−1	2.5	53	1.55	6.27
11	−1	2.6	48	1.50	4.82
12	−1	2.6	42	1.63	2.38
13	−1	3.0	58	1.33	3.44
14	−1	3.1	42	1.67	3.20
15	−1	3.1	52	1.52	2.35
16	−1	3.3	47	1.64	2.56
17	−1	3.4	47	1.55	2.32
18	−1	3.5	56	1.41	2.68
19	0	2.3	51	1.64	0.98
20	0	2.4	45	1.66	5.16
21	0	2.4	58	1.55	2.29
22	0	2.8	54	1.60	3.08
23	0	2.9	49	1.57	6.77
24	0	3.2	44	1.75	3.11
25	0	3.4	53	1.61	3.62

4.2. Data Processing

4.2.1. Weld Bead Scanning

After obtaining the LKW experimental sample, the next challenge is how to obtain the weld width. In this work, the entire weld width was chosen to be analyzed rather than a certain cross section. The weldment surface was scanned using a scanner and then the width information was extracted from the captured image. The scanner is a product of Shanghai Microtek Trade Co., Ltd., Shanghai, China, whose model is MRS-2400A48U. A series of images with a pixel density of 600 dpi were collected. Table 3 illustrates part of weld bead and the corresponding image processing results of No. 1 to No. 5 specimens.

Table 3. Weld bead of No. 1 to No. 5 specimens.

No.	Weld Appearance	Image Processing
1		
2		
3		
4		
5		

4.2.2. Image Processing and Feature Extracting

Welding defects are likely to occur at the beginning and end of welding, and the appearance of the weld bead will fluctuate considerably. To improve the prediction accuracy of the weld width, only the stable part of the weld bead was selected for analysis in this work. Perpendicular to the welding direction, the Matlab program (R2017a, MathWorks Inc., Natick, MA, USA) was applied to extract the number of pixels (NP) between the highest and lowest white pixels. The scanned image has a pixel density (PD) of 600 dpi, so the data obtained here can be converted into length in mm according to the following equation

$$WW = NP \div PD \times UC \tag{11}$$

where UC represents the unit conversion between inches and millimeters, equal to 25.4. In addition, WW indicates weld width.

For each weld bead, a set of discretized data of weld width is obtained. The minimum resolution of the scanned image is a pixel, corresponding to the weld width of 0.0423 mm according to Equation (11). For example, the scatter plot of discrete weld width of No. 1 weld bead along the length direction is shown in Figure 5a. Then the probability of weld width is calculated, and the maximum likelihood estimation method is adopted to obtain the mean and variance. A frequency histogram and fitting normal distribution density function of the No. 1 weld bead are shown in Figure 5b. The mean and variance of the normal distribution fitting will be used to construct the SKM.

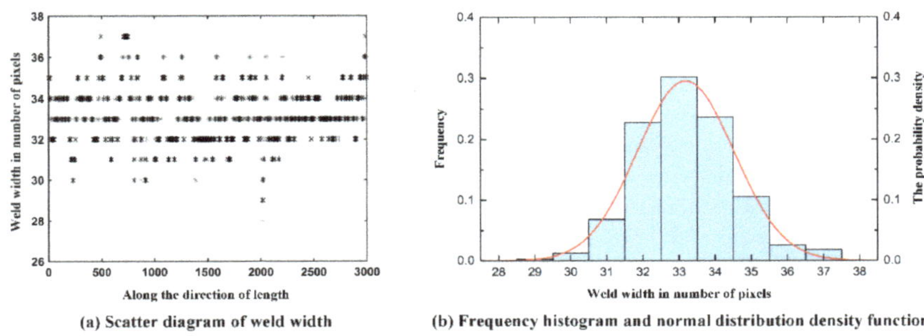

(a) Scatter diagram of weld width

(b) Frequency histogram and normal distribution density function

Figure 5. Discretized data of weld width.

4.3. Prediction Performance of the Stochastic Kriging Model

The Matlab program was used to construct SKM. The regression model is with polynomials of order 0. The parameters (i.e.,β, θ, σ^2) that need to be estimated correspond to (1.502, [0.516, 8.124, 0.010]$'$, 0.015). According to the difference of focal position, the three-dimensional surface of the SKM is shown in Figure 6.

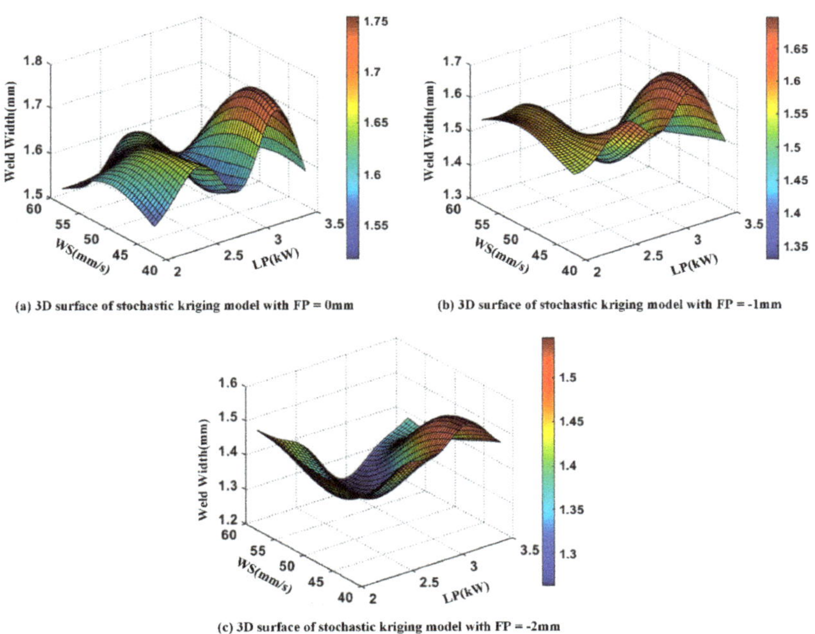

(a) 3D surface of stochastic kriging model with FP = 0mm

(b) 3D surface of stochastic kriging model with FP = -1mm

(c) 3D surface of stochastic kriging model with FP = -2mm

Figure 6. The stochastic kriging model for weld width prediction.

To verify the effectiveness of the SKM, the kriging model was also constructed with the same sample points, and the 3D surface of the kriging model is shown in Figure 7. To ensure a fair comparison, the estimated parameters of the two models have the same initial value and range. The range of parameters is 0.01 to 20, and their initial value is 0.1. The parameter estimates (i.e., β, θ, σ^2) of kriging model correspond to (-0.138, 0.800, 0.015).

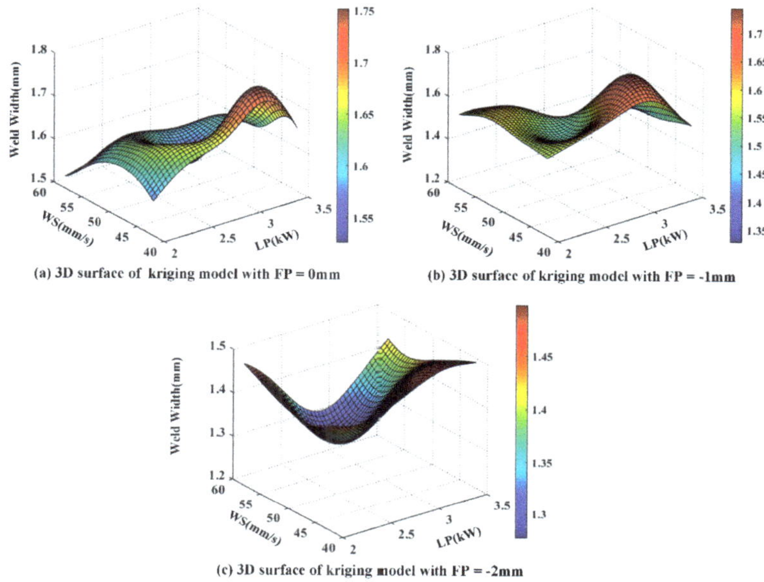

(a) 3D surface of kriging model with FP = 0mm

(b) 3D surface of kriging model with FP = -1mm

(c) 3D surface of kriging model with FP = -2mm

Figure 7. The kriging model for weld width prediction.

Two validation methods are introduced to compare the prediction performance of the weld width between the SKM and kriging model. First, 5 additional test points were generated randomly based on the same experimental and data processing methods to obtain the corresponding experimental values. The actual values of weld width at the test points are shown in Table 4. Figure 8 plots the relative prediction errors of the kriging and the SKMs. Both of the models have the largest prediction bias at No. 2 point, which are 9.88% and 11.46% respectively. The minimum prediction deviation appears at No. 3 point, corresponding to 0.47% and 3.60%. Obviously, the prediction accuracies of the SKM and kriging model have the same trend, and the SKM performs better due to considering the input uncertainty.

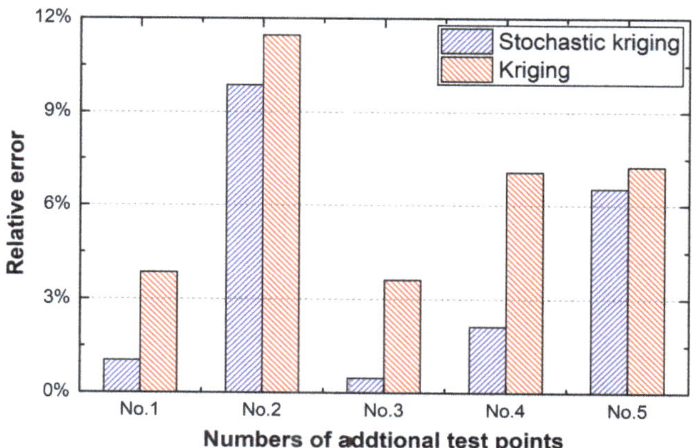

Figure 8. Relative prediction errors of the stochastic kriging model and the kriging model.

Table 4. Weld bead of additional test points.

No.	FP (mm)	LP (kW)	WS (mm/s)	Weld Width (mm)	Variance (10^{-6})
1	−2	3.2	58	1.32	1.22
2	−2	3.3	42	1.69	1.71
3	−1	2.7	53	1.45	4.55
4	−1	3.0	51	1.46	2.39
5	0	2.5	55	1.74	3.29

Furthermore, k-fold cross-validation is applied to further verify the prediction performance of SKM. k-fold cross-validation can make full use of existing data as both training and verification points, and its mathematical form is extremely simple [28]. The specific procedure to get the variance of k-fold cross-validation is as follows:

Step 1: Randomly divide the experimental data set $\Phi_n = \{(x_1, y_1), (x_2, y_2), \ldots, (x_n, y_n)\}$ into non-repeating and equal-length k groups $G_k = \{M_1, M_2, \ldots, M_k\}$, where n is the number of sample points and y_i is the response of x_i.

Step 2: Remove one group M_i from the set G_k, then construct approximate model by using the remaining set $G_{k-1} = \{M_1, \ldots, M_{i-1}, M_{i+1}, \ldots, M_k\}$.

Step 3: Predict the response values at the removal set Mi by using the model constructed at Step 2, and calculate the root mean square error ($RMSE$) by

$$RMSE_i = \sqrt{\sum_{j=1}^{m} (\hat{y}_{ij} - y_{ij})^2} \tag{12}$$

where y_{ij} is the jth true response at group M_i, while $\hat{y}_{ij}$ represents the corresponding approximate model prediction. m is the number of data in M_i.

Step 4: Repeat Step 2 to Step 3 k times, until the whole data set is considered.

In this work, the 25 sample points were randomly divided into five groups, and each of them consists five sample points. The order of groups was $G_k = \{(13, 23, 22, 2, 3), (24, 18, 16, 6, 21), (20, 12, 5, 10, 9), (11, 7, 1, 4, 14), (25, 8, 19, 17, 15)\}$, where the number corresponded to the No. in Table 2. The comparison results between SKM and kriging model are summarized in Figure 9. It is obvious that stochastic kriging provides better prediction performance. To better illustrate the problem and avoid contingency, k-fold cross-validation was repeated several times, and the same conclusion was drawn. All repeated cross-validation shows that SKM has better performance in weld width prediction. Taking Figure 9 as an example, SKM provides smaller RMSE except for the third group.

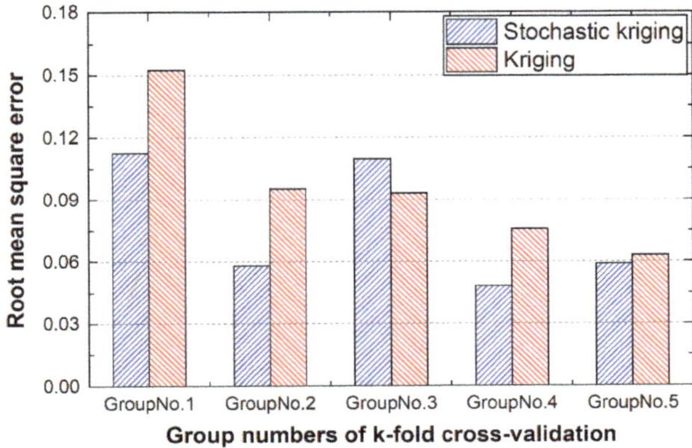

Figure 9. The comparison of RMSE between stochastic kriging model and kriging model.

5. Conclusions

In this work, the SKM has been used to predict the weld width of LKW. Unlike the general data acquisition method, this paper effectively obtains the properties of the whole weld by combining the weld bead scan with the image processing. Based on the estimated mean and variance of the normal distribution fitting of weld width, the SKM was constructed to fit the relationships between the laser power, the welding speed, the focal position and the weld width. Additional experiments and k-fold cross-validation are introduced to compare the prediction performance of weld width between SKM and kriging model. The comparison results indicate that the SKM is feasible and very prominent for predictions of the weld bead shape.

Author Contributions: Data curation, X.R.; Formal analysis, L.S.; Methodology, Q.Z., J.H. and L.C.; Supervision, Q.Z.; Validation, L.C.; Writing—original draft, X.R. and L.C.; Writing—review & editing, Q.Z. and L.C.

Funding: This research received no external funding.

Acknowledgments: This research has been supported by National Natural Science Foundation of China (NSFC) under Grant No. 51505182.

Conflicts of Interest: The authors declare no conflict of interest.

Nomenclature

DOE	Design of experiment
FP	Focal position
LKW	Laser keyhole welding
LP	Laser power
OLHS	Optimal Latin hypercube sampling
SKM	Stochastic kriging model
WPP	Welding process parameters
WS	Welding speed
WW	Weld width

References

1. Yuce, C.; Tutar, M.; Karpat, F.; Yavuz, N. The optimization of process parameters and microstructural characterization of fiber laser welded dissimilar HSLA and MART steel joints. *Metals* **2016**, *6*, 245. [CrossRef]
2. Cao, L.; Shao, X.; Jiang, P.; Zhou, Q.; Rong, Y.; Geng, S.; Mi, G. Effects of welding speed on microstructure and mechanical property of fiber laser welded dissimilar butt joints between AISI316L and EH36. *Metals* **2017**, *7*, 270. [CrossRef]
3. Murugan, N.; Gunaraj, V. Prediction and control of weld bead geometry and shape relationships in submerged arc welding of pipes. *J. Mater. Process. Technol.* **2005**, *168*, 478–487. [CrossRef]
4. Ridha Mohammed, G.; Ishak, M.; Ahmad, S.N.A.S.; Abdulhadi, H.A. Fiber laser welding of dissimilar 2205/304 stainless steel plates. *Metals* **2017**, *7*, 546. [CrossRef]
5. Loginova, I.; Khalil, A.; Pozdniakov, A.; Solonin, A.; Zolotorevskiy, V. Effect of pulse laser welding parameters and filler metal on microstructure and mechanical properties of Al-4.7 Mg-0.32 Mn-0.21 Sc-0.1 Zr Alloy. *Metals* **2017**, *7*, 564. [CrossRef]
6. Kim, I.; Son, J.; Kim, I.; Kim, J.; Kim, O. A study on relationship between process variables and bead penetration for robotic CO2 arc welding. *J. Mater. Process. Technol.* **2003**, *136*, 139–145. [CrossRef]
7. Fukuda, S.; Morita, H.; Yamauchi, Y.; Nagasawa, I.; Tsuji, S. Expert system for determining welding condition for a pressure vessel. *ISIJ Int.* **1990**, *30*, 150–154. [CrossRef]
8. Gunaraj, V.; Murugan, N. Application of response surface methodology for predicting weld bead quality in submerged arc welding of pipes. *J. Mater. Process. Technol.* **1999**, *88*, 266–275. [CrossRef]
9. Nagesh, D.; Datta, G. Prediction of weld bead geometry and penetration in shielded metal-arc welding using artificial neural networks. *J. Mater. Process. Technol.* **2002**, *123*, 303–312. [CrossRef]
10. Srivastava, S.; Garg, R. Process parameter optimization of gas metal arc welding on IS: 2062 mild steel using response surface methodology. *J. Manuf. Process.* **2017**, *25*, 296–305. [CrossRef]

11. Pal, S.; Pal, S.K.; Samantaray, A.K. Radial basis function neural network model based prediction of weld plate distortion due to pulsed metal inert gas welding. *Sci. Technol. Weld. Join.* **2007**, *12*, 725–731. [CrossRef]
12. Jiang, P.; Wang, C.; Zhou, Q.; Shao, X.; Shu, L.; Li, X. Optimization of laser welding process parameters of stainless steel 316L using FEM, Kriging and NSGA-II. *Adv. Eng. Softw.* **2016**, *99*, 147–160. [CrossRef]
13. Ankenman, B.; Nelson, B.L.; Staum, J. Stochastic kriging for simulation metamodeling. *Oper. Res.* **2010**, *58*, 371–382. [CrossRef]
14. Chen, X.; Ankenman, B.E.; Nelson, B.L. The effects of common random numbers on stochastic kriging metamodels. *ACM Trans. Model. Comput. Simul. (TOMACS)* **2012**, *22*, 7. [CrossRef]
15. Xie, W.; Nelson, B.; Staum, J. The influence of correlation functions on stochastic kriging metamodels. In Proceedings of the 2010 Winter Simulation Conference, Baltimore, MD, USA, 5–8 December 2010; pp. 1067–1078.
16. Kamiński, B. A method for the updating of stochastic kriging metamodels. *Eur. J. Oper. Res.* **2015**, *247*, 859–866. [CrossRef]
17. Liu, M.; Staum, J. Stochastic kriging for efficient nested simulation of expected shortfall. *J. Risk* **2010**, *12*, 3. [CrossRef]
18. El-Batahgy, A.-M. Effect of laser welding parameters on fusion zone shape and solidification structure of austenitic stainless steels. *Mater. Lett.* **1997**, *32*, 155–163. [CrossRef]
19. Cicală, E.; Duffet, G.; Andrzejewski, H.; Grevey, D.; Ignat, S. Hot cracking in Al–Mg–Si alloy laser welding-operating parameters and their effects. *Mater. Sci. Eng. A* **2005**, *395*, 1–9. [CrossRef]
20. Cao, L.; Yang, Y.; Jiang, P.; Zhou, Q.; Mi, G.; Gao, Z.; Rong, Y.; Wang, C. Optimization of processing parameters of AISI 316L laser welding influenced by external magnetic field combining RBFNN and GA. *Results Phys.* **2017**, *7*, 1329–1338. [CrossRef]
21. Zhou, Q.; Rong, Y.; Shao, X.; Jiang, P.; Gao, Z.; Cao, L. Optimization of laser brazing onto galvanized steel based on ensemble of metamodels. *J. Intell. Manuf.* **2016**, 1–15. [CrossRef]
22. Jiang, P.; Cao, L.; Zhou, Q.; Gao, Z.; Rong, Y.; Shao, X. Optimization of welding process parameters by combining Kriging surrogate with particle swarm optimization algorithm. *Int. J. Adv. Manuf. Technol.* **2016**, *86*, 2473–2483. [CrossRef]
23. Michalska, J.; Sozańska, M. Qualitative and quantitative analysis of σ and χ phases in 2205 duplex stainless steel. *Mater. Charact.* **2006**, *56*, 355–362. [CrossRef]
24. Lophaven, S.; Nielsen, H.; Søndergaard, J. *DACE-A MATLAB Kriging Toolbox—Version 2.0.(2002)*; Technical University of Denmark: Lyngby, Denmark, 2002.
25. Seberry, J. Orthogonal designs. In *Orthogonal Designs: Hadamard Matrices, Quadratic Forms and Algebras*; Springer: Cham, Switzerland, 2017; pp. 1–5.
26. Montgomery, D.C.; Runger, G.C.; Hubele, N.F. *Engineering Statistics*; John Wiley & Sons: Hoboken, NJ, USA, 2009.
27. Park, J.-S. Optimal Latin-hypercube designs for computer experiments. *J. Stat. Plan. Inference* **1994**, *39*, 95–111. [CrossRef]
28. Refaeilzadeh, P.; Tang, L.; Liu, H. Cross-validation. In *Encyclopedia of Database Systems*; Springer: Cham, Switzerland, 2009; pp. 532–538.

Article

3D Multiphysical Modelling of Fluid Dynamics and Mass Transfer in Laser Welding of Dissimilar Materials

Jiazhou Wu, Hua Zhang *, Yan Feng and Bingbing Luo

Key Laboratory of Robot and Welding Automation of Jiangxi, School of Mechanical and Electrical Engineering, Nanchang University, Nanchang 330031, China; woojz12@126.com (J.W.); fengyan76@ncu.edu.cn (Y.F.); lbingbing11@126.com (B.L.)
* Correspondence: huazhang_ncu@126.com; Tel.: +86-791-8396-9630

Received: 15 May 2018; Accepted: 6 June 2018; Published: 11 June 2018

Abstract: A three-dimensional multiphysical transient model was developed to investigate keyhole formation, weld pool dynamics, and mass transfer in laser welding of dissimilar materials. The coupling of heat transfer, fluid flow, keyhole free surface evolution, and solute diffusion between dissimilar metals was simulated. The adaptive heat source model was used to trace the change of keyhole shape, and the Rayleigh scattering of the laser beam was considered. The keyhole wall was calculated using the fluid volume equation, primarily considering the recoil pressure induced by metal evaporation, surface tension, and hydrostatic pressure. Fluid flow, diffusion, and keyhole formation were considered simultaneously in mass transport processes. Welding experiments of 304L stainless steel and industrial pure titanium TA2 were performed to verify the simulation results. It is shown that spatters are shaped during the welding process. The thickness of the intermetallic reaction layer between the two metals and the diffusion of elements in the weld are calculated, which are important criteria for welding quality. The simulation results correspond well with the experimental results.

Keywords: laser welding; keyhole; weld pool behavior; mass transfer; dissimilar metal

1. Introduction

Dissimilar metal welding often confronts the problems of brittle intermetallic compounds, residuals stresses, and crack formations, which lead to lower joint performance. Thethickness of an intermetallic reaction layer is an important factor affecting the formation of intermetallic compounds, besides the differences in the physical and chemical properties of the materials. As long as the thickness of the intermetallic reaction layer is maintained in the proper range, the welded workpiece will have good mechanical properties. If the thickness is too large, it is possible for new intermetallic compounds to be formed, which makes the welded joint brittle and hard and reduces plasticity and ductility. Therefore, precise control of the heat source is highly important in the welding process. Compared with the traditional welding method, laser welding has several advantages such as precise energy control, high density due to minimal laser beam diameter, and narrow heat affected zones. The rapid melting and solidification of metal can reduce the formation of intermetallic compounds in laser welding of dissimilar materials, which improves the mechanical properties of the welded joint. Laser welding is an important joining method for dissimilar metals.

The behaviors of keyhole and weld pool have an important influence on welding quality in deep penetration laser welding, which has been widely confirmed through theoretical and experimental research. Early studies primarily focused on the analysis of temperature and heat flow fields in similar metal welding processes. Rai et al. [1,2] proposed a numerical model for keyhole mode laser welding to calculate the temperature and velocity fields, weld geometry, and solidification for different materials. Their results suggested that convection was the primary mechanism for heat transfer; however, the upper surface of the workpiece was assumed to be flat, and the shape of the keyhole was not the actual vapor-liquid interface, as this model did not consider the effect of the recoil pressure induced by metal evaporation. In recent years, research on keyhole formation, molten pool dynamics, and welding defects such as the formation of porosity and spatter has become a heavily researched topic. Na et al. [3,4] and Pang et al. [5–7] obtained the shape of a real-time keyhole, the model of which used a ray tracing method to simulate multiple laser reflections along the keyhole wall. Their study suggested that the formation of the keyhole was a result of the interaction of the recoil pressure caused by metal vaporization, the surface tension, and hydrostatic pressure. In addition, the formation of pores was due to the collapse of the keyhole. Recently, Hua et al. [8,9] performed a simulation study on the mechanism of spatter formation at full penetration condition. Their results suggested that the shear force generated by the high-speed plasma movement was the main cause of spatter formation, which was in full agreement with the experimental results of Chen et al. [10].

Although laser welding processes of the dissimilar metals and same metals are somewhat similar, laser welding of dissimilar materials has its own individual features, such as the formation of intermetallic compounds. Hu et al. [11] developed a 3D heat and mass transfer model of steel-nickel laser welding to analyze the influence of convection on the temperature field and mass transfer, and their results suggested that strong convection had an important impact on heat and mass transfer. Moreover, concentration gradients and sufficient time also contributed to homogeneous welds. For the study of convection in dissimilar laser welding, Yu et al. [12] have obtained the same conclusions as Hu et al. [11], and they noted that a thickness of intermetallic reaction layer could be decreased if the heat input was reduced to a reasonable range. Esfahani et al. [13] have conducted a dynamic analysis of the flow field based on a turbulence model in the weld pool. Their research also confirmed that the mixing of dissimilar metal materials was the result of convection in the molten pool, and the increase of convective intensity was caused due to the increase of heat. Tomashchuk et al. [14] proposed a simulation model for electron beam dissimilar welding via interlayers based on COMSOL soft, and the influence of the change of welding parameters on the thickness of the diffusion layer was studied.

In summary, several researchers have conducted further theoretical studies of the coupling behaviors of keyhole and molten pool in similar metal laser welding [1–10]; according to simulation methods. However, the numerical analysis of laser welding of dissimilar metal is mainly focused on the study of temperature and heat flow field [10–16]. Although several researchers have investigated the influence of the thickness of the intermetallic reaction layer between two metals on the mechanical properties of the workpiece, studies that have examined the behavior of the keyhole in laser welding of dissimilar materials were rarely reported. Additionally, numerical simulations regarding the laser welding of dissimilar materials have yet to be developed to simultaneously consider the coupling of keyhole formation, heat transfer, and mass transfer. The keyhole formation is an important feature of laser welding, and its dynamic behavior has an important influence on the welding quality.

In this study, a 3D multiphysics transient model considering both keyhole behavior and mass transfer is proposed and is verified by welding experiments. The simulation considers major physical factors such as evaporation, recoil pressure, surface tension, and Rayleigh scattering of the laser beam. Convection, diffusion and keyhole formation are considered simultaneously in the mass transport process between the two metals. The spatter formation mechanism, keyhole and molten pool dynamics, and influence of process parameters on the thickness of the intermetallic reaction layer are analyzed.

2. Experimental Procedure

To validate the result of numerical model, 304L stainless steel and pure titanium TA2 with dimensions of 60 mm × 40 mm × 0.8 mm were used in laser overlap welding experiments, and the chemical compositions of the materials are shown in Table 1. In this experiment, 304L stainless steel is placed on top, as shown in Figure 1, since titanium can very easily absorb oxygen and hydrogen in air at high temperature, which makes the weld joint brittle; and crack easily. Moreover, there is little difference in melting point between pure titanium and stainless steel, which does not affect the formation of the molten pool. A fiber laser (IPG YLS-2000, IPG Photonics Corporation, Oxford, MA, USA, maximum power: 2 KW, wavelength: 1.07 μm) is used, and continuous laser welding is performed using a KUKA (KUKA Roboter GmbH, Augsburg, Germany) robot movement. During the laser welding process, the side shielding gas is argon, and velocity of gas flow is 15 L/min. To analyze the influence of different parameters on the welding quality, the range of laser power is 500 W–520 W, and the range of welding speed is 3.2 m/min–3.6 m/min. Laser welding experimental and simulation parameters are shown in Table 2.

Table 1. Chemical compositions of 304L and TA2 (wt. %).

304L							
C	Si	Mn	P	S	Ni	Cr	Fe
$\leq$0.03	$\leq$1.00	$\leq$2.00	$\leq$0.035	$\leq$0.03	8.00~11.00	18.00~20.00	balance

TA2						
Fe	C	N	H	O	Si	Ti
$\leq$0.30	$\leq$0.10	$\leq$0.05	$\leq$0.015	$\leq$0.25	$\leq$0.015	balance

Table 2. Welding experiment and simulation parameters.

Power (W)	Welding Speed (m/min)	Beam Defocus (mm)	Gas Flux Rate (L/min)
500	3.2	0	15
500	3.6	0	15
520	3.6	0	15

The 304L stainless steel and TA2 were cleaned with acetone and dried before welding to avoid the effect of oil and water on the welding joint. Spatter formation and melted pool behavior were observed using a high-speed camera during the welding process. After welding, metallographic specimens were made. Optical microscope (OM), scanning electron microscope (SEM), and energy dispersive spectrometer (EDS) were used to verify the validity of the model.

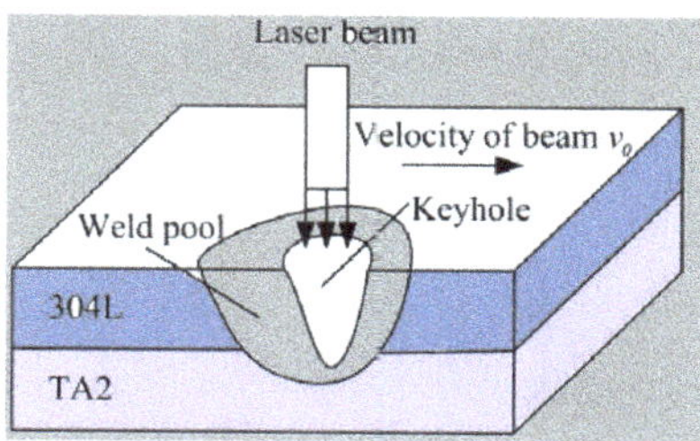

Figure 1. Schematic diagram of laser overlap welding.

3. Model Description

The calculated zone of the 3D laser welding model is shown in Figure 1, including the plasma and workpiece zone. Taking into account the physical symmetry in the welding and the calculation speed of computer, half of the welding zone is selected as the calculation domain. Laser welding is a complex process of heat and mass transfer, and in order to simplify the calculation, the model is assumed to be as follows:

(1)   The effect of side shielding gas on the behavior of keyhole and molten pool is ignored.

(2)   The calculated fluid is Newtonian and incompressible and is in local thermal equilibrium; furthermore, this fluid satisfies the basic equations of fluid motion.

(3)   The temperature-dependent thermo-physical parameters are calculated, derived from the JMatPro software (Release Version 7.0.0, Sente Software Ltd., Guildford, UK).

(4)   In the simulation, only iron and titanium components are considered, and other alloy elements are ignored.

3.1. Laser Heat Source Model

In fiber laser welding, the fundamental mode Gaussian beam is the main output mode, in which the spot is circular, and the distribution of light intensity theoretically satisfies the Gauss function, as shown in Equation (1).

$$q(r) = \frac{3q_{max}}{\pi r_{eff}^2} \exp\left(-\frac{3r^2}{r_{eff}^2}\right) \tag{1}$$

where q_{max} is the maximum heat flux density of the laser beam, r is the heat source radius, r_{eff} is the waist radius, and $q(r)$ is the heat flux density at radius r.

In the process of laser transmission, the beam has a certain degree of divergence, and the power tends to decrease with an increase in weld depth. In addition, the metal vapor plasma in the keyhole hinders laser energy transmission. At present, the beam quality factor M^2 proposed by Siegman et al. [17] is generally used to evaluate the quality of the laser beam, as shown in Equation (2).

$$r(z) = r_{eff}\sqrt{1 + (\frac{z - z_0}{z_R})} \tag{2}$$

where z_0 is the longitudinal coordinate of the waist, z_R is the Rayleigh constant, and z is the longitudinal coordinate of the heat source.

In laser continuous welding, the radius r in Equation (1) can be expressed as

$$r = \sqrt{(x - v_0t)^2 + y^2} \tag{3}$$

where v_0 is the welding speed, and t is the welding time.

During laser welding, the amount of defocus is set to zero, and the cells in the interface and inside of the keyhole are heated by the beam, while the value of heat flux density is calculated using Equation (1). As mentioned above, the shape of the heat source changes with the shape of keyhole. When the depth of the keyhole increases, the effective radius of the heat source increases. However, the peak energy decreases. It is worth noting that Equation (1) shows the surface heat flux density, not the volume density. It is necessary to transform Equation (1) into the body heat flux according to Zhou et al. [18].

3.2. Governing Equation

With the above assumptions, numerical analysis of laser welding of dissimilar metal is carried out by solving equations such as mass, momentum, energy and species transport, as expressed in Equations (4)–(6).

The mass conservation equation is described as

$$\frac{\partial}{\partial t}(\rho) + \nabla \cdot (\rho\overline{V}) + S_{MA} = 0 \tag{4}$$

where ρ is density, $\overline{V}$ is velocity vector, and S_{MA} is mass source.

The momentum conservation equation is described as

$$\frac{\partial}{\partial t}\left(\overline{V}\right) + \overline{V}\cdot\nabla\overline{V} = -\frac{1}{\rho}\nabla P + \mu\nabla^2\overline{V} + S_{MT} + g \tag{5}$$

where p is pressure, g is gravity constant, μ is dynamic viscosity, and S_{MT} is momentum source.

The energy conservation equation is described as

$$\frac{\partial h}{\partial t} + \overline{V}\cdot\nabla h = k\nabla^2 T + S_E \tag{6}$$

where h is enthalpy, k is coefficient of thermal conductivity, T is temperature, and S_E is energy source.

The mass fraction of each species can be predicted, through the solution of a convection and diffusion equation for the species, as expressed in Equation (7).

$$\frac{\partial}{\partial t}\left(\rho Y_i\right) + \nabla\cdot\left(\rho\overline{V}Y_i\right) = -\nabla\overrightarrow{J_i} \tag{7}$$

where Y_i is the mass fraction of species i, and J_i is the diffusion flux of species i. In this study, it is assumed that the molten pool fluid is laminar, implying that J_i can be expressed as

$$\overrightarrow{J_i} = -\rho D_{im}\nabla Y_i - D_{Ti}\frac{\nabla T}{T} \tag{8}$$

where D_{im} is the mass diffusion coefficient for species i in the mixture, and D_{Ti} is the thermal diffusion coefficient. According to the above assumptions, the diffusion between titanium and iron is studied, and other alloy elements are not considered.

The mass source term S_{MA} in Equation (4) mainly considers the gas-liquid mass transfer due to the evaporation or solidification of the material. The momentum source S_{MT} in Equation (5) mainly includes the recoil pressure due to the violent evaporation of liquid metal, the momentum loss caused by the solidification of liquid metal, and the Boussinesq buoyancy due to the change of the density of the melted metal. The energy source S_E in Equation (6) primarily includes the laser heat source, and the energy loss caused by the radiation and evaporation of the liquid metal on the keyhole wall.

In deep penetration laser welding, the formation of the keyhole is an important feature different from laser conductive welding. On the keyhole free surface, the recoil pressure is the driving force of the keyhole formation. However, surface tension prevents keyhole formation. A general expression for calculating the recoil pressure is as follows [18]:

$$P_r = AB_0/\sqrt{T_v}\exp(-U/T_w) \tag{9}$$

where P_r is the recoil pressure, A is the adjustment coefficient, which is generally 0.55, B_0 is the evaporation constant, T_w is the temperature of the keyhole surface, and U is a constant related to the material.

It is assumed that the recoil pressure P_r always vertically acts on the keyhole wall, which is decomposed into axial components during calculation, and the components P_x along the x-axis is defined as follows:

$$P_x = \frac{n_x}{\sqrt{n_x^2 + n_y^2 + n_z^2}}\cdot P_r \tag{10}$$

where n_x, n_y, and n_z are, respectively, the volume fraction gradient components in each axial direction. Using the same calculation method, the other two axial recoil pressures P_y and P_z can also be calculated.

During the welding process, the recoil pressure and heat source act on the keyhole surface, meaning that the calculation of the keyhole shape is very important. The keyhole surface is calculated by solving the volume fraction Equation (11) [19], as shown.

$$\frac{\partial F}{\partial t} + \overline{V} \cdot \nabla F = 0 \tag{11}$$

where F is the volume fraction of cells.

During keyhole formation, mass transfer due to evaporation and recondensation is studied, as shown in [20].

$$S_{lv} = -m_{lv} \tag{12}$$

$$S_{vl} = m_{lv} \tag{13}$$

where S_{lv} and S_{vl} are the mass source in Equation (4); m_{lv} is cell mass, and a negative value means recondensation.

In simulation, the momentum sink due to solidification is calculated using enthalpy-porosity technique; the latent heat is described using properly changing the specific heat, and the value of momentum sink S_{mush} is approximated as follows [21]:

$$S_{mush} = \frac{(1 - f_l)^2}{f_l^3 + \varphi} A_{mush} \cdot \overline{V} \tag{14}$$

where f_l is the liquid volume fraction of cells, φ is a tiny number to prevent the denominator from being zero, and A_{mush} is a porous medium constant.

Mass transport between dissimilar metals is calculated by solving the mass transport Equation (7); therefore, the mixed phase is composed of iron and titanium in the simulation model. In the mixture, density ρ, coefficient of thermal conductivity k, specific heat c_p, and dynamic viscosity μ are averaged and defined as follows:

$$\rho = \alpha \rho_1 + (1 - \alpha)\rho_2 \tag{15}$$

$$k = \alpha k_1 + (1 - \alpha)k_2 \tag{16}$$

$$c_p = \alpha c_{p1} + (1 - \alpha)c_{p2} \tag{17}$$

$$\mu = \alpha \mu_1 + (1 - \alpha)\mu_2 \tag{18}$$

where α is the mass fraction of the species, subscript 1 represents the first species, and subscript 2 represents the second species.

3.3. Boundary Conditions

As mentioned above, the laser heat source acts on the keyhole surface to heat the workpiece. Meanwhile, energy loss such as gas convection, radiation and evaporation are considered, and the energy balance equation is expressed as

$$k\frac{\partial T}{\partial n} = q - h_c(T - T_\infty) - \sigma\varepsilon(T^4 - T_\infty^4) - WH_{lv} \tag{19}$$

where n is the vector normal to the liquid-gas interface, h_c is the convection heat transfer coefficient, T_∞ is the ambient temperature, σ is the Boltzmann constant, ε is the emissivity of radiation, W is the speed of evaporation, and H_{lv} is the latent heat of evaporation.

For other boundaries such as the rear, left, right and bottom of the workpiece, only energy loss due to convection and radiation is considered, and expressed as follows:

$$k\frac{\partial T}{\partial n} = h_c(T - T_\infty) - \sigma\varepsilon(T^4 - T_\infty^4) \tag{20}$$

For momentum boundary, the surface tension P_s is defined as

$$P_s = \kappa \gamma \tag{21}$$

where κ is the surface curvature, and γ is the temperature-dependent surface tension coefficient.

3.4. Numerical Method

The developed 3D model is solved using a commercial software package named Fluent, and single-machine parallel computing is used in order to improve the calculating speed. The computed domain with dimensions of 6 mm $\times$ 2.8 mm $\times$ 2.4 mm is meshed, and the regular hexagon grid is adopted to facilitate the conversion of surface data and volume data in the simulation with a cell size of 0.08 mm. In this model, the mixed phase in species transport model is composed of iron and titanium. Moreover, the phase in multiphase flow model Volume of Fluid (VOF) is composed of plasma and the mixed phase established in the previous step.

The governing equation is discretized using a finite volume method; user-defined functions(UDFs) such as initializing of the calculation domain, source terms, boundary conditions, and material property definitions are written using C code, and pressure implicit with splitting of operators(PISO) algorithm is used to solve discrete equations. In the PISO algorithm, the pressure correction equation is solved twice; therefore, compared with other algorithms, the PISO algorithm has good convergence and high efficiency.

4. Results and Discussion

To improvewelding quality, the formation of excessive intermetallic compounds must be suppressed. Meanwhile, the burning loss of alloying elements in the welding process should be prevented in laser overlap welding. Therefore, precise control of the heat source has a significant impact on joint performance, and the selection of welding parameters is crucial in dissimilar metal welding. In the simulation and experiment, the following welding process parameters, as shown in Table 2, are used to analyze the effects of different parameters on mass transfer between the two metals.

The physical property values used in the calculation are shown in Table 3. Other important parameters used in the simulation are listed in Table 4.

Table 3. Physical properties of 304L and TA2 [1,22].

Physical Properties	304L	TA2
Density (kg/m^3)	7000	4110
Specific heat (J/(kg·K))	712	594
Heat conductivity (W/(m·K))	29	40
Dynamic viscosity (N·s/m^2)	0.007	0.005
Boiling point (K)	3100	3315
Surface tension (N/m)	1.4	1.65
Surface tension temperature coefficient (N (m·K))	-4.9×10^{-4}	-2.6×10^{-4}
Coefficient of thermal expansion (/K)	1.96×10^{-5}	1.1×10^{-5}
Melting latent (J/kg)	2.47×10^5	3.89×10^5
Evaporation latent (J/kg)	6.34×10^6	8.88×10^6

Table 4. Data used in the simulation.

Nomenclature	Value
Laser beam radius at focus (mm)	0.2
Planck constant (J·s)	5.67×10^{-8}
Stefan-Boltzmann constant (W/(m^2·K^4))	1.38×10^{-23}
Ambient temperature (K)	300
Density of plasma (kg/m^3)	0.06
Specific heat of plasma (J/(kg·K))	610
Heat conductivity of plasma (W/(m·K))	3.72
Convective heat transfer coefficient (W/(m^2·K))	60
Radiation Emissivity	0.4
Gas constant (J/(kg·mol))	8.3×10^3

4.1. Keyhole Formation and Weld Pool Dynamics

Figure 2 shows the calculated temperature distribution, keyhole formation and velocity fields in the laser welding of dissimilar materials. At the very beginning of laser welding, when the welding time is less than 3.05 ms, the top surface of 304L stainless steel is heated and metal is melted. When the temperature of the melted metal is higher than the boiling point of the material, violent vaporization occurs. Meanwhile, the metal vapor plasma is formed under the action of the laser. As shown in Figure 2, the keyhole is formed under the action of the recoil pressure due to evaporation, surface tension, and hydrostatic pressure. The recoil pressure causes the keyhole to deepen downwards, while the surface tension prevents the keyhole from moving downwards. As shown in Figure 2c,d, the dynamic and energy balance are achieved on the keyhole wall, and the maximum depth of the keyhole is obtained. These phenomena are consistent with previous research results in similar metal welding, such as Na et al. [3,4], Pang et al. [5–7], Zhao et al. [20] and Tan et al. [21]. Moreover, it can also be seen from Figure 2c,d, as a result of the recoil pressure, that the keyhole also appears in the lower region. Therefore, the lower metal may be squeezed into the upper region. Thus, it is possible that elements are mixed, and intermetallic compounds are formed. However, the mixing of elements and the formation of intermetallic compounds have a significant influence on welding quality [11]. Therefore, keyhole behavior directly affects joint performance in dissimilar metal welding. Moreover, laser heat is transmitted through the keyhole wall to melt the welded metal such that the keyhole shape also affects heat transfer.

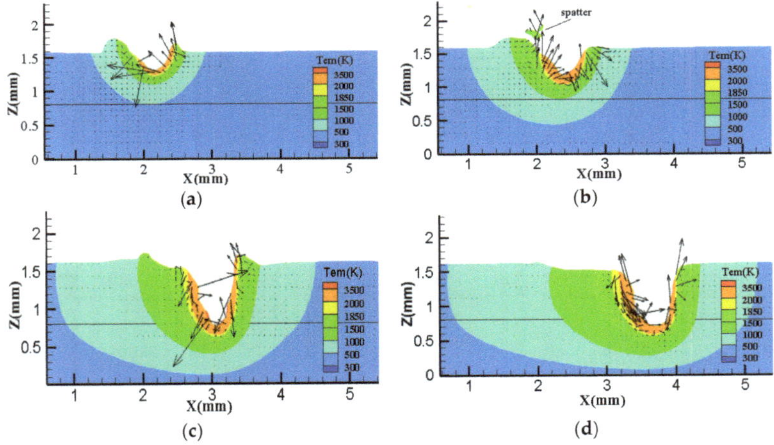

Figure 2. Longitudinal section views of the calculated temperature and velocity fields in laser welding (laser power 500 W, welding velocity 3.6 m/min): (**a**) t = 3.05 ms; (**b**) t = 7.20 ms; (**c**) t = 18.20 ms; (**d**) t = 28.35 ms.

It can be seen in Figure 2b that spatter is formed in the welding process, which may damage the laser equipment due to spatter into the light path and may also affect the welding quality. In essence, laser welding is the process of rapid heating and rapid cooling of the material, so that the weld pool width is relatively small. Due to the role of recoil pressure, the molten pool near the keyhole wall has an upward trend, as shown in Figure 2b. According to the previous study by Hua et al. [8,9], when the stagnation pressure of the melted metal is higher than the surface pressure, spatter may appear. The shear force and recoil pressure are the driving force of spatter formation, which has been experimentally confirmed (Figure 3).

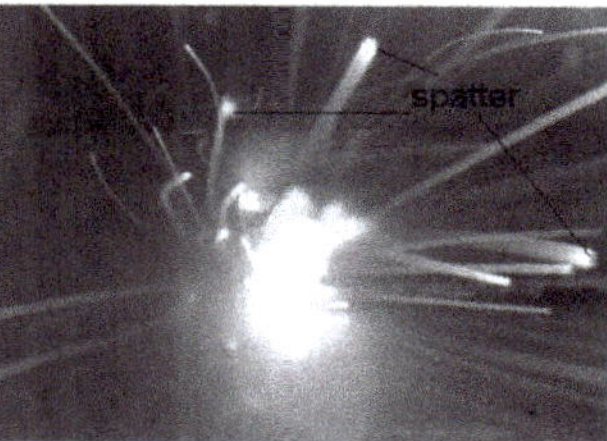

Figure 3. High-speed photographs of spatter formation during laser welding.

The velocity vector indicates the flow state of melted metal in the welding pool, which is an important factor affecting heat and mass transfer. On the surface of 304L stainless steel, the melted metal flows outwards from the keyhole center due to the Marangoni force, as shown in Figure 4a. Meanwhile, the melted metal in the front of the keyhole flows along the keyhole to the rear of the weld pool, which is resolidified to form the weld bead. In the depth direction of the weld bead (z-axis), as shown in Figures 2 and 4b, the molten pool flow is analyzed from three regions, such as the keyhole wall, near the keyhole, and near the solid phase. The keyhole wall surface is affected by the shear force due to the high-speed movement of the metal vapor plasma, the recoil pressure, and the surface tension, thereby increasing its complexity. Similar to surface tension, the shear force is the tangential force at the gas-liquid interface, the size of which is determined by the velocity of metal vapor plasma inside the keyhole [8,9]. The direction of the recoil pressure is the normal direction of the gas-liquid interface, the size of which is a function of the gas-liquid interface temperature, as shown in Equation (9). The melted metal near the keyhole flows downwards from the top, and the melted metal near the solid phase flows upwards from the welding pool bottom. The simulation results are consistent with previous theoretical calculations by Na et al. [3,4], Pang et al. [5–7] and Zhao et al. [20] and agree with experimental results by Katayama et al. [23]. Moreover, it can be clearly seen from Figures 2 and 5 that the molten pool flow affects the heat transfer and also has an important influence on the mixing of the elements in the weld zone and the formation of the metal reaction layer in dissimilar metal welding, which is a key factor affecting joint performance.

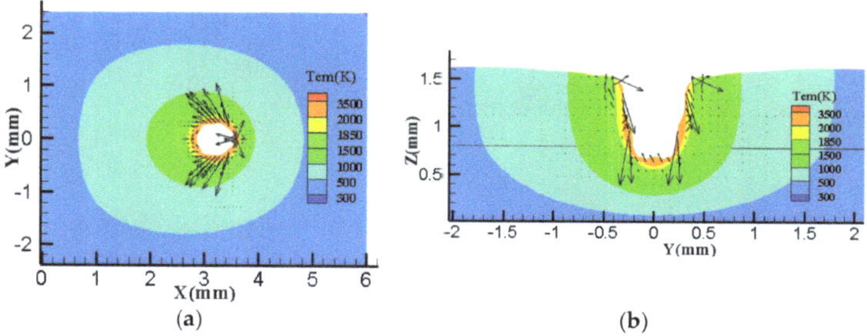

Figure 4. Calculated temperature and velocity fields in laser welding at time t = 28.35 ms (laser power 500 W, welding velocity 3.6 m/min): (**a**) top view; (**b**) cross-section view.

4.2. Mass Transfer

Fluid flow and diffusion are two basic methods of mass transport, which have been considered by solving Equation (7). Furthermore, keyhole formation is also an important factor affecting melted metal transport, as shown in Figures 5–7, and it cannot be ignored. In this study, the effects of fluid flow, diffusion and keyhole formation on the formation of intermetallic reactive layers are considered simultaneously, and the concentration distribution of elementsis studied in the weld. Using the process parameters in Table 2, simulation and welding experiments are performed to analyze the effects of flow, diffusion and keyhole formation on the formation of intermetallic reaction layers and the concentration distribution of elements. Meanwhile, the change of parameters is also considered, such as laser power and welding velocity.

Figure 5a,b shows the effects of different welding speeds on mass transport in dissimilar metal welding with a laser power of 500 W. Compared with Figure 5b, it can be seen from Figure 5a that the keyhole depth and the thickness of the intermetallic reaction layer increase slightly at x = 3.6 mm, and the concentration of Ti in the weld beam is also higher. This is due to the increase in laser line energy caused by a decrease in welding speed. Therefore, the mass transport of dissimilar metals increases, resulting in an increase in the thickness of the intermetallic reaction layer. As stated by Hu et al. [11] and Yu et al. [12], if the thickness of the intermetallic reaction layer is excessively large, intermetallic compounds may be produced to attenuate joint performance [24–28]. In addition, excessive Ti in the weld beam is also not conducive to welding quality.

The influence mechanism of laser power change on mass transport is similar to that of welding velocity change, as shown in Figure 5b,c. When the laser power increases, the welding line energy increases, and therefore the keyhole depth, the thickness of the reaction layer between the metals and the concentration of Ti in the weld zone will increase slightly, which is the same as the effect of welding speed decrease on mass transport. The effective variation range of power and welding velocity is very small, which means that laser welding is an important approach for dissimilar metal welding because of its precise control of energy.

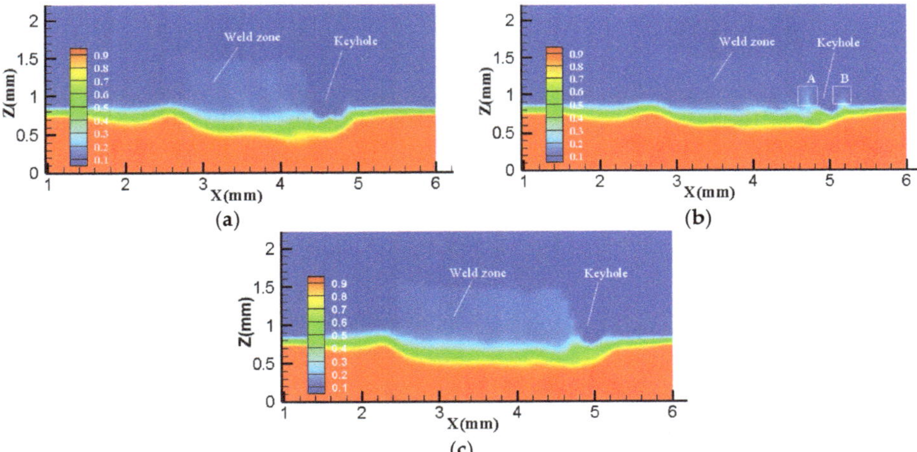

Figure 5. Longitudinal section views of the calculated concentration profile of Ti for different process parameters: (**a**) laser power 500 W, and welding velocity 3.2 m/min; (**b**) laser power 500 W, and welding velocity 3.6 m/min; (**c**) laser power 520 W, and welding velocity 3.6 m/min.

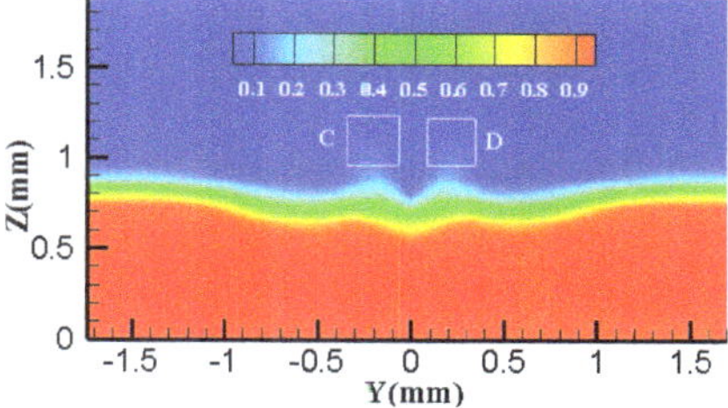

Figure 6. Cross section views of the calculated concentration profile of Ti at time t = 50.15 ms (laser power 500 W, and welding velocity 3.6 m/min).

In addition to fluid flow and diffusion, keyhole formation has an important influence on the mass transport in dissimilar metal welding. In the welding process, the diameter of the laser beam is very small, and the action time with the metal is extremely short, meaning that the amount of melted metal is very small. When the recoil pressure acts on the keyhole surface in the lower region, the melted metal in the lower region will be pressed into the upper region, as shown in regions A and B in Figure 5b, and regions C and D in Figure 6a. Moreover, with an increase in keyhole depth in the lower region, the concentration of Ti increases in the upper weld zone, as shown in Figure 5a,c. Furthermore, during the actual welding process, when the welding conditions change suddenly, or the welding process is unstable, the melted metal from the lower region in the upper region may be re-solidified, and form weld defects, which have been confirmed in welding experiments by Yu et al. [11] and Yang et al. [24].

4.3. Validation of Simulation Results

Figure 7 shows SEM image of the welding joint of 304L stainless steel and pure titanium TA2, with a laser power of 500 W and the welding speed of 3.6 m/min. The concentration distribution of elements in different positions of welded joints was obtained through EDS to verify the validity of the calculation results.

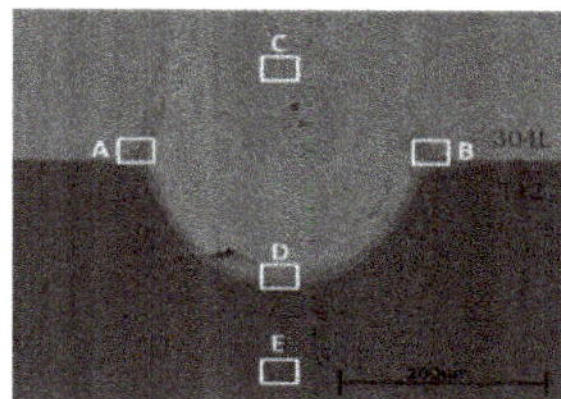

Figure 7. Scanning electron microscope (SEM) image of the welding joint of 304L stainless steel and TA2.

The experimental element distributions measured through EDS are shown in Figures 8–10 and Table 5. As can be seen from Figures 8 and 9 and Table 5, the titanium (Ti) concentration in the weld zone is about 11.23 at. % (region C), while increasing rapidly to 40.71 at. % (region D) at the interface regions between 304L stainless steel and pure titanium TA2, and finally reaches 100 at. % (region E) in the lower metal region (pure titanium). This is because the melted Ti enters the molten pool zone through the convection and diffusion occurring in the welding process. As shown in Figures 8 and 9 and Table 5, the Fe concentration in region E is 0 at. %, while the Ti concentration in region C is 11.23 at. %. This indicates that the convection and diffusion of the molten pool has an important influence on mass transport. Moreover, as can be seen from Figure 5, the simulated Ti concentration in the weld zone is found to be 11.75 at. %, which is consistent with the experimental results. Furthermore, the thickness of the intermetallic reaction layer has an important effect on the properties of the joint in dissimilar metal welding. As shown in Table 5 and Figure 9, in the interface regions between 304L stainless steel and pure titanium TA2, the Ti (40.71 at. %), and Fe (37.73 at. %) concentrations are very large and very close, possibly due to the fact that intermetallic compounds will be generated [11,24,28]. At x = 3.6 mm in Figure 5b, when the Ti concentration is between 50 at. % and 42 at. %, the thickness of the intermetallic reaction layer is about 30 μm, which is consistent with the experimental results. The difference from the experimental data may arise from the influence of alloy elements such as Cr and Ni on mass transfer in 304L stainless steel ignored in the simulation. The simulation results have important theoretical significance.

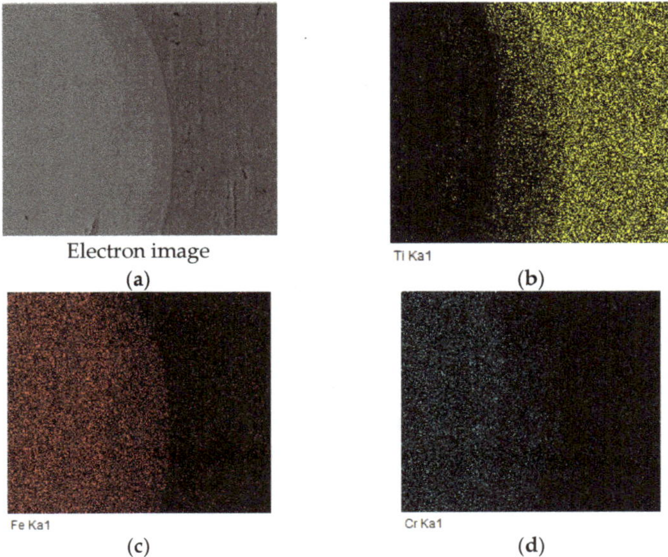

Figure 8. Distribution of elements near the interface (**a**) Electron image, (**b**) Ti, (**c**) Fe, and (**d**) Cr.

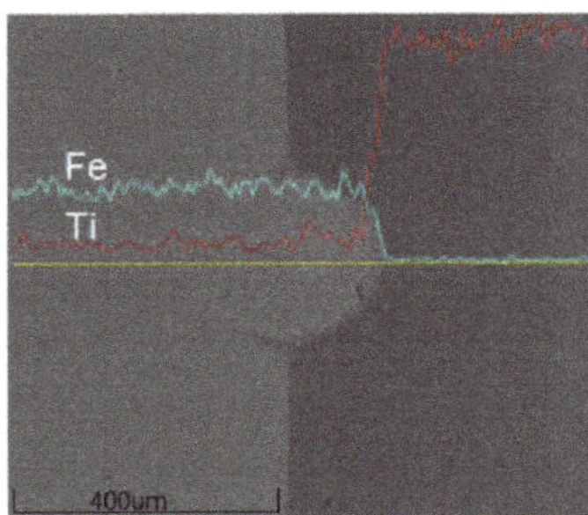

Figure 9. Distribution of elements Ti and Fe along the axis line of weld.

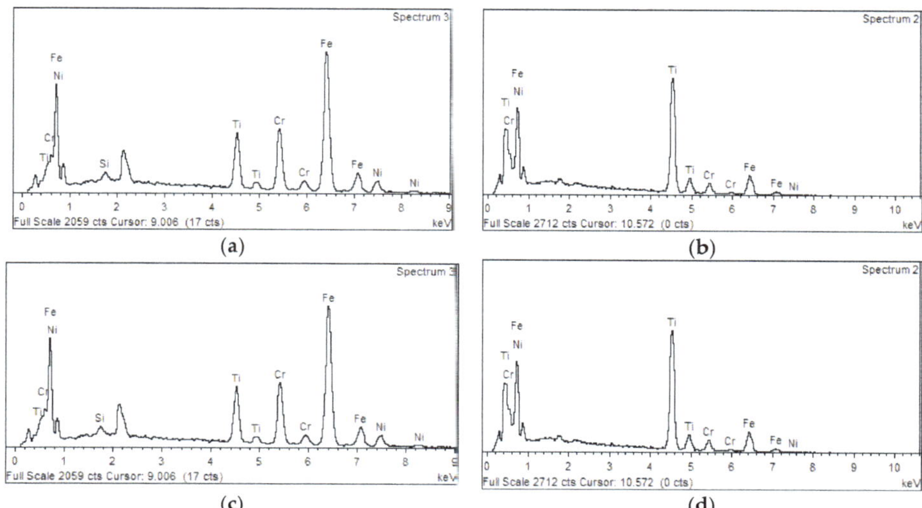

Figure 10. Energy dispersive spectrometer(EDS) spectra of (**a**) region A, (**b**)region B, (**c**) region C, and (**d**) region D.

Table 5. Experimental atomic fraction of Fe and Ti for various location in Figure 7 (at. %).

Element	A	B	C	D	E
Fe	44.45	45.39	62.50	37.73	0
Ti	33.22	41.82	11.23	40.71	100

When the laser keyhole is formed in the lower region, the melted metal in the lower region is squeezed into the upper region due to recoil pressure, as shown in regions A and B in Figure 7. It can be clearly seen from Table 5 and Figure 10 that the concentration values of Ti in regions A and B are very high, and the concentration of titanium are 33.22 at. % and 41.82 at. %, respectively, which verifies the accuracy of the calculated results in Figure 6. The keyhole formation has an important effect on mass transfer in laser welding of dissimilar materials. The calculated results are in good agreement with the experimental results.

5. Conclusions

A 3D multiphysics transient model of laser welding of dissimilar materials is proposed, and the dynamic behaviors of the keyhole and molten pool are analyzed. Moreover, mass transport between dissimilar metals is calculated, which is consistent with the experimental results. The primary points of this study are as follows.

(1) Recoil pressure is the driving force for keyhole formation. The laser beam heats the workpiece through the keyhole wall, and the flow of the molten pool has an important effect on energy transmission.

(2) Fluid flow and diffusion are two important mechanisms of mass transport. As the laser line energy increases, the thickness of the intermetallic reaction layer and the diffusion of elements in the weld will increase. Accurate control of laser energy is the key to reduce the formation of intermetallic compounds.

(3)   In the premise of ensuring connection strength and avoiding the burning of alloying elements, the depth of the keyhole in the lower region should be controlled accurately, and the melted metal in the lower region should also be prevented from entering the upper region in large quantities.

Author Contributions: H.Z. and J.W. conceived and designed experiments; Y.F. analyzed the data; B.L. performed the experiments and sampled; J.W. performed the simulation and wrote the paper.

Acknowledgments: This work was supported by the National High Technology Research and Development Program of China (No. 2013AA041003) and the National Natural Science Foundation of China (No. 51665039).

Conflicts of Interest: The authors declare no conflict of interest.

Symbol

Symbols	Meaning	Symbols	Meaning
q_{max}	Maximum heat flux density of the laser beam	P_x	Recoil pressure components along the x-axis
r	Radius	n_x	Volume fraction gradient along the x-axis
r_{eff}	Waist radius	F	Volume fraction of cells
$q(r)$	Heat flux density at radius r	S_{lv}	Mass source
M^2	Beam quality factor	S_{vl}	Mass source
z_0	Longitudinal coordinate of the waist	m_{lv}	Cell mass
z_R	Rayleigh constant	f_l	Liquid volume fraction of cells
v_0	Welding speed	φ	A tiny number
t	Welding time	A_{mush}	Porous medium constant
ρ	Density	c_p	Specific heat
$\overline{V}$	Velocity vector	α	Mass fraction of the species
S_{MA}	Mass source	n	Vector normal
p	Pressure	h_c	Convection heat transfer coefficient
g	Gravity constant	T_∞	Ambient temperature
μ	Dynamic viscosity	σ	Boltzmann constant
S_{MT}	Momentum source	ε	Emissivity of radiation
h	Enthalpy	W	Speed of evaporation
k	Coefficient of thermal conductivity	H_{lv}	Latent heat of evaporation
T	Temperature	P_s	Surface tension
S_E	Energy source	κ	Surface curvature
Y_i	Mass fraction of species i	γ	Temperature-dependent surface tension coefficient
J_i	Diffusion flux of species i	A	Adjustment coefficient
D_{im}	Mass diffusion coefficient for species i	B_0	Evaporation constant
D_{Ti}	Thermal diffusion coefficient	T_w	Temperature of the keyhole surface
P_r	Recoil pressure	U	Constant related to the material

References

1.   Rai, R.; Elmer, J.W.; Palmer, T.A.; DebRoy, T. Heat transfer and fluid flow during keyhole mode laser welding of tantalum, Ti-6AI-4V, 304L stainless and vanadium. *J. Phys. D Appl. Phys.* **2007**, *40*, 5753–5766. [CrossRef]
2.   Rai, R.; Burgardt, P.; Milewski, J.O.; DebRoy, T. Heat transfer and fluid flow during electron beam welding of 21Cr-6Ni-9Mn steel and Ti-6AI-4V alloy. *J. Phys. D Appl. Phys.* **2009**, *42*, 025503. [CrossRef]
3.   Cho, W.I.; Na, S.J.; Thomy, C. Numerical simulation of molten pool dynamics in high power disk laser welding. *J. Mater. Process. Technol.* **2012**, *212*, 262–275. [CrossRef]
4.   Cho, J.H.; Na, S.J. Implementation of real-time multiple reflection and Fresnel absorption of laser beam in keyhole. *J. Phys. D Appl. Phys.* **2006**, *39*, 5372–5378. [CrossRef]
5.   Pang, S.Y.; Chen, L.L.; Zhou, J.X. A three dimensional sharp interface model for self consistent keyhole and weld pool dynamics in deep penetration laser welding. *J. Phys. D Appl. Phys.* **2011**, *44*, 025301. [CrossRef]
6.   Pang, S.Y.; Chen, W.D.; Zhou, J.X. Self consistent modeling of keyhole and weld pool dynamics in tandem dual beam laser welding of aluminum alloy. *J. Mater. Process. Technol.* **2015**, *217*, 131–143. [CrossRef]
7.   Pang, S.Y.; Chen, X.; Zhou, J.X. 3D transient multiphase model for keyhole, vapor plume, and weld pool dynamics in laser welding including the ambient pressure effect. *Opt. Lasers Eng.* **2015**, *74*, 47–58. [CrossRef]

8. Wu, D.Y.; Hua, X.M.; Fang, L. Understanding of spatter formation in fiber laser welding of 5083 aluminum alloy. *Int. J. Heat Mass Transfer.* **2017**, *113*, 730–740. [CrossRef]
9. Wu, D.Y.; Hua, X.M.; Huang, L.J. Numerical simulation of spatter formation during fiber laser welding of 5083 aluminum alloy at full penetration condition. *Opt. Laser Technol.* **2018**, *100*, 157–164. [CrossRef]
10. Li, S.C.; Chen, G.Y.; Katayama, S.J. Relation between spatter formation and dynamic molten pool during high-power deep-penetration laser welding. *Appl. Surf. Sci.* **2014**, *303*, 481–488. [CrossRef]
11. Hu, Y.W.; He, X.L.; Yu, G. Heat and mass transfer in dissimilar welding of stainless and nickel. *Appl. Surf. Sci.* **2012**, *258*, 5914–5922. [CrossRef]
12. Zhao, S.S.; Yu, G.; H, X.L. Numerical simulation and experimental investigation of laser overlap welding of Ti6Ai4V and 42CrMo. *J. Mater. Process. Technol.* **2011**, *211*, 530–537. [CrossRef]
13. Esfahani, N.M.R.; Coupland, J.; Marimuthu, S. Numerical simulation of alloy composition in dissimilar laser welding. *J. Mater. Process. Technol.* **2015**, *224*, 135–142. [CrossRef]
14. Tomashchuk, I.; Sallamand, P.; Jouvard, J.M. Multiphysical modeling of dissimilar welding via interlayer. *J. Mater. Process. Technol.* **2011**, *211*, 1796–1803. [CrossRef]
15. Iseav, V.I.; Cherepanov, A.N.; Shapeev, V.P. Numerical study of heat models of laser welding of dissimilar metals with an intermediate insert. *Int. J. Heat Mass Transfer.* **2016**, *99*, 711–720. [CrossRef]
16. Tomashchuk, I.; Sallamand, J.M.; Jouvard, J.M. The simulation of morphology of dissimilar copper-steel electron beam welds using level set method. *Comput. Mater. Sci.* **2010**, *48*, 827–836. [CrossRef]
17. Siegman, A.E. Defining, measuring, and optimizing laser beam quality. *SPIE* **1990**, *1224*, 2–13.
18. Zhou, J.; Tsai, H.L. Investigation of mixing and diffusion processes in hybrid spot laser-MIG keyhole welding. *J. Phys. D Appl. Phys.* **2009**, *42*, 095502. [CrossRef]
19. Semak, V.; Matsunawa, A. The role of recoil pressure in energy balance during laser materials processing. *J. Phys. D Appl. Phys.* **1997**, *30*, 2541–22252. [CrossRef]
20. Zhao, H.Y.; Niu, W.C.; Zhang, B. Modelling of keyhole dynamics and porosity formation considering the adaptive keyhole shape and three-phase coupling during deep-penetration laser welding. *J. Phys. D Appl. Phys.* **2011**, *44*, 485302. [CrossRef]
21. Tan, W.D.; Shin, Y.C. Analysis of multi-phase interaction and its effects on keyhole dynamics with a multi-physics numerical model. *J. Phys. D Appl. Phys.* **2014**, *47*, 345501. [CrossRef]
22. Cho, J.H.; Farson, D.F.; Hollis, K.J. Numerical analysis of weld pool oscillation in laser welding. *J. Mech. Sci. Technol.* **2015**, *29*, 1715–1722. [CrossRef]
23. Naito, Y.; Katayama, S.; Matsunawa, A. Keyhole behavior and liquid flow in molten pool during laser arc hybrid welding. *Proc. SPIE.* **2003**, *3888*, 34–45. [CrossRef]
24. Yang, J.; Li, Y.L.; Zhang, H. Microstructure and mechanical properties of pulsed laser welded Al/steel dissimilar joint. *Trans. Nonferr. Met. Soc. China* **2016**, *26*, 994–1002. [CrossRef]
25. Zhou, K.; Cai, L.L. Online Nugget Diameter Control System for Resistance Spot Welding. *Int. J. Adv. Manuf. Technol.* **2013**, *68*, 2571–2588. [CrossRef]
26. Cui, L.; Chen, B.X.; Qian, W. Microstructures and mechanical properties of dissimilar Al/Steel butt joints produced by autogenous laser keyhole welding. *Metals* **2017**, *7*, 492. [CrossRef]
27. Xue, Z.Q.; Hu, S.S.; Lee, D.K. Molten pool characterization of laser lap welded copper and aluminum. *J. Phys. D Appl. Phys.* **2013**, *46*, 495501. [CrossRef]
28. Kobayashi, S.; Yakou, T. Control of intermetallic compound layers at interface between steel and aluminum by diffusion-treatment. *Mater. Sci. Eng. A* **2002**, *338*, 44–53. [CrossRef]

Article

Study on the Size Effects of H-Shaped Fusion Zone of Fiber Laser Welded AZ31 Joint

Guang-Feng Lu, Lin-Jie Zhang *, Yi Pei, Jie Ning and Jian-Xun Zhang

State key laboratory of mechanical behavior for materials, Xi'an Jiaotong University, Xi'an 710049, China;
luguangfeng@stu.xjtu.edu.cn (G.-F.L.); peiyi@xjtu.edu.cn (Y.P.); ningjie2013@stu.xjtu.edu.cn (J.N.);
jxzhang@mail.xjtu.edu.cn (J.-X.Z.)
* Correspondence: zhanglinjie@mail.xjtu.edu.cn; Tel.: +86-29-8266-3115

Received: 25 January 2018; Accepted: 15 March 2018; Published: 21 March 2018

Abstract: There are two kinds of typical cross-section profiles for the fusion zone (FZ) of a laser welded thin section joint, i.e., a V-shaped cross-section and an H-shaped cross-section. Previous researches indicated that tensile strength of the V-shaped joint was lower than that of the H-shaped one due to the greater heterogeneity of strain distribution on the V-shaped joint during tensile process. In this work, impacts of the aspect ratio of FZ on the mechanical properties of laser welded thin section joints with an H-shaped cross-section profile were investigated. Welding conditions corresponding to two typical H-shaped joints (i.e., $W_{narrower}$ with a narrower FZ, and W_{wider} with a wider FZ) were decided through a laser welding orthogonal experimental plan. Then, the microstructure and properties of both joints were examined and compared. The results show that the tensile strength of joint $W_{narrower}$ and joint W_{wider} was about 72% and 80.9% that of the base metal, respectively. Both joints fractured in the FZ during tensile processes. The low-cycle fatigue life of the base metal, the joint $W_{narrower}$ and the joint W_{wider} were 3377.5 cycles, 2825 cycles and 3155.3 cycles, respectively. By using high-speed imaging, it was found that the fatigue crack of joint $W_{narrower}$ initiated and propagated inside the fusion zone, while the fatigue crack of the joint W_{wider} initiated at the edge of the base metal and propagated for a distance within the base metal before entering into the fusion zone. This work promoted our understanding about the influence of the weld bead shape on the properties of laser welded thin section joints.

Keywords: magnesium alloy thin sheet; fiber laser welding; microstructure; mechanical properties; H-shaped fusion zone

1. Introduction

Impacts of cross-section profiles of fusion zone (FZ) on the service performance of laser-welded joints have been studied by many researchers. Benyounis et al. [1] developed linear and quadratic polynomial equations which described the correlation between weld bead geometry and welding parameters including laser power, welding speed and defocusing amount. Hann et al. [2] presented a simple physical model to predict melt depth and width by using the concept of mean surface enthalpy which was derived from both material parameters and laser parameters. Li [3] investigated the effect of parameters on weld bead geometry in high-power laser welding of thick plate. The results showed that the width and depth of the weld seam decreased with the increasing of the welding speed. As a result, the geometry of the weld cross-section changed from a big-head shape to a needle-like shape. Kim et al. [4] proposed a simple scaling law that could predict the penetration depth by considering heat flow characteristics and multiple reflections. Volpp et al. [5] presented an analytical model to calculate the keyhole geometry using a ray-tracing method. Ayoola et al. [6] found that the weld bead geometry can be regulated by changing the spatial and temporal distribution of laser energy, such as power density, interaction time, and energy density. Tchoffo Ngoula [7] studied the influence of weld toe geometry on the fatigue life of cruciform welded joints. The result showed that the flank angle has a significant influence on the calculated fatigue life: the fatigue life decreases with increasing flank angle. Matsuoka [8] found that the weld geometry could affect the thermal deformation of thin stainless steel sheet in laser micro-welding and the smallest deformation could be achieved by forming a proper weld geometry. Furthermore, Okamoto [9] ensured that the smallest deformation was achieved around the maximum aspect ratio of the weld bead regardless of specimen thickness and spot diameter. Liu et al. [10] investigated the correlations between the microstructure and typical weld shapes of SUS201 including [peanut-shaped welds (PWs), nail-shaped welds (NWs), and wedge-shaped welds (WWs). The result showed that the three weld morphologies presented different microstructures and distributions of alloying elements.

Fiber laser welding has many advantages, such as high energy density, low heat input, small welding deformation, and flexibility [11–16]. Compared with Nd:YAG lasers, a fiber laser beam has a higher power density and larger Rayleigh length. Therefore, fiber lasers are more suitable for welding thin-walled metals [12,14,16–20]. Casalino et al. [18] studied the laser offset welding of AZ31B magnesium alloy to 316 stainless steel. The result showed that a thin intermetallic compounds was formed in the FZ and the ultimate tensile strength exceeded the value of 100 MPa. Campanelli et al. [20] used fiber laser welding to produce Ti6Al4V butt joints with a thickness of 2 mm and found that the increase of the welding speed led to restricted HAZ and FZ and a conic-shaped bead. In this work, fiber laser welding was used to conduct the experiment. In recent years, laser-arc hybrid welding also attracted researchers' interest [12,17,21]. Zhang et al. [12] compared the microstructure and properties of MIG welding joint and laser-MIG hybrid welding joints. In laser full-penetration welding of thin workpieces, there are two kinds of typical cross-section profiles of the fusion zone (FZ): V-shaped and H-shaped [22]. Previous studies revealed that the weld joint with a V-shaped FZ showed a lower tensile strength than that of a joint with an H-shaped FZ due to that the heterogeneity of strain distribution on the former was greater than that on the latter [23]. However, investigation on effects of aspect ratio of FZ on the mechanical properties of laser welded thin section joints with H-shaped cross-section was rather limited.

Magnesium alloys have been widely used in aviation, aerospace, transportation, electronics, and other industrial sectors, since magnesium alloys have high specific strength, high elastic modulus, and good vibration resistance, and also possess good electrical conductivity, thermal conductivity, electromagnetic shielding performance, and good recyclability. In this paper, the influences of the aspect ratio of FZ on the mechanical properties of laser-welded 1-mm thick AZ31 magnesium alloy joints with H-shaped FZ were investigated. The results of this study have certain guiding significance on optimizing the service performance of laser welded thin-wall AZ31 structures.

2. Materials and Methods

2.1. Materials and Welding Test

This research used 1-mm AZ31 magnesium alloy sheet and the base metal (BM) is composed of equiaxed grains, as shown in Figure 1. Its chemical composition has been shown in Table 1. Before the welding, the plates had been burnished by sandpaper and cleaned with acetone to remove the oxide film and oil on the surface. The welding equipment was an IPG YLS-4000 laser device produced by IPG Photonics Corporation (Oxford, MA, USA) and high-purity argon was used as the shielding gas in the LBW process. The welding direction was perpendicular to the rolling direction of the base metal. The rolling direction of the plate was determined according to the surface morphology of the plate. A set of orthogonal experiments were conducted from which two weld joints with quite different weld widths were chosen to compare the microstructure and properties. The parameters are shown in Table 2.

Figure 1. Microstructure of BM (base metal).

Table 1. Chemical composition of AZ31 magnesium alloy (wt %).

Al	Zn	Mn	Si	Fe	Cu	Ca	Be	Mg
3.19%	0.81%	0.34%	0.02%	0.005%	0.05%	0.04%	0.1%	Balance

Table 2. L25 (5^3) orthogonal test table.

Welded Joint Number	Welding Power/W	Welding Speed/m/min	Defocusing Amount/mm
1	500	2	−2
2	500	3.5	0
3	500	5	2
4	500	6.5	4
5	500	8	6
6	1000	2	0
7	1000	3.5	2
8	1000	5	4
9	1000	6.5	6
10	1000	8	−2
11	1500	2	2
12	1500	3.5	4
13	1500	5	6
14	1500	6.5	−2
15	1500	8	0
16	2000	2	4
17	2000	3.5	6
18	2000	5	−2
19	2000	6.5	0
20	2000	8	2
21	2500	2	6
22	2500	3.5	−2
23	2500	5	0
24	2500	6.5	2
25	2500	8	4

2.2. Microstructure Examination and Microhardness Test

The metallographic samples were polished and then etched with a kind of corrosive (4.2 g picric acid, 10 mL acetic acid, 10 mL diluted water and 70 mL ethanol) to reveal the microstructure. The microstructure of the weld joints was observed under a Nikon MA200 metallographic microscope (Nikon, Tokyo, Japan). The microhardness of the weld joints was measured using a Vickers hardness tester with the load of 0.49 N and a dwell time of 15 s.

2.3. Mechanical Property Test

The dimensions of tensile test specimen and fatigue test specimen are shown in Figure 2. The tensile test was carried out using a CSS-88100 universal testing machine (SINOMACH, Beijing, China) with a stretching rate 1 mm/min. The sinusoidal tensile-tensile load was employed in the fatigue test with the aid of an INSTRON fatigue testing machine. The maximum transient load was 1.44 KN and the minimum transient load was 0.144 KN. The stress ratio was $R = \sigma_{min}/\sigma_{max} = 0.1$ and the frequency was 15 HZ. The evolution of the fatigue crack were recorded by using a high speed camera and the shooting rate was 24 fps, as shown in Figure 2. The tensile fracture morphology and fatigue fracture morphology were observed with SEM (JEOL, Tokyo, Japan). Before the fatigue test, the specimens were polished smooth with a metallographic procedure.

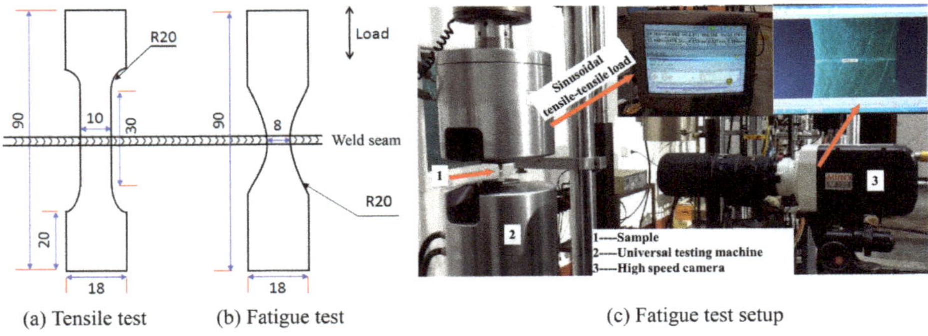

Figure 2. Details of the mechanical properties tests: (**a**) dimension of the tensile test specimens, (**b**) dimension of the fatigue test specimens, and (**c**) fatigue test setup.

3. Results and Discussion

3.1. Effect of Parameters on Weld Shape

Figure 3 shows the upper surface morphology of the welded joints achieved through the orthogonal experiments. Figure 4 shows the cross-sections of the welded joints. In order to investigate the influence of different aspect ratio on the property of welded joints of magnesium alloy, two welded joints with different aspect ratios and good shape quality were chosen. The welded joint 15 (Figure 3o) and welded joint 18 (Figure 3r) had a relatively good forming quality. Additionally, they presented different aspect ratios as showed in the Figure 4o,r. Thus, these two joints were chosen as the research objects. Joint 15, with a narrower fusion zone, was recorded as $W_{narrower}$ and joint 18 was recorded as W_{wider}.

3.2. Microstructure

Figure 5 shows the cross-section of welded joints $W_{narrower}$ and W_{wider}. The average weld width of joint $W_{narrower}$ is about 0.52 mm. However, the average weld width of W_{wider} is about 0.92 mm,

1.77 times of that of joint $W_{narrower}$. The aspect ratios of the two joints are 1.92 and 1.09, respectively. The difference was mainly caused by the different defocusing amount during welding process.

Figure 6c,f present the fusion boundary of joint $W_{narrower}$ and W_{wider}, respectively. On the base metal side are the equiaxed grains and they did not show any obvious growth compared with the microstructure of the BM (Figure 1). Some equiaxed grains next to the fusion boundary precipitated some fine intermetallic phases. On the fusion zone side is the transition zone. This zone is composed of columnar grains as shown in Figure 6d,g. This is mainly because of the large temperature gradient around the fusion boundary which resulted from the high energy density of laser and good thermal conductivity of magnesium alloys. The center of the fusion zone is composed of equiaxed grains as shown in Figure 6e,h. The temperature gradient of the center of fusion zone was relatively smooth causing the formation of equiaxed crystals. During the welding process, many particles precipitated inside the fusion zone which are composed of Mg, Al, and Zn [24]. It is mainly because that in the solidification process, the alloying elements dissolved in the base metal diffused and gathered together which caused the formation of precipitations. Due to a larger defocusing amount, both the columnar grain zone and equiaxed grain zone of the joint W_{wider} are wider than that of joint $W_{narrower}$. The size of grains of the two welded joints are similar. Compared with the base metal, the equiaxed grains in the fusion zone are obviously larger.

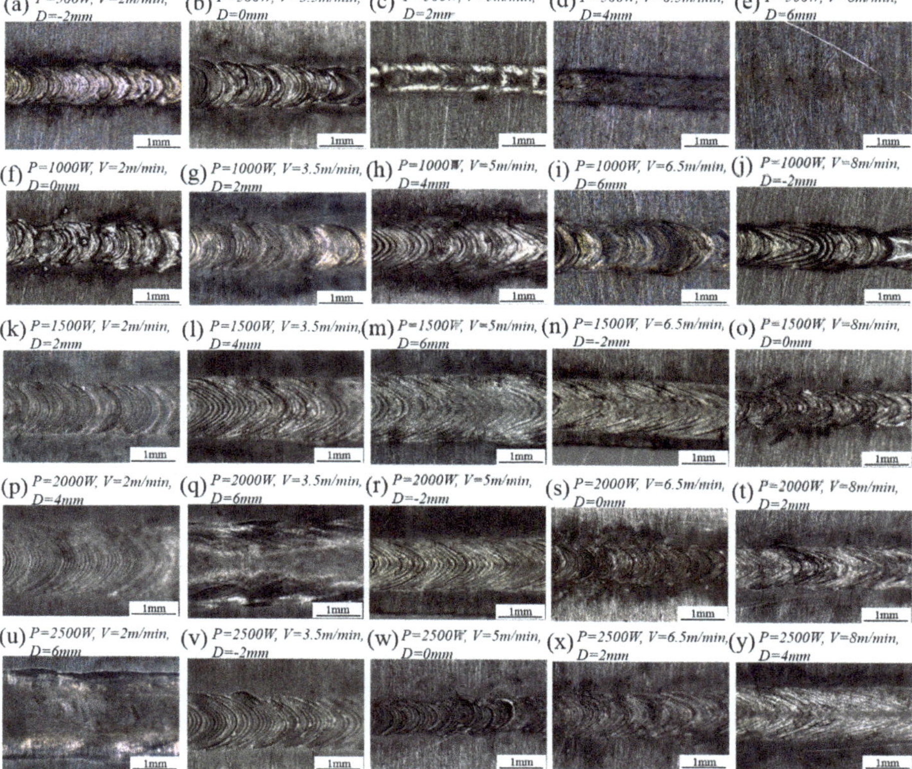

Figure 3. The upper surface morphologies and corresponding welding parameters of the welded joints achieved through the orthogonal experiments: (**a**–**y**) the upper surface morphologies and welding parameters of welded joints 1–25 in Table 2, respectively.

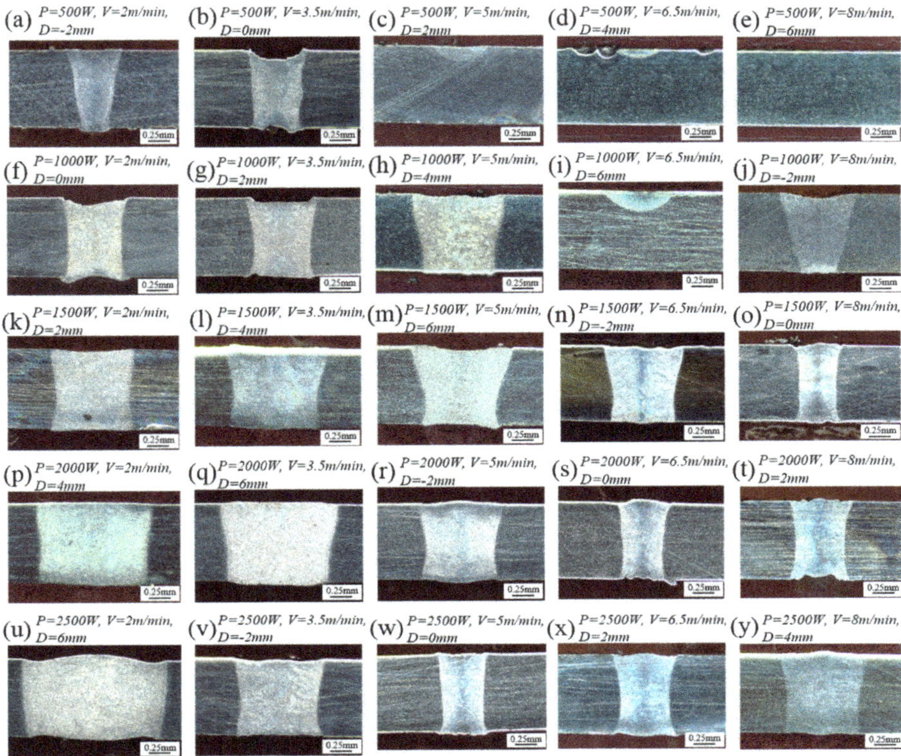

Figure 4. The cross-section morphologies and corresponding welding parameters of the welded joints achieved through the orthogonal experiments: (**a–y**) the cross-section morphologies and welding parameters of welded joints 1–25 in Table 2, respectively.

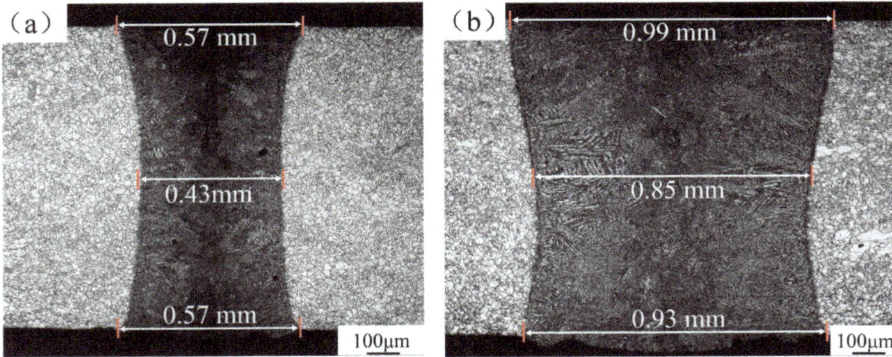

Figure 5. The width of the fusion zone of welded joints: (**a**) $W_{narrower}$ (P = 1500 W, V = 8 m/min, f = 0 mm) and (**b**) W_{wider} (P = 2000 W, V = 5 m/min, f = 2 mm).

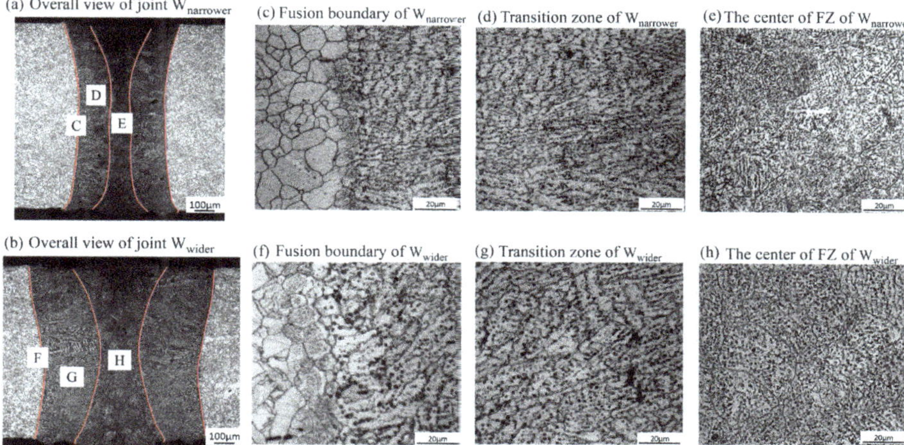

Figure 6. The microstructure of welded joints: (**a**, **b**) overall view of joint $W_{narrower}$ and W_{wider}; and (**c–h**) high-resolution images of positions C–H in panel (**a**, **b**).

3.3. Microhardness

Figure 7 shows the microhardness distribution profile along the center line in the direction of the plate thickness. It can be seen from Figure 7a that the microhardness around the FZ is similar to the microhardness far away from the FZ. The hardness of this material can be affected by both the grain size and the precipitated phases. Although the grain size of the fusion zones is larger than the base metal, many precipitated particles occurred in the FZ. According to the Orowan hardening mechanism [25], the hardness can benefits from the small particles of intermetallic compounds. In addition, there is no obvious difference between the microhardness distribution of the two welded joints, which can be attributed to the similar microstructure of the FZs.

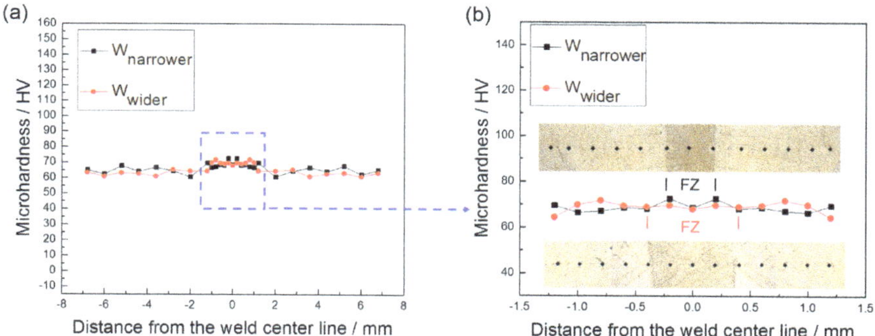

Figure 7. The microhardness (HV) distribution profile along the center line in the direction of the plate thickness: (**a**) overview of the microhardness distribution profile, and (**b**) the microhardness distribution around the FZ.

3.4. Tensile Test

Figure 8 shows the tensile test results of the base metal and the two joints. The tensile strengths of $W_{narrower}$ and W_{wider} are very close to that of base metal, which is about 97.1% and 97.8% that of the base metal, respectively. The high tensile strength of the welded joint can attribute to the precipitated phases. It can be seen from Figure 6 that most precipitated phases occurred in the grain and distributed uniformly. These phases can compensate the decrease of mechanical properties caused by the coarse grain size in the FZ by hinting dislocation movement. The elongation rate of the welded joint W_{wider} is about 80.9% that of the base metal, slightly larger than that of the welded joint $W_{narrower}$, which is up to 72% that of base metal. This might because that the joint W_{wider} has a larger weld width and can endure a greater deformation [26]. Both joints fractured in the FZ during tensile processes.

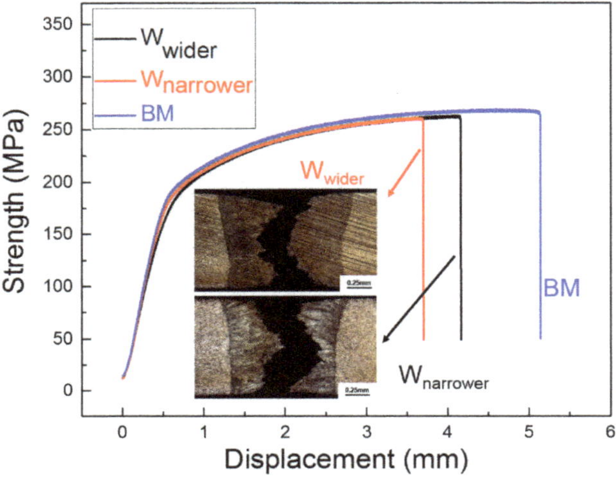

Figure 8. Tensile test results of joint $W_{narrower}$, joint W_{wider}, and BM.

Figure 9 shows the tensile fractures of base metal and two joints. The fracture of the base metal is composed of many small and dense dimples. However, the fracture of joint $W_{narrower}$ shows a cleavage fracture pattern. The fracture of joint W_{wider} shows characteristics of both of these two fracture patterns.

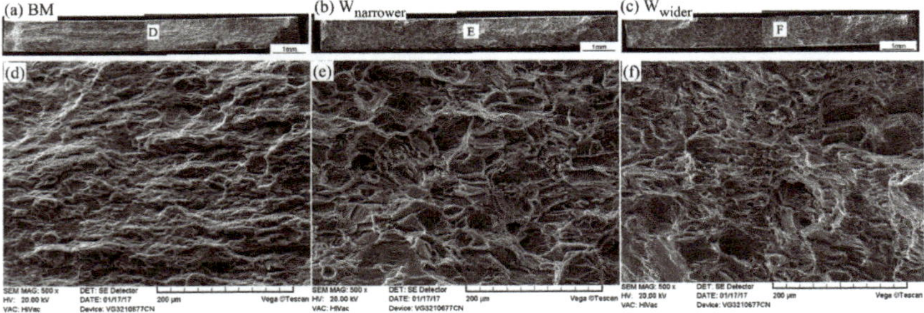

Figure 9. SEM (scanning electron microscope) images of the tensile fractures: (**a–c**) the overall view of BM, joint $W_{narrower}$, and joint W_{wider} respectively; (**d–f**) the high resolution images of region D–F in panel (**a–c**).

3.5. Fatigue Test

Figure 10 shows the crack evolution in the low-cycle fatigue test of base metal. The white dashed line shows the fatigue crack path. The crack of the base metal specimen initiated at the notch of the edge (Figure 10a) because of a great stress concentration. Subsequently, the crack propagated perpendicular to the loading direction. After a period of about 162.8 s, the crack propagated 13% the neck width of the specimen (Figure 10b) and from this time, the propagation rate became significantly larger. After 3.2 s, the crack propagated to 33% the whole crack path (Figure 10c). In the last 0.21 s, the crack propagated rapidly and the specimen broke completely (Figure 10d). The fatigue life of the base metal specimen was 3377.5 cycles.

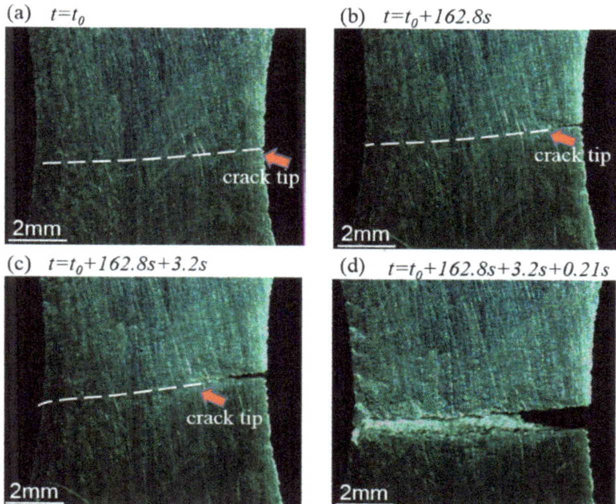

Figure 10. Crack evolution in the low-cycle fatigue test of base metal: (**a**–**d**) the status of the crack propagation at different times.

Figure 11 shows the crack evolution in the low-cycle fatigue test of the welded joint $W_{narrower}$. The fatigue crack initiated at the edge of the specimen (Figure 11a). It took 90 s to propagate through 14% of the whole path (Figure 11b) and during subsequent 0.83 s, the crack propagated rapidly and the specimen broke suddenly (Figure 11c). Compared with the base metal specimen, the fatigue crack propagation rate of joint $W_{narrower}$ was relatively larger and steadily increasing without mutation. It took less time for the fatigue crack propagation stage of joint $W_{narrower}$. Figure 12 shows the fracture path of the welded joint $W_{narrower}$ after the low-cycle fatigue test. It can be seen from the picture that the fatigue crack initiated and propagated inside the fusion zone. The fatigue life of joint $W_{narrower}$ was 2825 cycles, which reached 83.6% that of the base metal.

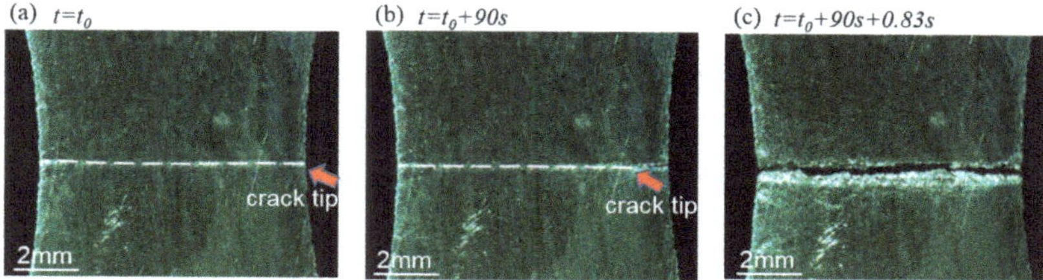

Figure 11. Crack evolution in the low-cycle fatigue test of the welded joint $W_{narrower}$: (**a**–**c**) the status of the crack propagation at different times.

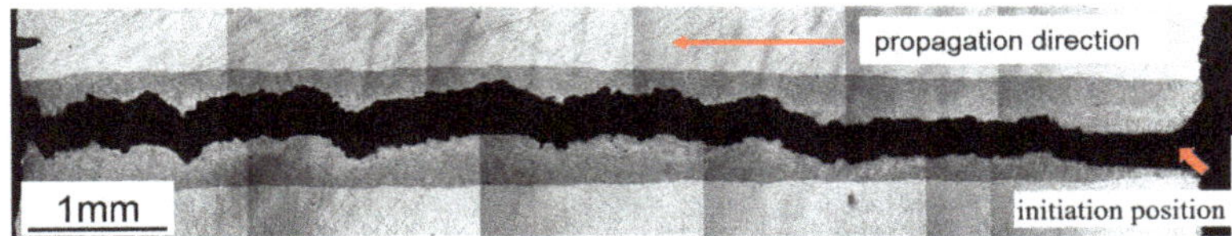

Figure 12. The fracture path on the welded joint $W_{narrower}$ after low-cycle fatigue testing.

Figure 13 shows the crack evolution in the low-cycle fatigue test of joint W_{wider}. Different from the fatigue failure behavior of joint $W_{narrower}$, the crack of joint W_{wider} initiated at the edge of the base metal about 1 mm away from the fusion boundary (Figure 13a). This can be attributed to the better tensile ductility showed from the tensile test. The crack propagated slowly perpendicular to the loading direction for 98 s (Figure 13b), and then the propagation rate became larger. During the subsequent 0.67 s, the crack changed direction and propagated toward the fusion zone (Figure 13c). Then, it propagated perpendicular to the loading direction inside the FZ as the white dotted line shows in Figure 13c and finally caused the fracture of this specimen (Figure 13d). The fatigue life of joint W_{wider} was 3155.3 cycles, which was 93.4% that of the base metal, obviously superior to joint $W_{narrower}$. Figure 14 shows the fatigue crack path of joint W_{wider}. It can be seen that the fatigue crack propagated through the BM first and then through the fusion zone.

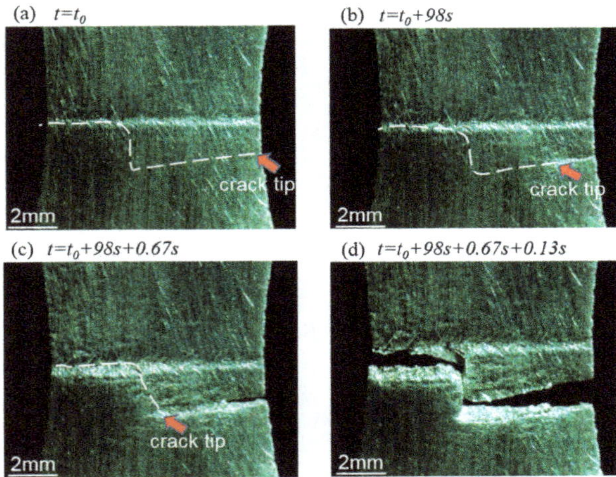

Figure 13. Crack evolution in the low-cycle fatigue test of the welded joint W_{wider}: (**a–d**) the status of the crack propagation at different times.

Figure 14. The fracture path on the weld joint W_{wider} after low-cycle fatigue testing.

Figures 15–17 show the evolution of fatigue cracks of BM, joint $W_{narrower}$, and joint W_{wider}. It can be seen that the initiation and propagation stage of joint $W_{narrower}$ and W_{wider} was relatively shorter than that of the BM and the fatigue crack of the joint $W_{narrower}$ propagated fastest among three specimens. The final rupture stage was very short compared with the initiation and propagation stage.

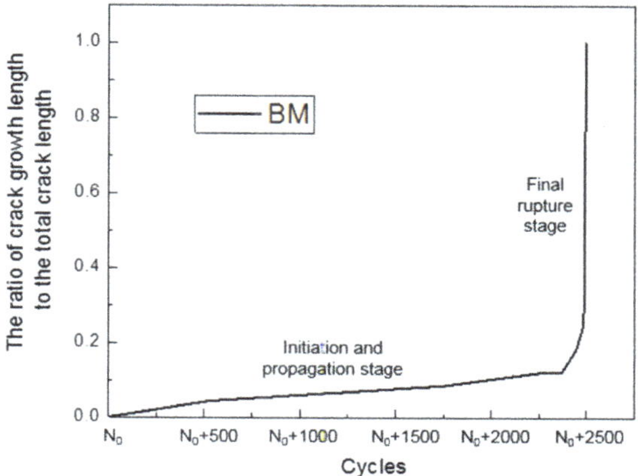

Figure 15. The evolution of fatigue cracks of BM in terms of cycles (N_0 represents where the crack of BM initiated).

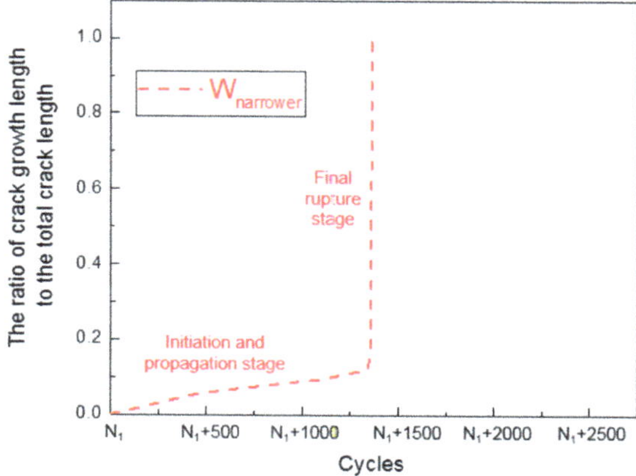

Figure 16. The evolution of fatigue cracks of joint $W_{narrower}$ in terms of cycles (N_1 represents where the crack of joint $W_{narrower}$ initiated).

Figure 18 presents the SEM images of the fatigue fracture. The fatigue crack initiation and stable propagation zones of the three samples is shown in the Figure 18a–c and marked with E, G, and J, respectively. The other regions are unstable crack propagation and final rupture zones. Figure 18e, Figure 18g,j are the SEM images with higher magnification for zones E, G, and J. It shows that the stable propagation zones of the three samples are characterized by a lamellar structure and the fracture surface is relatively flat. Additionally, it is revealed in the overall morphologies that the crack initiation and stable propagation zones of BM are larger than that of joint $W_{narrower}$, which indicate that the stable crack propagation stage of FZ is short than that of the BM. Figure 18d,f are the SEM images with higher magnification for zones D and F. It is revealed that the unstable crack propagation zone and final rupture zone of BM are characterized by small and shallow dimples. However, that of FZ is characterized by a tearing ridge which is similar to the fracture morphology observed in the tensile test. This indicates that the FZ has a lower capability of preventing the propagation of the fatigue crack than the BM. Figure 18h,i are the SEM images with higher magnification for zone H and zone I. The zone I is on the base metal part of crack path of joint W_{wider} and shows identical morphology with the BM. Zone H is on the fusion zone part of the crack path of joint W_{wider} and presents similar morphology with joint $W_{narrower}$.

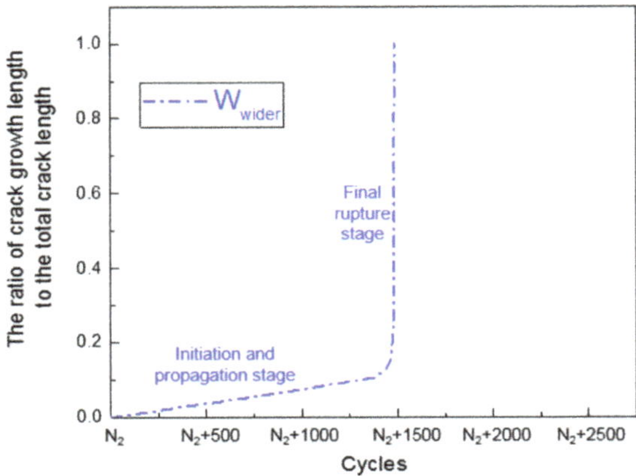

Figure 17. The evolution of fatigue cracks of joint W_{wider} in terms of cycles (N_2 represents where the crack of joint W_{wider} initiated).

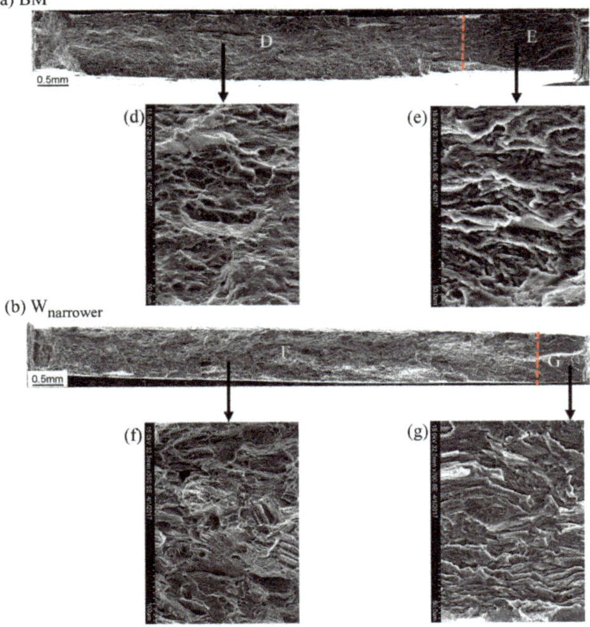

Figure 18. *Cont.*

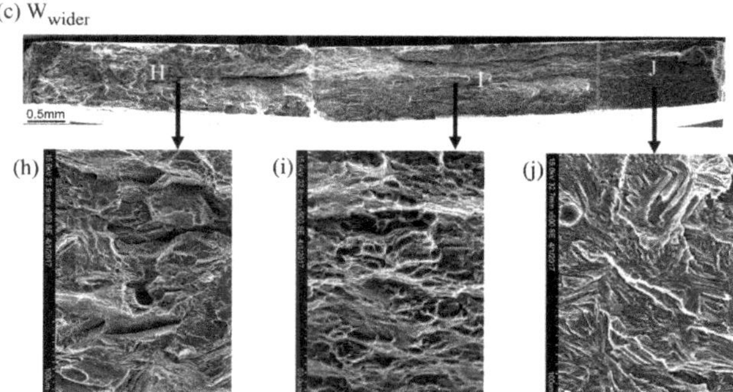

Figure 18. The SEM images of fatigue fracture: (**a–c**) overall view of BM, joint $W_{narrower}$ and W_{wider}; and (**d–j**) typical morphologies of region D–J in panel (**a–c**).

4. Conclusions

In this study, tensile tests and fatigue tests were repeated only two times. We have to recognize that this is an obvious limitation of the experimental work. We will pay attention to it in future studies. The major conclusions are as follows:

1. The fusion zone of the welded joint is composed of two different zones. The transition zone is composed of columnar grains and are distributed on both sides of the fusion zone. The center of the fusion zone is composed of equiaxed grains. Many precipitated phases were found on the fusion zone. The equiaxed grains next to the fusion boundary on the base metal side did not show obvious growth. The fusion zones of the two welded joints show similar microhardness.
2. The tensile strengths of both two joints exceed 90% that of base metal. The elongation rate of the joint W_{wider} is about 80.9% that of the base metal slightly larger than that of the joint $W_{narrower}$ which is up to 72% that of the base metal. The fracture of the base metal is composed of many small and dense dimples. However, the fracture of joint $W_{narrower}$ shows a cleavage fracture pattern. The fracture of joint W_{wider} shows characteristics of both of these two fracture patterns.
3. The low-cycle fatigue life of the base metal, the joint $W_{narrower}$ and the joint W_{wider} are 3377.5 cycles, 2825 cycles and 3155.3 cycles, respectively. The initiation and propagation of the fatigue crack of joint $W_{narrower}$ is inside the fusion zone. However, the fatigue crack of the joint W_{wider} initiated at the edge of the base metal and propagated for a distance on the base metal before propagating inside the fusion zone.

In summary, the microstructure and microhardness distribution of the joints W_{wider} and $W_{narrower}$ were similar to each other. The ductility and fatigue life of the joint W_{wider} was slightly larger than that of the joint $W_{narrower}$. The result of this research can provide reference for optimizing the service performance of laser welded AZ31 thin-walled structures used in the fields of aerospace, automobile, light railway traffic, etc. Certainly, to achieve high-quality products of thin-walled AZ31 structures, there are still many topics should be investigated in the future, such as the size effects of the H-shaped fusion zone of AZ31 joint on its corrosion resistance, flame resistance, high-cycle fatigue properties, and so on.

Acknowledgments: The authors wish to thank Xue-Wu Wang from School of Information Science and Engineering of East China University of Science and Technology for his valuable comments and advices. This work was supported by National Natural Science Foundation of China (grant no. 61773165 and grant no. 51275391).

Author Contributions: Lin-jie Zhang and Jie Ning conceived and designed the experiments; Yi Pei and Jian-Xun Zhang analysed the experimental data; Guang-Feng Lu did the experiments and wrote the paper.

References

1. Benyounis, K.Y.; Olabi, A.G.; Hashmi, M.S.J. Effect of laser welding parameters on the heat input and weld-bead profile. *J. Mater. Process. Technol.* **2005**, *164*, 978–985. [CrossRef]
2. Hann, D.B.; Iammi, J.; Folkes, J. A simple methodology for predicting laser-weld properties from material and laser parameters. *J. Phys. D Appl. Phys.* **2011**, *44*, 445401. [CrossRef]
3. Li, S.; Chen, G.; Zhou, C. Effects of welding parameters on weld geometry during high-power laser welding of thick plate. *Int. J. Adv. Manuf. Technol.* **2015**, *79*, 177–182. [CrossRef]
4. Kim, J.; Ki, H. Scaling law for penetration depth in laser welding. *J. Mater. Process. Technol.* **2014**, *214*, 2908–2914. [CrossRef]
5. Volpp, J.; Vollertsen, F. Analytical Modeling of the Keyhole Including Multiple Reflections for Analysis of the Influence of Different Laser Intensity Distributions on Keyhole Geometry. *Phys. Procedia* **2013**, *41*, 453–461. [CrossRef]
6. Ayoola, W.A.; Suder, W.J.; Williams, S.W. Parameters controlling weld bead profile in conduction laser welding. *J. Mater. Process. Technol.* **2017**, *249*, 522–530. [CrossRef]
7. Ngoula, D.T.; Beier, H.T.; Vormwald, M. Fatigue crack growth in cruciform welded joints: Influence of residual stresses and of the weld toe geometry. *Int. J. Fatigue* **2016**, *101*, 253–262. [CrossRef]
8. Matsuoka, S.; Okamoto, Y.; Okada, A. Influence of Weld Bead Geometry on Thermal Deformation in Laser Micro-Welding. *Procedia CIRP* **2013**, *6*, 492–497. [CrossRef]
9. Okamoto, Y.; Matsuoka, S.; Otowa, T.; Okada, A. Influence of Bead Geometry on Weld Distortion in Laser Micro-welding of Thin Stainless Steel Sheet with High-speed Scanning. *Int. J. Electr. Mach.* **2017**, *20*, 9–15. [CrossRef]
10. Liu, S.; Mi, G.; Yan, F.; Wang, C.; Jiang, P. Correlation of high power laser welding parameters with real weld geometry and microstructure. *Opt. Laser Technol.* **2017**, *94*, 59–67. [CrossRef]
11. Zhang, M.; Chen, G.; Zhou, Y.; Li, S. Direct observation of keyhole characteristics in deep penetration laser welding with a 10 kW fiber laser. *Opt. Express* **2013**, *21*, 19997–20004. [CrossRef] [PubMed]
12. Zhang, L.J.; Bai, Q.L.; Ning, J.; Wang, A.; Yang, J.N.; Yin, X.Q.; Zhang, J.X. A comparative study on the microstructure and properties of copper joint between MIG welding and laser-MIG hybrid welding. *Mater. Des.* **2016**, *110*, 35–50. [CrossRef]
13. Pang, S.; Chen, X.; Shao, X.; Gong, S.; Xiao, J. Dynamics of vapor plume in transient keyhole during laser welding of stainless steel: Local evaporation, plume swing and gas entrapment into porosity. *Opt. Lasers Eng.* **2016**, *82*, 28–40. [CrossRef]
14. Zhou, L.; Li, Z.Y.; Song, X.G.; Tan, C.W.; He, Z.Z.; Huang, Y.X.; Feng, J.C. Influence of laser offset on laser welding-brazing of Al/brass dissimilar alloys. *J. Alloy. Compd.* **2017**, *717*, 78–92. [CrossRef]
15. Pang, S.; Chen, X.; Li, W.; Shao, X.; Gong, S. Efficient multiple time scale method for modeling compressible vapor plume dynamics inside transient keyhole during fiber laser welding. *Opt. Laser Technol.* **2016**, *77*, 203–214. [CrossRef]
16. Guo, W.; Crowther, D.; Francis, J.A.; Thompson, A.; Liu, Z.; Li, L. Microstructure and mechanical properties of laser welded S960 high strength steel. *Mater. Des.* **2015**, *85*, 534–548. [CrossRef]
17. Leo, P.; Renna, G.; Casalino, G.; Olabi, A.G. Effect of power distribution on the weld quality during hybrid laser welding of an Al–Mg alloy. *Opt. Laser Technol.* **2015**, *73*, 118–126. [CrossRef]
18. Casalino, G.; Guglielmi, P.; Lorusso, V.D.; Mortello, M.; Peyre, P.; Sorgente, D. Laser offset welding of AZ31B magnesium alloy to 316 stainless steel. *J. Mater. Process. Technol.* **2017**, *242*, 49–59. [CrossRef]
19. Casalino, G.; Mortello, M.; Campanelli, S.L. Ytterbium fiber laser welding of Ti6Al4V alloy. *J. Manuf. Process.* **2015**, *20*, 250–256. [CrossRef]
20. Campanelli, S.L.; Casalino, G.; Mortello, M.; Angelastro, A.; Ludovico, A.D. Microstructural Characteristics and Mechanical Properties of Ti6Al4V Alloy Fiber Laser Welds. *Procedia CIRP* **2015**, *33*, 428–433. [CrossRef]
21. Casalino, G.; Campanelli, S.L.; Ludovico, A.D. Laser-arc hybrid welding of wrought to selective laser molten stainless steel. *Int. J. Adv. Manuf. Technol.* **2013**, *68*, 209–216. [CrossRef]
22. Krasnoperov, M.Y.; Pieters, R.R.G.M.; Richardson, I.M. Weld pool geometry during keyhole laser welding of thin steel sheets. *Sci. Technol. Weld. Join.* **2014**, *9*, 501–506. [CrossRef]

23. Gao, X.L.; Zhang, L.J.; Liu, J.; Zhang, J.X. Effects of weld cross-section profiles and microstructure on properties of pulsed Nd:YAG laser welding of Ti6Al4V sheet. *Int. J. Adv. Manuf. Technol.* **2014**, *72*, 895–903. [CrossRef]

24. Quan, Y.J.; Chen, Z.H.; Gong, X.S.; Yu, Z.H. Effects of heat input on microstructure and tensile properties of laser welded magnesium alloy AZ31. *Mater. Charact.* **2008**, *59*, 1491–1497. [CrossRef]

25. Wang, X.; Wang, K. Microstructure and properties of friction stir butt-welded AZ31 magnesium alloy. *Mater. Sci. Eng. A* **2006**, *431*, 114–117.

26. Ning, J.; Zhang, L.J.; Bai, Q.L.; Yin, X.Q.; Niu, J.; Zhang, J.X. Comparison of the microstructure and mechanical performance of 2A97 Al-Li alloy joints between autogenous and non-autogenous laser welding. *Mater. Des.* **2017**, *120*, 144–156. [CrossRef]

Article

Microstructure and Properties of Hybrid Laser Arc Welded Joints (Laser Beam-MAG) in Thermo-Mechanical Control Processed S700MC Steel

Jacek Górka [1,*] and Sebastian Stano [2]

1   Welding Department, Silesian University of Technology, Konarskiego 18A, 44-100 Gliwice, Poland
2   Welding Institute in Gliwice, Bl. Czeslawa 16-18, 44-100 Gliwice, Poland; sebastian.stano@is.gliwice.pl
*   Correspondence: jacek.gorka@polsl.pl; Tel.: +48-32-237-1445

Received: 27 December 2017; Accepted: 9 February 2018; Published: 15 February 2018

Abstract: The article presents the microstructure and properties of joints welded using the Hybrid Laser Arc Welding (HLAW) method laser beam-Metal Active Gas (MAG). The joints were made of 10-mm-thick steel S700MC subjected to the Thermo-Mechanical Control Process (TMCP) and characterised by a high yield point. In addition, the welding process involved the use of solid wire GMn4Ni1.5CrMo having a diameter of 1.2 mm. Non-destructive tests involving the joints made it possible to classify the joints as representing quality level B in accordance with the ISO 12932 standard. Destructive tests of the joints revealed that the joints were characterised by tensile strength similar to that of the base material. The hybrid welding (laser beam-MAG) of steel S700MC enabled the obtainment of good plastic properties of welded joints. In each area of the welded joints, the toughness values satisfied the criteria related to the minimum allowed toughness value. Tests involving the use of a transmission electron microscope and performed in the weld area revealed the decay of the precipitation hardening effect (i.e., the lack of precipitates having a size of several nm) and the presence of coagulated titanium-niobium precipitates having a size of 100 nm, restricting the growth of recrystallised austenite grains, as well as of spherical stable TiO precipitates (200 nm) responsible for the nucleation of ferrite inside austenite grains (significantly improving the plastic properties of joints). The tests demonstrated that it is possible to make welded joints satisfying quality-related requirements referred to in ISO 15614-14.

Keywords: steel S700MC; hybrid welding; HLAW; laser beam; MAG

---

1. Introduction

The use of advanced technologies in metallurgical processing as well as a new look at the significance and the role of alloying elements used in steels have enabled the fabrication of various groups of steels characterised by vast ranges of mechanical and plastic properties [1]. The development of new steel grades, particularly high-strength low alloys (HSLA) having ferritic, ferritic-pearlitic, ferritic-bainitic, bainitic, or tempered martensitic structures, has made it possible to significantly reduce the weight of structural elements and that of entire structures [2–4]. The reduction of thicknesses of steels manufactured in thermo-mechanical control processes (TMCP) (dictated by the needs of the automotive, ship-building, and petroleum industries) without compromising previously obtained performance characteristics has made it possible to achieve significant savings resulting from lower material processing and transport-related costs [5–11]. Advanced TMCP steels must satisfy not only strength-related but also environmental and social criteria [12]. In the case of the arc welding and laser welding of thermomechanically rolled steels, joints with insufficient impact strength are obtained, but these deficiencies can be eliminated during the hybrid welding process [13–15]. During welding, the material's microstructure, which provides high strength and plastic properties and is achieved by thermo-mechanical rolling, is destroyed. In weld metal, the structure of cast metal with properties are based on the used welding method (welding parameters) and additional material. Therefore, there is a need for the assessment of the welding method and parameters' impact on welded joints from steels, which parameters are a result of TMCP properties, especially when unconventional heat sources, such as a laser or laser-arc hybrids, are used [16–19].

The development of the above-named steels entails research works concerning various technologies enabling the joining of such steels [20–22]. Increasingly often, metals and their alloys are joined using highly efficient welding processes enabling the obtainment of high-quality welded joints accompanied by high joining process efficiency and less labour. The above-presented approach has resulted in the improvement of existing welding methods and, among other things, in the development of hybrid welding (in the late 1990s). However, only recently has the hybrid welding process seen growing popularity in, e.g., the shipbuilding or automotive industries, and it is gradually replacing laser and arc welding.

The hybrid laser arc welding (HLAW) technology combines two conventional welding methods. This process involves the simultaneous use of a heat source in the form of a laser radiation beam and an electric arc. According to the PN-EN ISO 15614-14 standard, a welding process can be referred to as hybrid where two coupled heat sources are used to form one common weld pool (Figure 1). The combination of two independent welding methods into one hybrid process results in the synergic effect of two heat sources. Consequently, the hybrid welding process is characterised by advantages typical of both methods. In addition, the above-presented combination reduces or eliminates limitations and disadvantages related to the use of only one heat source. Controversially, the beginning of the development of the hybrid technology (combining two independent heat sources) in welding engineering is seen by many as the announcement (in 1972) made by a group of engineers from the Philips Research Laboratory, Eindhoven, Holland (led by W.G. Essers and A. C. Liefkens) concerning the development of a new torch combining functions of plasma welding (PAW) and Gas Metal Arc Welding (GMAW). Initial attempts involving the combination of the laser method with the arc process were conducted at the Imperial College in London in the 1970s. A group of scientists supervised by William Steen demonstrated unquestionable advantages resulting from the combination of a plasma arc with a CO_2 laser beam. The welding rate increased by 50–100%, whereas the penetration depth increased by 20% when compared with that obtained using one heat source, i.e., a laser beam. The process of hybrid welding can involve the use of two independent heat sources with two independent technological heads. However, the foregoing requires precise positioning and the use of a simple system enabling the synchronised activation of both heat sources. In addition, it is possible to use special heads dedicated to the hybrid process that ensure the appropriate positioning of the two heat sources. In the HLAW, method it is possible to use nearly any industrial laser, yet the recent significant development of solid-state lasers (disc and fibre lasers) has made them the most popular sources of laser radiation in the hybrid method. In addition, a plasma cloud formed during the hybrid-welding process is more transparent for the electromagnetic wave emitted by the above-named lasers (approximately 1 µm), nearly entirely regardless of the type of shielding gas used in the arc method. Arc methods usually used in the HLAW method are those where the electrode is simultaneously filler metal fed to the welding area in a continuous manner (Metal Inert Gas (MIG), Metal Active Gas (MAG)). The filler metal makes it possible to adjust the chemical composition of the weld through supplying appropriate alloying elements to the weld pool and ensures the proper course of the welding process when a gap between joined elements is present (in terms of laser welding, there should be no gap between elements to be joined). The laser beam enables the obtainment of deep penetration using low linear energy, stabilises the arc, and improves the thermal efficiency of the process. The electrode wire ensures the complete filling of the weld groove gap and the formation of excess weld metal. The process of hybrid welding can be particularly useful in large-size industrial-scale production, primarily because of a higher acceptable tolerance when preparing the elements to be welded, the possibility of joining sheets in one run, and a lower accuracy when positioning the sheets to be joined [23–31]. HLAW is a relatively new welding process which causes specific thermal conditions that result in a change of the crystallization mode and precipitation type and size in a welded joint. The aforementioned conditions have a substantial impact on the properties of thermo-mechanically rolled steel joints. Determination of HLAW welded joints' properties enables the application of contemporary, highly efficient welding technologies in an industrial environment. Industrial demand for research in the field of joining contemporary steels with recent welding processes was the basis for collaboration between the Silesian University of Technology and the Welding Institute.

In this study, the welding of novel hot-rolled 700 MPa tensile strength Nb-Ti-V microalloyed steels of 10 mm thickness was carried out by HLAW. The joint microstructure, hardness, strength, and impact toughness were examined, and face and root bend and radiographic tests were performed.

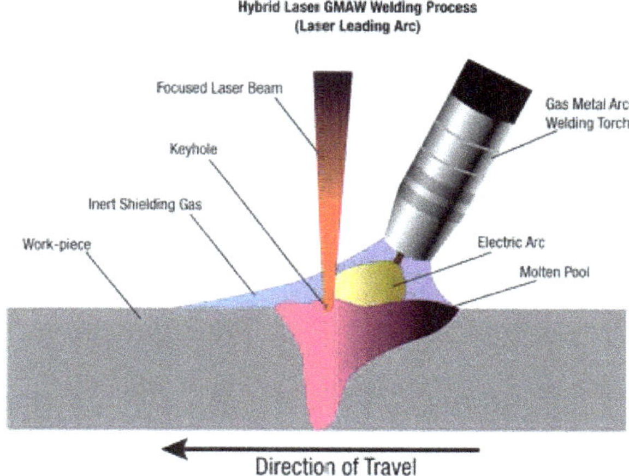

Figure 1. Schematic diagram of the hybrid welding process [26].

2. Experimental Section

The research-related tests aimed to identify the properties of butt hybrid welded (laser beam-MAG) joints made of 10-mm-thick steel S700MC using a copper strip and solid wire GMn4Ni1.5CrMo having a diameter of 1.2 mm. The chemical composition and the properties of the steel and weld deposit are presented in Tables 1 and 2, whereas the steel microstructure is presented in Figure 2.

Table 1. The real chemical composition performed using Optical Emission Spectroscopy (OES) and the mechanical properties of the original S700MC steel material.

Chemical Composition, wt %											
C	Si	Mn	P	S	Al$_{tot.}$	Nb	V	Ti	B	Mo.	Ce **
0.056	0.16	1.18	0.01	0.005	0.027	0.044	0.006	0.12	0.002	0.0150	0.33

Mechanical Properties				
Tensile Strength Rm, MPa	Yield Point Re, MPa	Elongation A$_5$, %	Hardness HV	Impact strength, J/cm^2 ($-20\,^\circ$C)
822	768	19	280	135

* Total amount of Nb, V, and Ti should amount to a maximum of 0.22%. ** Ce: carbon equivalent.

Table 2. Chemical composition and mechanical properties of filler metal GMn4Ni1.5CrMo.

Chemical Composition, wt %						
C	Mn	Si	Cr	Ni	Mo	Ti
0.1	1.8	0.7	0.3	2.0	0.55	0.07

Mechanical Properties			
Tensile Strength R_m, MPa	Yield Point R_e, MPa	Elongation A_5, %	Impact Strength, J/cm^2 ($-40\,^\circ$C)
900	810	18	55

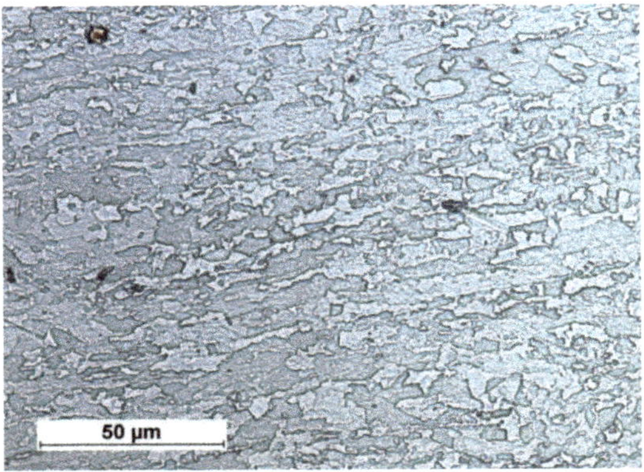

Figure 2. Microstructure of bainitic-ferritic steel S700MC with visible effects of plastic deformation.

The tests of thin foils performed using a transmission electron microscope revealed that the hardening of steel S700MC was primarily caused by dispersive (Ti,Nb)(C,N)-type precipitates (a few nm in size) formed in the steel ferrite during cooling (Figure 3). The growth of recrystallised austenite grains in the steel was reduced by spherical (Nb,Ti)C and (Ti,Nb)(C,N) carbide precipitates having diameters of between 10 and 50 nm (Figure 4).

Element	Mass Concentration, %	Atomic Concentration, %
N K	10.77	32.72
Ti K	56.27	49.96
Mn K	1.74	1.34
Fe K	5.48	4.18
Nb (K)	25.72	11.77

(a)

Figure 3. *Cont.*

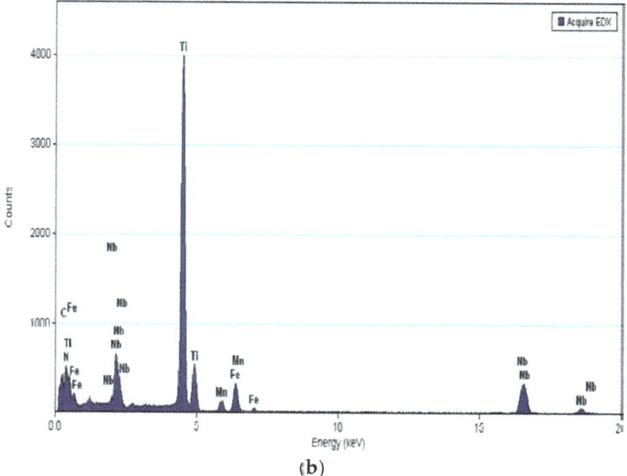

(b)

Figure 3. Precipitation of carbonitride (Ti,Nb)(C,N) with numerous small dispersive precipitates corresponding to the hardening of the steel. (**a**) recipitation of carbonitride (Ti,Nb)(C,N); (**b**) large particle-related EDX spectrum.

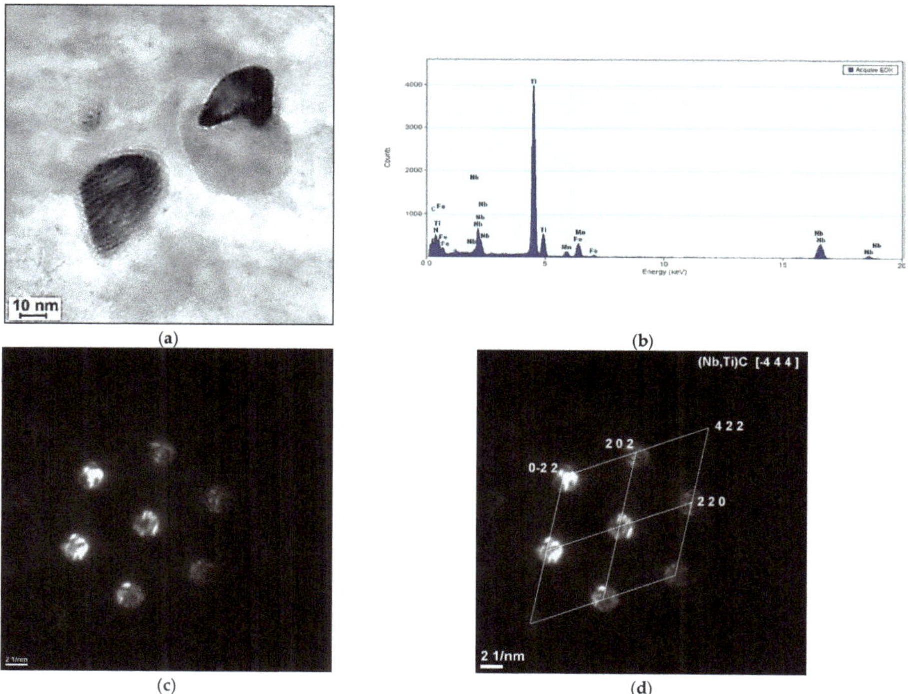

(a)

(b)

(c)

(d)

Figure 4. Dispersive precipitation of carbides (Ti,Nb)C in S700MC steel limiting the growth of recrystallized austenite grains. (**a**) dispersive precipitation of carbides (Ti,Nb)C; (**b**) EDX spectrum; (**c**) diffraction pattern; (**d**) diffraction pattern solution.

2.1. Welding Process

The welded joints were made at Welding Institute in Gliwice using a robotic station (Figure 5). The tests were performed using a TruLaser Robot (TRUMPF, Stuttgart, Germany) 5120 cell-equipped with a TruDisk 12002 disc laser (TRUMPF, Stuttgart, Germany) having a power of 12,000 W (wavelength λ = 1030 nm) and an EWM Phoenix 452 RC PULS (EWM AG, Mündersbach, Germany) synergic power source. The hybrid welding head (laser beam-MAG) was fixed on the robot's wrist. The focal length of the laser optics collimator was f_c = 200 mm, whereas the focal length of the focusing lens amounted to f_{foc} = 400 mm. The diameter of the optical fibre supplying energy from the laser to the robot was d_{fiber} = 0.4 mm. The above-presented arrangement of the optics made it possible to obtain a laser beam focus diameter d_{foc} = 0.8 mm (Figure 6). The electrode extension was l = 18 mm. The shielding gas used in the tests was mixture M21 (18% CO_2 + 82% Ar), whereas the gas flow rate amounted to 18 dm^3/min. The filler metal was solid wire GMn4Ni1.5CrMo having a diameter of 1.2 mm. The welding process was carried out in one pass. The electrode was inclined in relation to the welded surface at angle α = 65°, whereas the distance between the electrode tip and the laser beam was a = 2 mm (Figure 6). The parameters used when making the joint (adjusted on the basis of preliminary tests [32,33]) are presented in Table 3.

Figure 5. Hybrid welding station (laser beam-Metal Active Gas (MAG)).

Table 3. Parameters of the hybrid welding of 10-mm-thick steel S700MC.

Pre-Weld Metal Preparation	Welding Sequence
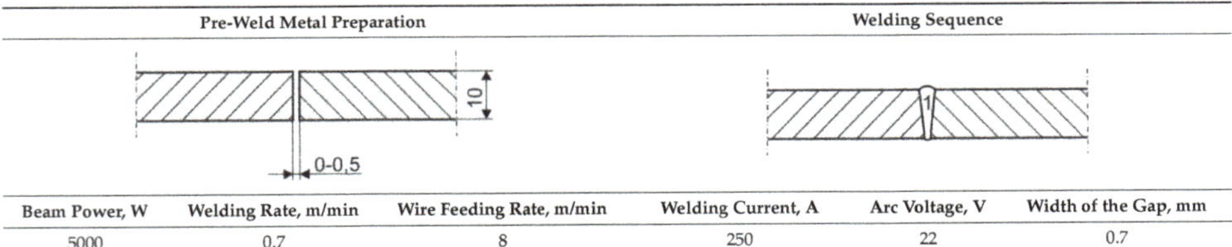	

Beam Power, W	Welding Rate, m/min	Wire Feeding Rate, m/min	Welding Current, A	Arc Voltage, V	Width of the Gap, mm
5000	0.7	8	250	22	0.7

Shielding gas: C18 ferromix; filler metal wire: GMn4Ni1.5CrMo having a diameter of 1.2 mm; welding performed using a copper strip: a rectangular groove; dimension of welded joints: 350 mm × 150 mm × 10 mm; fixing with metalwork clamps; a copper washer on the side of the weld ridge; positioning of the laser beam in the axis of the welding groove.

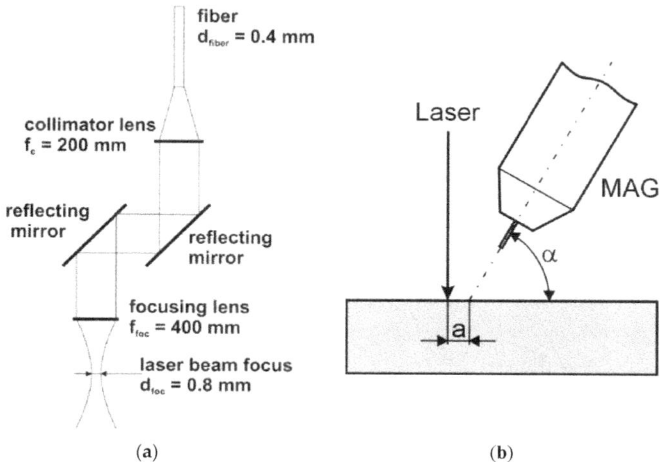

Figure 6. Diagram of the laser optical system and the position of the laser beam and electric arc in the hybrid technology. (**a**) the laser optical system; (**b**) position of the laser beam and electric arc.

2.2. Tests of Welded Joints

The test welded joint was subjected to the following non-destructive tests:

- visual tests performed on the basis of the requirements specified in the PN-EN ISO 17637:2011 standard;
- magnetic particle tests performed following the guidelines referred to in the PN-EN ISO 3059:2005, PN-EN ISO 9934-2:2003, and PN-EN ISO 9934-3:2003 standards. The necessary contrast was obtained using white contrast paint MR 72. The tests were performed using magnetic powder suspension MR 76S (MR International, Fränkisch-Crumbach, Germany) and a yoke electromagnet;
- radiographic tests based on the PN-EN 1435 standard performed using a CERAM 235 X-ray tube (Balteau NDT, Hermalle-sous-argenteau, Belgium) with the X-ray beam having a diameter $d = 2$ mm, a voltage $U = 180$ kV, a current $I = 3$ mA, and intensifying screens OW of -0.15 mm. The test results were recorded using an AGFA C5 photographic plate with an exposure time $t = 2.3$ min and a focal length $f = 700$ mm. Images were assessed using a 13FEEN wire-type image quality indicator.

Following the non-destructive tests, the welded joint was subjected to the following destructive tests:

- tensile tests performed in accordance with PN-EN ISO 6892-1:2010 using a ZWICK/ROELL Z 330RED (Zwick Roell, Ulm, Germany) testing machine and specimens sampled in accordance with PN-EN ISO 4136:2011 (dimensions of the sample: 300 mm × 35 mm × 10 mm);
- a face bend test of the butt weld (FBB) and a root bend test of the butt weld (RBB) performed in accordance with the PN-EN ISO 5173:2010 standard (dimensions of the sample: 300 mm × 20 mm × 10 mm). The bend tests were performed using a ZWICK/ROELL Z 330RED testing machine (Zwick Roell, Ulm, Germany) with an additional module enabling the performance of bend tests involving the use of a bending mandrel having a diameter of 30 mm. The distance between the rollers was set at 60 mm. To identify the position of the weld axis, the faces of the specimens were etched using Adler's reagent;
- impact strength tests performed in accordance with PN-EN ISO 148-1:2010 using specimens with the V-notch and a ZWICK/ROELL RKP 450 impact testing machine (Zwick Roell, Ulm, Germany). The tests were conducted at a temperature of $-30\,°C$ (due to industrial requirements). Because of

the thickness of the plates being welded (10 mm) and the necessity of performing a preparatory mechanical treatment, the specimens were reduced in cross-section to 7.5 mm. The samples were extracted from the base metal, the heat-affected zone (HAZ), and the FL (fusion line), and the specimens were etched using Nital;

- macroscopic metallographic tests performed using an Olympus SZX9 light stereoscopic microscope (Olympus, Tokyo, Japan); the test specimens were etched using Adler's reagent (CHMES, Poznań, Poland);
- microscopic metallographic tests performed using a NIKON ECLIPSE MA100 light microscope (Nikon, Tokyo, Japan); the test specimens were etched using Nital;
- hardness measurements performed using a Vickers 401MVD hardness testing machine (Wilson Wolpert, Norwood, Massachusetts, USA) and a load of 1 kg;
- X-ray phase analysis performed using an X'Pert PRO diffractometer and an X'Celerator strip detector (PANalytical, Almelo, The Netherlands);
- tests of thin foils performed using a Titan 80–300 kV (FEI) high-resolution scanning transmission electron microscope (HR S/TEM, Thermo Fisher Scientific, Waltham, MA, USA) provided with an XFEG electron gun with the Schottky field emission characterised by enhanced brightness.

Samples for destructive testing were prepared by machining.

3. Results and Discussion

The visual tests and the magnetic particle tests of the welded joint did not reveal surface-breaking welding imperfections, such as cracks, porosity, incomplete fusion, or a lack of penetration (Figure 7). The radiographic tests did not reveal the presence of internal welding imperfections. The welded joint satisfied the requirements related to quality level B according to ISO 12932. The macroscopic metallographic tests did not reveal the presence of welding imperfections in the weld and HAZ (Figure 8). Excess penetration in the weld root was related to the shape and dimensions of the copper strip used during welding.

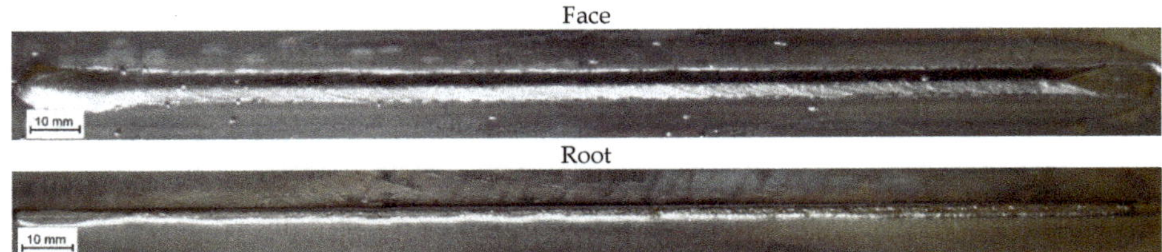

Figure 7. The face and the root of the hybrid welded joint (laser beam-MAG) in steel S700MC.

The microscopic metallographic tests revealed a bainitic-ferritic microstructure in the weld area. The heat-affected zone (HAZ) was characterised by variably sized grains, which could be ascribed to a significant heat input during the hybrid welding process (Figure 9). In addition, the microscopic tests revealed the probability of nitride precipitates' presence in the HAZ and in the base material; this is evidenced by their distinctive sharp shape (Figure 10). HLAW welding is characterized by relatively short cooling times $t_{8/5}$, which leads to the formation of martensite in the HAZ. However, as it is low-carbon martensite, it has no negative impact on plastic properties. In specific region of the HAZ with an increase of distance to the fusion line, ferrite content is increasing in lieu of bainite.

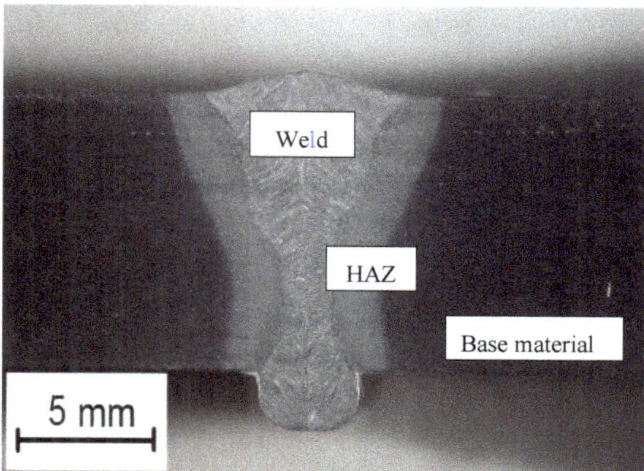

Figure 8. Macrostructure of the hybrid welded joint (laser beam-MAG) made in steel S700MC. HAZ: heat-affected zone.

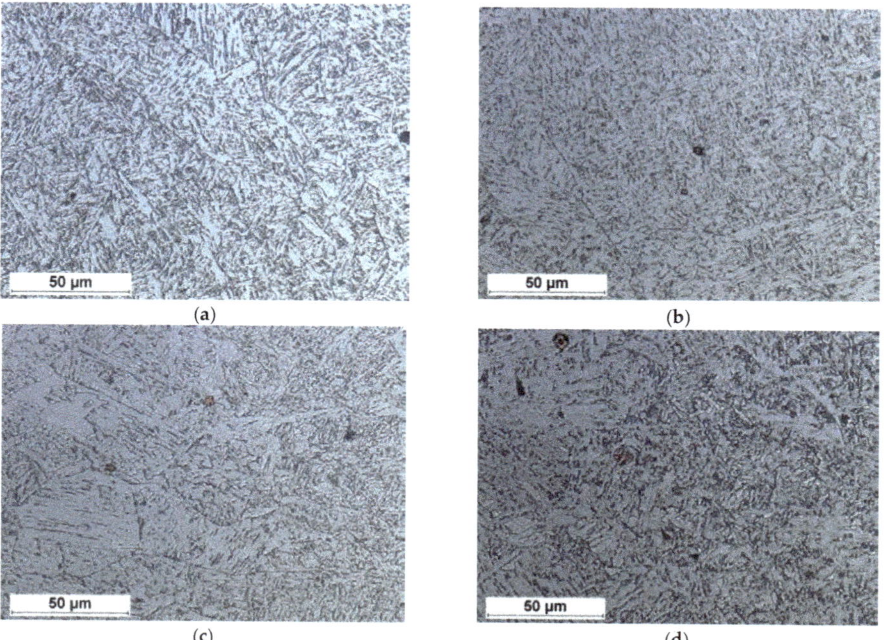

Figure 9. *Cont.*

Figure 9. Microstructure of the hybrid welded joint (laser beam-MAG). (**a**) microstructure of the upper part of the weld; (**b**) microstructure of the lower part of the weld; (**c**) microstructure of the upper part of the fusion line; (**d**) microstructure of the lower part of the fusion line; (**e**) microstructure of the coarse-grained part of the HAZ; (**f**) microstructure of the fine-grained part of the HAZ.

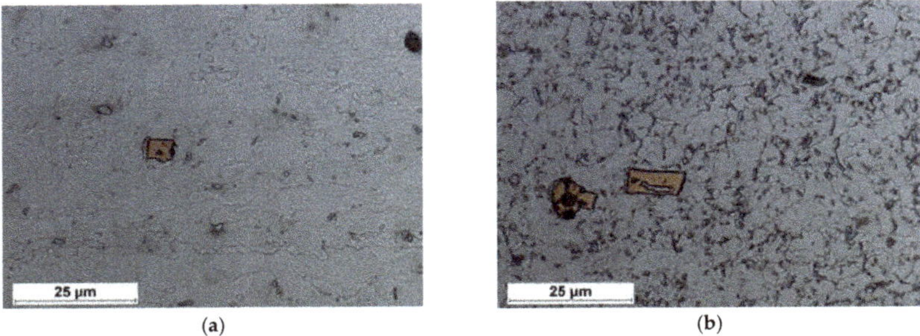

Figure 10. Nitride precipitates in the hybrid welded joint (laser-MAG) made of S700MC steel. (**a**) base material; (**b**) HAZ.

The analysis of the destructive test results of the hybrid butt welded joint (laser beam-MAG) revealed that the test joint satisfied the requirements of the ISO 15614-14 standard (Table 4). The hybrid welding process (laser beam-MAG) resulted in a slight decrease in tensile strength (to approximately 790 MPa) in relation to the hardness of the base material (820 MPa). The rupture took place in the HAZ (an area of slight grain growth). The above-named decrease in tensile strength was connected with the loss of properties obtained by steel S700MC through the thermo-mechanical control process. This is mainly related to the increase in the proportion of ferrite in the structure and the grain growth in this area. The bend test resulted in a bend angle of 180°, both during bending on the face and root side, which demonstrated the high plastic properties of the joint. The impact strength test performed at a temperature of -30 °C revealed satisfactory toughness values in the weld, fusion line, and the HAZ. In the weld area, the toughness amounted to 89 J/cm^2. In the fusion line area, the toughness decreased to approximately 50 J/cm^2. In the HAZ, the toughness amounted to approximately 40 J/cm^2, (samples for impact tests were cut across the welded joint). The toughness of the base material amounted to 50 J/cm^2.

Table 4. Strength and plastic properties of the hybrid weld joint (laser beam-MAG) made of S700MC steel.

Tensile Strength *		Bending *, Bend Angle, °		Impact Strength KCV **, J/cm2 (Test Temperature −30 °C)		
Rm, MPa	Area of Rupture	Face	Root	Weld	FL	HAZ
790	HAZ	180	180	89	51	42

FL: Fusion line; * average result of two measurements; ** average result of three measurements.

The hardness measurements concerning the hybrid welded joints made in steel S700MC revealed that the lowest hardness was characteristic of the heat-affected zone and amounted to approximately 227 HV, whereas the highest hardness was that of the base material and amounted, on average, to 280 HV. The difference between the hardness of the base material and that of the HAZ amounted to approximately 20%. The hardness value in the upper part of the weld was similar to that of the base material (280 H V1) and was higher than the value measured in the lower part of the weld (by approximately 8%). The foregoing could be attributed to the fact that the upper part of the weld contained more alloying elements, increasing the hardenability (nickel, chromium) supplied along with the filler metal (Figure 11).

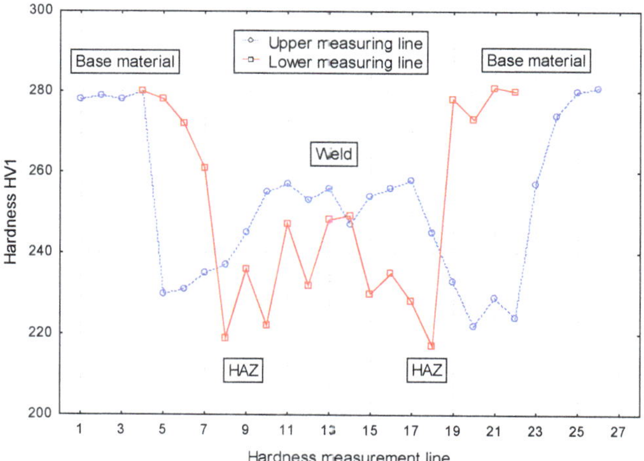

Figure 11. Hardness of the hybrid weld joint (laser beam-MAG) made of S700MC steel (position of measuring lines 2 mm from the top and bottom surface of the sheet).

The X-ray phase analysis revealed that the weld contained phase Feα and a slight amount of phase Feγ (Figure 12). The presence of phase Feγ could be attributed to the presence of austenitic alloying elements, e.g., Ni or C, in the weld deposit The analysis of the total intensity of X-radiation diffraction maxima from the lattice plane of phases Feα and Feγ of individual welded joints made it possible to determine that the content of retained austenite was restricted within the range of 3–6%.

The microscopic observations revealed that carbonitride precipitates (of several μm in size) in the weld area were dissolved entirely as a result of hybrid welding. The above-named observations were additionally confirmed by observations performed using the transmission electron microscope. The weld area was characterised by the decay of the precipitation hardening effect (lack of precipitates of several nm in size) and the presence of coagulated titanium-niobium precipitates of up to 500 nm (Figure 13), preventing the growth of recrystallised austenite and, consequently, improving the plastic properties of the weld. In addition, the weld area contained spherical and stable TiO precipitates (of 200 nm in size) responsible for the nucleation of ferrite inside austenite grains (Figure 14). The presence of the above-named ferrite translated into the high mechanical and plastic properties of the welded joints. This oxide is very stable even at high temperature and leads to ferrite formation inside coarse-grained HAZ grains.

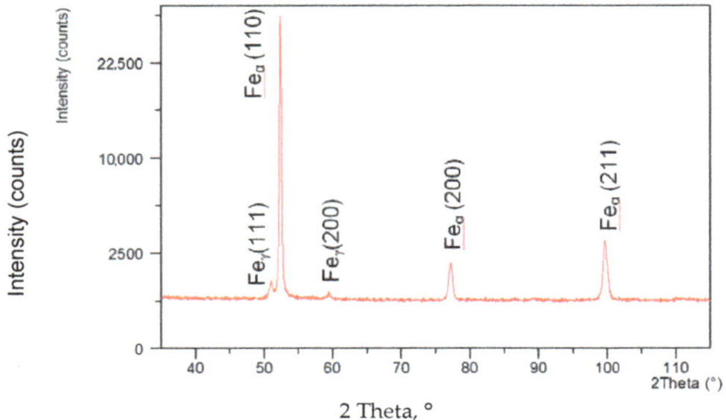

Figure 12. X-ray diffraction of the butt weld made in steel S700 MC.

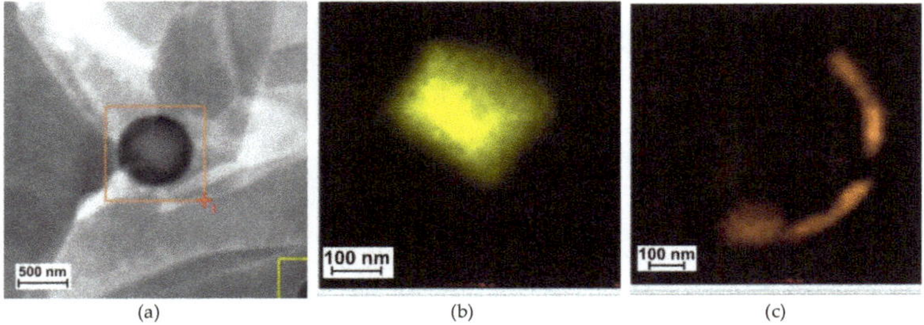

Figure 13. Titanium-niobium precipitates in the hybrid welded joint made of steel S700MC. (**a**) precipitate; (**b**) surface distribution of titanium; (**c**) surface distribution of niobium).

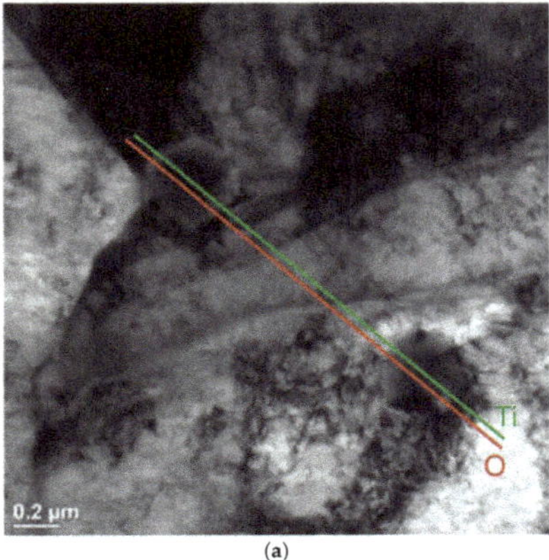

(**a**)

Figure 14. *Cont.*

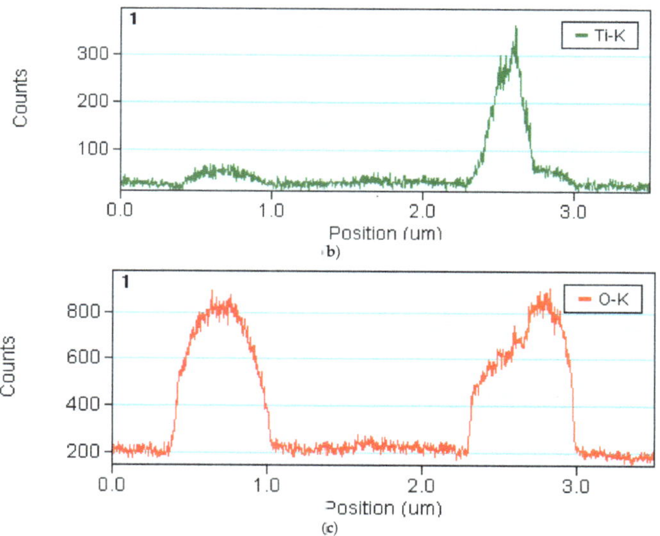

Figure 14. Spherical TiO precipitates in the hybrid welded joint made of steel S700MC. (**a**) precipitates; (**b**) linear analysis EDS of titanium concentration; (**c**) linear analysis EDS of oxygen concentration.

4. Conclusions

The tests of the hybrid welding (laser beam-MAG) of 10-mm-thick steel S700MC involving the use of a filler metal having the form of solid wire GMn4Ni1.5CrMo (1.2 mm in diameter) revealed the possibility of making welded joints satisfying the criteria formulated in the ISO 15614-14 standard. The test joints were characterised by a lack of welding imperfections as regards the shape, geometry, and discontinuity of the weld metal in the cross-section of the welded joint. The tensile strength of the welded joints was similar to that of the base material, whereas the plastic properties of the joints were satisfactory. The bend test performed both on the weld face and on the weld root side enabled the obtainment of a bend angle of 180°, whereas the toughness in the weld, fusion line, and in HAZ satisfied the criteria of the minimum yield point in relation to welded joints. The fusion line revealed a decrease in hardness, yet within a relatively narrow range and without compromising the operational properties of the welded joints. The microscopic tests of the welded joints revealed that the weld contained the typical bainitic-ferritic microstructure of dendritic nature. The heat-affected zone contained areas of variously sized grains, which was triggered by the thermal cycle effect. The HAZ and the base material revealed the presence of phases containing hardening microagents in the form of significant titanium nitride precipitates (indicated by their shape and the considerable content of titanium in the steel). The increase in the base material content in the weld was accompanied by the increase in the concentration of hardening microagents in the weld. The longer the time at which the material remained in the liquid state, the greater the amount of microagents which could dissolve in the matrix and re-precipitate (during cooling) or remain in the solution. The high temperature of the liquid metal pool could have resulted in the dissolution of even the most stable TiN particles. The cooling process did not provide appropriate conditions enabling the controlled re-precipitation of fine-dispersive carbides and carbonitrides (Ti,Nb) responsible for precipitation hardening. The welds made using the hybrid welding method were characterised by higher toughness resulting from the decay of the precipitation hardening effect (coagulation of precipitates) and the presence of spherical TiO precipitates responsible for the nucleation of ferrite inside austenite grains and, consequently, significantly improving the mechanical and plastic properties of the weld.

Acknowledgments: Work carried out as part of own research.

Author Contributions: Jacek Górka: assumptions, methodology, metallographic studies, analysis of test results, conclusions. Sebastian Stano: practical tests, optimization of welding parameters, destructive tests.

Conflicts of Interest: The authors declare no conflict of interest.

References

1. Flaxa, V.; Shaw, J. Material applications in ULSAB-AVC. *Steel Grips* **2003**, *1*, 255–261.
2. Opiela, M. Thermodynamic analysis of the precipitation of carbonitrides in microalloyed steels. *Mater. Tehnol.* **2015**, *49*, 395–401. [CrossRef]
3. Opiela, M. Elaboration of thermomechanical treatment conditions of Ti-V and Ti-Nb-V microalloyed forging steels. *Arch. Metall. Mater.* **2014**, *59*, 1181–1188. [CrossRef]
4. Grajcar, A. Thermodynamic analysis of precipitation processes in Nb-Ti-microalloyed Si-Al TRIP steel. *J. Therm. Anal. Calorim.* **2014**, *118*, 1011–1020. [CrossRef]
5. Mikia, C.; Homma, K.; Tominaga, T. High strength and high performance steels and their use in bridge structures. *J. Constr. Steel Res.* **2002**, *58*, 3–20. [CrossRef]
6. Lee, C.; Shin, H.; Park, K. Evaluation of high strength TMCP steel weld for use in cold regions. *J. Constr. Steel Res.* **2012**, *74*, 134–139. [CrossRef]
7. Ollilainen, V.; Hurmola, H.; Pontinen, H. Mechanical properties and machinability of a high-strength, medium-carbon, microalloyed steel. *J. Mater. Energy Syst.* **1984**, *5*, 222–232. [CrossRef]
8. Shipitsyn, S.; Babaskin, Y.; Kirchu, I.; Smolyakova, L.; Zolotar, N. Microalloyed steel for railroad wheels. *Steel Transl.* **2008**, *38*, 782–785. [CrossRef]
9. Rak, I.; Gliha, V.; Kocak, M. Weldability and toughness assessment of Ti-microalloyed offshore steel. *Metall. Mater. Trans.* **1997**, *28*, 199–206. [CrossRef]
10. Chang, K.; Lee, C.; Park, K.; Um, T. Experimental and numerical investigations on residual stresses in a multi-pass butt-welded high strength SM570-TMCP steel plate. *Int. J. Steel Struct.* **2011**, *11*, 315–324. [CrossRef]
11. Nishioka, K.; Ichikawa, K. Progress in termomechanical control of steel plates and their commercialization. *Sci. Technol. Adv. Mater.* **2012**, *13*, 023001. [CrossRef] [PubMed]
12. Grajcar, A.; Różański, M.; Kamińska, M.; Grzegorczyk, B. Effect of gas atmosphere on the non-metallic inclusions in laser-welded TRIP steel with Al and Si additions. *Mater. Tehnol.* **2016**, *50*, 945–950. [CrossRef]
13. Żuk, M.; Górka, J.; Czupryński, A.; Adamiak, M. Properties and structure of the weld joints of quench and tempered 4330V steel. *Metalurgija* **2016**, *55*, 613–616.
14. Górka, J. Microstructure and properties of the high-temperature (HAZ) of thermo-mechanically treated S700MC high-yield-strength steel. *Mater. Tehnol./Mater. Technol.* **2016**, *50*, 617–621. [CrossRef]
15. Górka, J. Welding Thermal Cycle-Triggered Precipitation Processes in Steel S700MC Subjected to the Thermo-Mechanical Control Processing. *Arch. Metall. Mater.* **2017**, *62*, 331–336. [CrossRef]
16. Burdzik, R.; Węgrzyn, T.; Konieczny, Ł.; Lisiecki, A. Research on influence of fatigue metal damage of the inner race of bearing on vibration in different frequencies. *Arch. Metall. Mater.* **2014**, *59*, 1275–1281. [CrossRef]
17. Kurc-Lisiecka, A.; Piwnik, J.; Lisiecki, A. Laser welding of new grade of advanced high strength steel STRENX 1100 MC. *Arch. Metall. Mater.* **2017**, *62*, 1651–1657. [CrossRef]
18. Górka, J.; Janicki, D.; Fidali, M.; Jamrozik, W. Thermographic Assessment of the HAZ Properties and Structure of Thermomechanically Treated Steel. *Int. J. Thermophys.* **2017**, *38*, 183. [CrossRef]
19. Janicki, D.; Górka, J.; Kwaśny, W.; Gołombek, K.; Kondracki, M.; Żuk, M. Diode laser surface alloying of armor steel with tungsten carbide. *Arch. Metall. Mater.* **2017**, *62*, 473–481. [CrossRef]
20. Grajcar, A.; Matter, P.; Stano, S.; Wilk, Z.; Różański, M. Microstructure and hardness profiles of bifocal laser-welded DP-HSLA steel overlap joints. *J. Mater. Eng. Perform.* **2017**, *26*, 1920–1928. [CrossRef]
21. Grajcar, A.; Morawiec, M.; Różański, M.; Stano, S. Twin-spot laser welding of advanced high-strength multiphase microstructure steel. *Opt. Laser Technol.* **2017**, *92*, 52–61. [CrossRef]
22. Kubiak, M.; Piekarska, W.; Stano, S.; Saternus, Z.; Domański, T. Numerical Prediction of Deformations in Laser Welded Sheets Made of X5CrNi18-10 Steel. *Arch. Metall. Mater.* **2015**, *60*, 1965–1972. [CrossRef]

23. Naito, Y.; Katayama, S.; Matsunawa, A. Keyhole behaviour and liquid flow in molten pool during laser-arc hybrid welding. In Proceedings of the International Congress on Laser Advanced Materials Processing (LAMP 2002), Osaka, Japan, 27–31 May 2002; Volume 4831, pp. 357–363.

24. Bagger, C.; Olsen, F. Comparison of plasma, metal inactive gas (MIG) and tungsten inactive gas (TIG) processes for laser hybrid welding. In Proceedings of the International Congress on Applications of Laser and Electro-Optics (ICALEO 2003), Jacksonville, FL, USA, 13–17 October 2003; pp. 11–20.

25. Orozco, N.J. Fully integrated hybrid-laser welding control process, Processes for Laser Hybrid Welding. In Proceedings of the International Congress on Applications of Laser and Electro-Optics (ICALEO 2003), Jacksonville, FL, USA, 13–17 October 2003; pp. 31–40.

26. Murakami, T.; Shin, M.H.; Nakata, K. Effect of welding direction on weld bead formation in high power fiber laser and MAG arc hybrid welding. *Trans. JWRI* **2010**, *39*, 175–177.

27. Banasik, M.; Urbańczyk, M. Laser + MAG Hybrid Welding of Various Joints. *Biul. Inst. Spaw.* **2017**. [CrossRef]

28. Banasik, M.; Urbańczyk, M. Laser + MAG Hybrid Welding of T-Joints. *Biul. Inst. Spaw.* **2017**. [CrossRef]

29. Banasik, M.; Turyk, E.; Urbańczyk, M. Spawanie hybrydowe laser + MAG elementów urządzeń dźwigowych wykonanych ze stali ulepszonej cieplnie S960QL. *Prz. Spaw.* **2017**, *89*. [CrossRef]

30. Wang, X.-N.; Zhang, S.-H.; Zhou, J.; Zhang, M.; Chen, C.-J.; Misra, R.D.K. Effect of heat input on microstructure and properties of hybrid fiber laser-arc weld joints of the 800 MPa hot-rolled Nb-Ti-Mo microalloyed steels. *Opt. Lasers Eng.* **2017**, *91*, 86–96. [CrossRef]

31. Sun, Q.; Di, H.-S.; Li, J.-C.; Wu, B.-Q.; Misra, R.D.K. A comparative study of the microstructure and properties of 800MPa microalloyed C-Mn steel welded joints by laser and gas metal arc welding. *Mater. Sci. Eng. A* **2016**, *669*, 150–158. [CrossRef]

32. Górka, J. *Własności i Struktura Złączy Spawanych stali Obrabianej Termomechanicznie o Wysokiej Granicy Plastyczności*; Wydawnictwo Politechniki Śląskiej: Gliwice, Poland, 2013.

33. Górka, J.; Stano, S. The structure and properties of filler metal-free laser beam welded joints in steel S700MC subjected to TMCP. In *Laser Technology 2016: Progress and Applications of Lasers*; SPIE: Bellingham, WA, USA, 2016; Volume 10159, pp. 183–190.

Article

Fiber Laser Welding of Dissimilar 2205/304 Stainless Steel Plates

Ghusoon Ridha Mohammed [1,2,*], Mahadzir Ishak [1], Syarifah Nur Aqida Syed Ahmad [1] and Hassan Abdulrssoul Abdulhadi [1,2]

[1]   Faculty of Mechanical Engineering, University Malaysia Pahang, Pekan 26600, Pahang, Malaysia; mahadzir@ump.edu.my (M.I.); aqida@ump.edu.my (S.N.A.S.A.); h.shamary@gmail.com (H.A.A.)
[2]   Institute of Technology, Middle Technical University Baghdad, Alzafaranya 10074, Iraq
*   Correspondence: ghusoon_ridha@yahoo.com; Tel.: +60-129-457-480

Received: 27 September 2017; Accepted: 27 November 2017; Published: 6 December 2017

Abstract: In this study, an attempt on pulsed-fiber laser welding on an austenitic-duplex stainless steel butt joint configuration was investigated. The influence of various welding parameters, such as beam diameter, peak power, pulse repetition rate, and pulse width on the weld beads geometry was studied by checking the width and depth of the welds after each round of welding parameters combination. The weld bead dimensions and microstructural progression of the weld joints were observed microscopically. Finally, the full penetration specimens were subjected to tensile tests, which were coupled with the analysis of the fracture surfaces. From the results, combination of the selected weld parameters resulted in robust weldments with similar features to those of duplex and austenitic weld metals. The weld depth and width were found to increase proportionally to the laser power. Furthermore, the weld bead geometry was found to be positively affected by the pulse width. Microstructural studies revealed the presence of dendritic and fine grain structures within the weld zone at low peak power, while ferritic microstructures were found on the sides of the weld metal near the SS 304 and austenitic-ferritic microstructure beside the duplex 2205 boundary. Regarding the micro-hardness tests, there was an improvement when compared to the hardness of duplex and austenitic stainless steels base metals. Additionally, the tensile strength of the fiber laser welded joints was found to be higher when compared to the tensile strength of the base metals (duplex and austenitic) in all of the joints.

Keywords: fiber laser welding; dissimilar material; stainless steel; microstructure; hardness; tensile strength

---

1. Introduction

In the automotive, oil, aerospace, and nuclear power industries, the welding of dissimilar materials is a strategic area of interest [1,2]. Owing to the differences the metallurgical and physico-mechanical behaviour of joined dissimilar materials (DM), they have attracted significant interest when compared to the joints of similar materials. Components with adequate ductility, high strength, good corrosion resistance, and desirable electrical and thermal conductivities have been achieved through DM welding [3,4]. Several DM welding techniques have been reported, including Welding-Brazing [5–7], Arc [8–10], Friction [11], Friction stir [12,13], Ultrasonic [14,15], Explosive welding [16], Laser [17–19], and Hybrid Laser-Arc welding [20,21].

There is an increased risk of encountering problems with DM during welding and in the resultant weldments. The noted problems with DM welding include (a) carbon migration from materials higher carbon content to the materials with less carbon content; (b) the variation in the thermal expansion coefficient of the materials, triggering a deviation of the thermal residual stress over disparate parts of the weldments; and, (c) the challenges in the execution of the post-welding heat treatment, especially in compounds where one of the joined materials is sensitive to an unwanted deposition at elevated temperatures [22]. Consequently, the solution largely depends on high-energy-density techniques, including electron beam and laser beam welding methods, which result in excessive industrial concerns [23].

The laser welding technique, both in continuous and pulse modes, gained interest in the 1960s. The laser welding requires a low heat input during welding, which results in a higher capability of combining heat-susceptible parts with low distortions. The small focusing zone of laser welding is the main reason for its wide application in cropping narrow welding zones with low heat involvement. Additionally, the welded beads from laser welding processes are created with limited residual stress and deformation [24]. To create a high-quality joint between two different materials with different physico-mechanical and metallurgical characteristics while using the laser welding technique, there must be a suitable combination of welding variables [25]. For Arc welding procedures, fiber laser welding (FLW) is one of the extensively recommended techniques due to its precise benefits such as higher efficacy, absence of flux obligation, low defects, low maintenance cost and compact size, good appearance, and simplicity of mechanization [26].

Austenitic stainless steels (ASS) account for about 70% of stainless steel groups. They have a face-centred cubic structure, with individual characteristics, such as weldability, corrosion resistance, toughness, and ductility [27]. Duplex stainless steels (DSS) were developed to accommodate strong bonding and higher resistance to the pitting zones, and stress corrosion cracking to support 300-series ASS [28,29]. When compared to ASS, DSS have greater product strength and stress-corrosion cracking resistance to chlorides. This distinctive feature of the DSS is the reason for their application in many industries such as chemical process plant piping, structures in marine environments, biomedical applications, oil and gas transmission lines, and nuclear power plants [30–36]. DSS is composed of a two-phase microstructure, showing an approximately similar ferrite and austenite balance [37]. Their exclusive properties are completely based on the equal stability exhibited by ferrite and austenite. However, with the duplex steels, it is advised to use a low welding joint energy [38–40].

DM welding is used in many industries for the combination of different stainless steels or with other materials. The reasons for this combination of these dissimilar materials are economical and evolutional. To create a weld joint with the benefits of the dissimilar materials and good service performance, it is necessary to select a suitable welding technique and a proper filler metal [41]. Since laser welding techniques does not involve the use of filler metals, it is therefore obvious that the process parameters play a key role in achieving the desired weld joints. The mechanical properties of stainless steel have strongly been reported to rely on the microstructure of the joint. It is necessary, therefore, to study the relationship of the microstructure with their mechanical properties. Berretta [42] had investigated the pulsed Nd:YAG laser welding of AISI 304–AISI 420 stainless steels, while Anawa [43] experimented on the optimization of the tensile strength of ferritic/austenitic laser-welded components. Similarly, another study has experimented on low carbon and austenitic stainless steels to determine the major parameters that influence the mechanical properties and microstructure of a laser welded joint steel [44]. The study reported that a variation of the laser beam position can be effective in controlling the composition of the fusion zone. Anawa and Olabi [45] statistically studied the dissimilar LBW of stainless steel and low carbon steel and reported a significant influence of the laser power and welding speed on the weld bead characteristics. Torkamany et al. [46] studied low carbon-austenitic stainless steel for the influence of dissimilar laser welding on its mechanical performance and microstructure. The study revealed that the transition of laser welding mode from conduction to keyhole had a significant influence on the size of the fusion zone and dilution percentage of low carbon steel. Even though several works have been reported on the application of laser welding [18,44,47], limited studies are available in the area of fiber laser techniques of dissimilar materials (DM), such as carbon steel with aluminium alloys [44], titanium alloy with steel [48], wrought to powder stainless steel [49], stainless steel to aluminium [50], magnesium alloy to stainless steel [51], and aluminium with titanium [52].

In summary, many studies have reported interesting findings on the potential of welding stainless steel. However most of these studies have failed to simultaneously examine the effect of fiber laser welding on dissimilar duplex-austenitic stainless steel. Therefore, this study aims to experimentally investigate the influence of fiber laser parameters on the mechanical and metallurgical characteristics of dissimilar 2205-304 stainless steel plates.

In this study, a fiber laser welding technique was employed to join two dissimilar metals plates (AISI 304 and AISI 2205) of 2 mm thickness. The joined metals were characterized for the tensile test, microhardness, macrostructure, microstructure, and phase composition, using an optical microscope (OM) and scanning electron microscopy (SEM) coupled with an energy dispersive spectrometer (EDS).

2. Materials and Methods

2.1. Base Materials

AISI 2205 duplex and AISI 304 austenitic stainless steels were the metals used in this study, with their chemical composition shown in Table 1. Before welding, the metal plates were shaped to sizes, with L × W × T dimensions of 75 × 50 × 2 mm, respectively.

Table 1. Chemical composition of the base metals (wt %).

Elements	C	Cr	Ni	Mn	Si	P	S	Mo	Nb
AISI 2205	0.03	22.07	4.8	1.15	0.53	0.003	0.005	3.65	0.002
AISI 304	0.07	18.3	8.6	1.18	0.47	0.003	0.005	-	0.002

The ASS is made up of randomly linking equiaxed austenite crystals (γ), whereas the DSS entailed of grainy and lengthened particles of ferrite (α) and austenite (γ), respectively, as shown in Figure 1a,b. The experimental setup is shown in Figure 2.

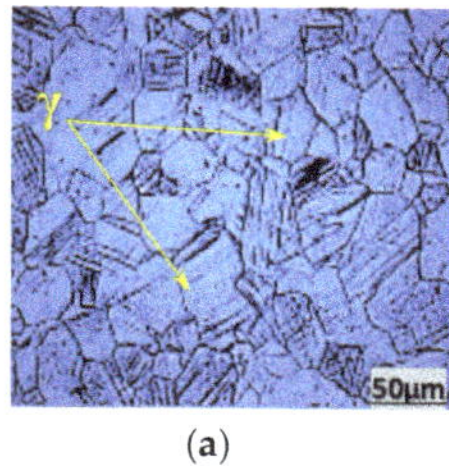
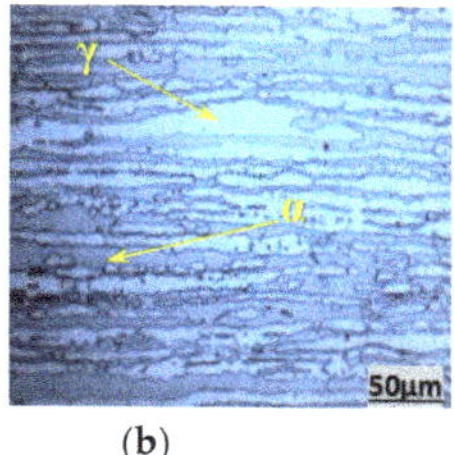

(a)

(b)

Figure 1. The microstructure of (**a**) Austenitic stainless steels (ASS) 304; (**b**) Duplex stainless steels (DSS) 2205.

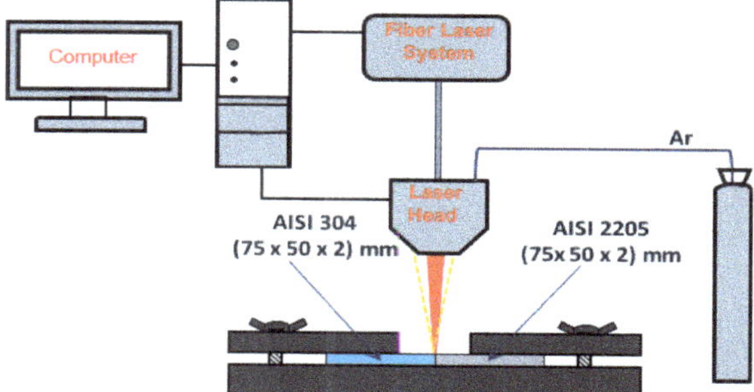

Figure 2. The experimental setup.

2.2. Laser Welding

The laser welding studies were conducted using a pulse-mode with an output power of 2 kW. The fiber laser welding (FLW) was performed at an angle of 10° from a vertical position, whereas the laser was fixed at the joint border. Table 2 showed the varied parameters (peak power PP, beam diameter BD, pulse width PW, and pulse repetition rate PRR) that were used during the welding process. Argon gas (99.99% purity) at a flow rate of 20 L/min was used as the shielding gas.

Table 2. The fiber laser welding parameters.

No.	PP (kW)	BD (mm)	PW (ms)	PRR (Hz)
1	1	0.3	5	15
2	1.2	0.3	5	15
3	1.4	0.3	5	15
4	1.6	0.3	5	15
5	1.8	0.3	5	15
6	1.4	0.24	5	15
7	1.4	0.3	5	15
8	1.4	0.35	5	15
9	1.4	0.4	5	15
10	1.4	0.45	5	15
11	1.4	0.5	5	15
12	1.4	0.35	3	15
13	1.4	0.35	4	15
14	1.4	0.35	5	15
15	1.4	0.35	6	15
16	1.4	0.35	7	15
17	1.4	0.35	8	15
18	1.4	0.3	5	8
19	1.4	0.35	5	10
20	1.4	0.3	5	12
21	1.4	0.3	5	15
22	1.4	0.3	5	18
23	1.4	0.3	5	20

2.3. Metallography

The weld geometry was studied at several weld zones to observe the changes in the bead profile, and weld joint microstructures through low-intensity optical microscopy (MT8000 Series Metallurgical Microscopes, Roanoke, NC, USA) and scanning electron microscopy (SEM) (HITACHI Table Top Microscope TM3030Plus, Schaumburg, IL, USA). Energy dispersive X-ray analysis (EDS, HITACHI, Roanoke, NC, USA) technique has been utilized for the qualitative analysis of the samples. Metallographic investigations were conducted on a "compound region" of 30 mm × 10 mm × 2 mm, which involved all of the weldment regions (base metal, heat-affected, and fusion zones). The samples were etched with modified Kalling's solution (5 mg $CuCl_2$ in 100 mL ethanol and 100 mL HCl) for a period of 20 s. The compound weldment regions were polished with alumina and distilled water, using emery sheets of SiC of varied sand sizes ranging from 240 to 1200. The polishing process followed regular metallographic techniques to achieve a perfect mirror finishing of approximately 1 μm.

2.4. Mechanical Tests

Figure 3 shows the weld bead dimensions (W: width of weld bead, D: depth of penetration), while the micro-hardness was measured at 0.5 mm from the top surface of the welded joint. Vickers hardness analysis technique was applied to study the mechanical behavior by using a hardness tester, Tukon 1202 (Instran ITW Test & Measurement Co., Shanghai, China). Hardness measurements were studied across the interface. Standard 500 gf load was smeared for a residing time of 10 s and the quantities were measured at consistent intervals of 0.20 mm. The tensile tests were performed following the American Society for Testing and Materials (ASTM E8M-04 Standard). Sub-sized specimens were prepared due to the small plate thickness. The samples for the tensile tests were transversely cut to the weld direction and were machined to the dimensions that are presented in Figure 4.

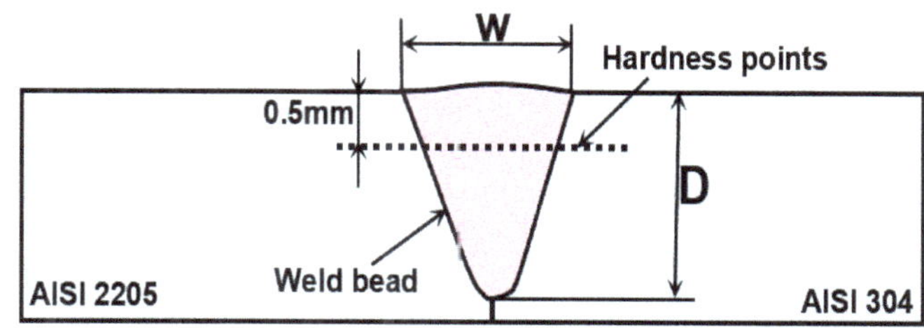

Figure 3. Weld bead profile and the micro-hardness position across it.

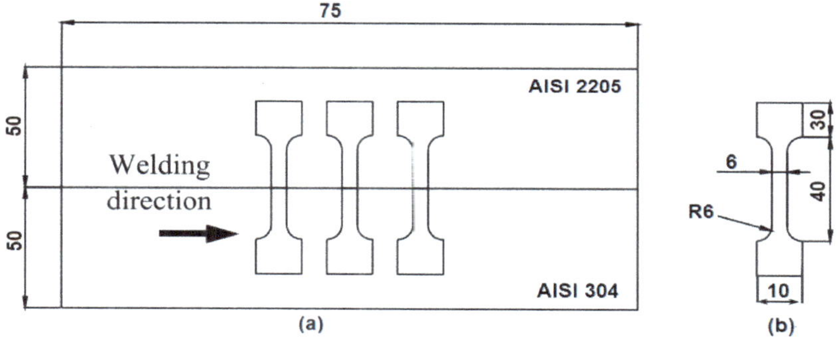

Figure 4. Sketch shows the geometry of (**a**) the American Society for Testing and Materials (ASTM) E8 standard tensile strength of the laser-welded samples and the base metals; (**b**) the Sub-sized specimen (all dimensions in mm).

3. Results and Discussion

3.1. The Influence of Weld Parameters on the Weld Profile

3.1.1. Influence of Peak Power on Weld Geometries

The peak power exhibited a continuous effect on the heat input in each individual pulse (Tests **1–5**). A linear increase in the weld depth and width was observed at varied peak power rates. The weld depth was significantly influenced by the increase in peak power when compared to the weld width (see Figure 5a). A maximum depth and width value of 2 mm and 0.9 mm, respectively, at full penetration was recorded at a peak power of 1.8 kW.

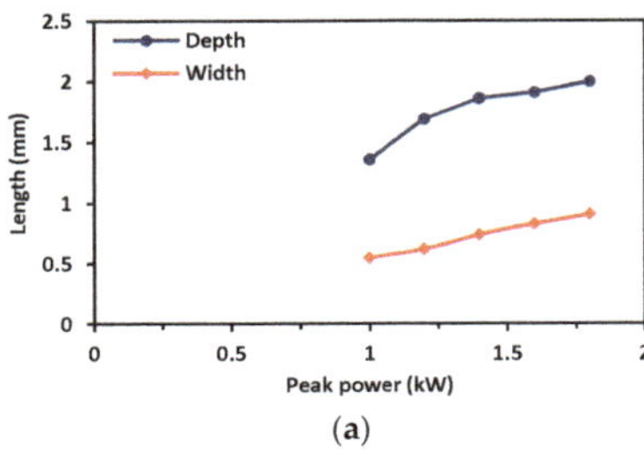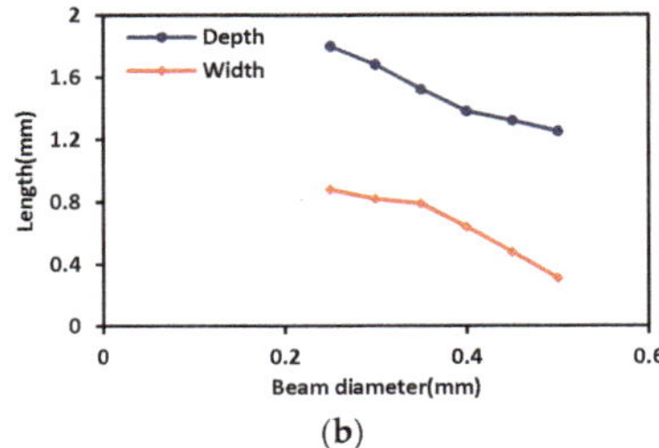

Figure 5. Effects of (**a**) peak power and; (**b**) laser diameter on weld width and depth.

3.1.2. Laser Beam Diameter Effects on Weld Geometries

Figure 5b shows a decrease in the weld penetration depth and weld width with the laser beam diameter (Tests **6–11**). The decrease in the weld width was due to the decline in the applied laser ray focal point. However, a decrease in the beam diameter caused an increase in the laser density at the same laser power, which effectively increased the weld depth. The optimum values were the obtained at a laser beam diameter of 0.35 mm, which fashioned almost a weld depth and width of 1.38 mm and 0.79 mm, respectively.

3.1.3. Influence of Pulse Width on the Weld Geometries

The pulse width demonstrated a bilateral influence on the weld bead dimensions (Tests **12–17**). Figure 6a showed that the best weld dimensions can be achieved at the optimum value, depending on the materials involved and the thickness of the joined parts. The weld depth reached the maximum value with increasing pulse length due to the increased system pulse overlapping. Furthermore, there was a decrease in the weld depth as the pulse width exceeded the optimal value, but, on the weld width, there was no significant effect. The decrease in the laser power is attributed the effect of the increase in the pulse length and energy decline. An optimal pulse width of 5 ms at 1.4 kW was achieved, resulting in a weld depth of 0.79 mm and weld width of 1.73 mm.

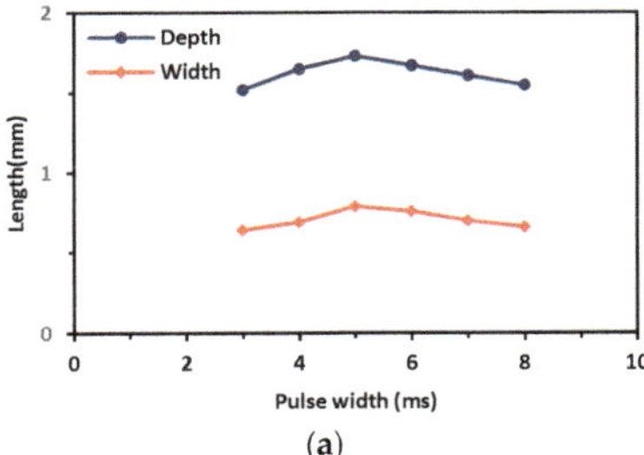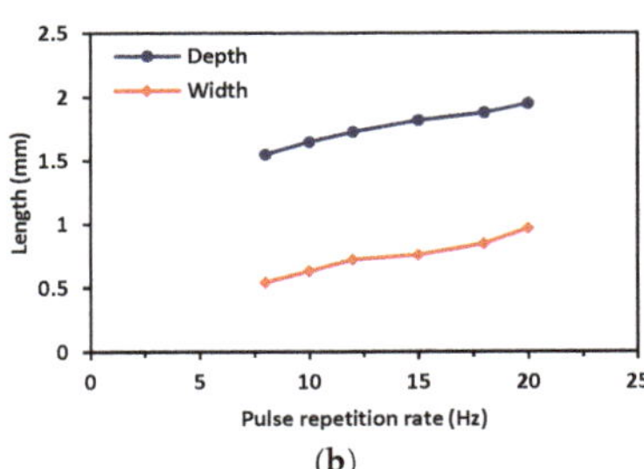

Figure 6. Influence of (**a**) pulse width and; (**b**) pulse repetition rate on the depth and width of the weld bead.

3.1.4. Influence of Pulse Repetition Rate on the Weld Geometries

Figure 6b showed an increasing pattern of the weld width and depth, with an increasing pulse repetition rate (Tests **18–23**). This result is largely due to the enhanced pulse overlapping because of an increase in the pulse repetition rate. The pulse overlapping resulted in an increased pulse density, which increased the weld depth. The optimal pulse repetition rate value was obtained at 20 Hz; thereby, producing a weld depth of 1.95 mm and weld width of 0.97 mm.

3.2. Metallography and Visual Examination

Macro and Microstructure of the Welded Joint

A top view of the fiber-laser welded dissimilar joints is illustrated in Figure 7, showing the different zones (fusion zone (FZ), HAZ) of the welded seam for the joined plates. The Macrograph of the weld geometry of the welding tests that were studied during the analysis is presented in Figure 8. As shown in the Figures, several variables (dimensions of the welds) and their effects were determined by the presence of a persistent variable. Samples **1–5** were used to scrutinize the behavior of the peak power, while Samples **6–11** were studied to reflect the effect of laser beam diameter. Samples **12–17** were used to show the properties of the pulse width, while Samples **18–23** were studied for the effects of the pulse repetition rate.

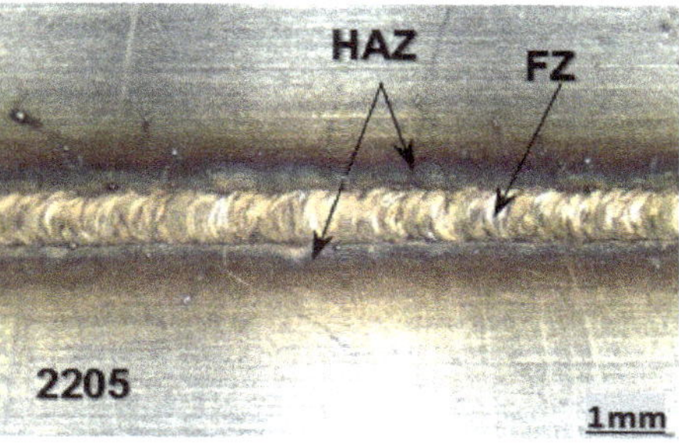

Figure 7. Plan view of the seam welding.

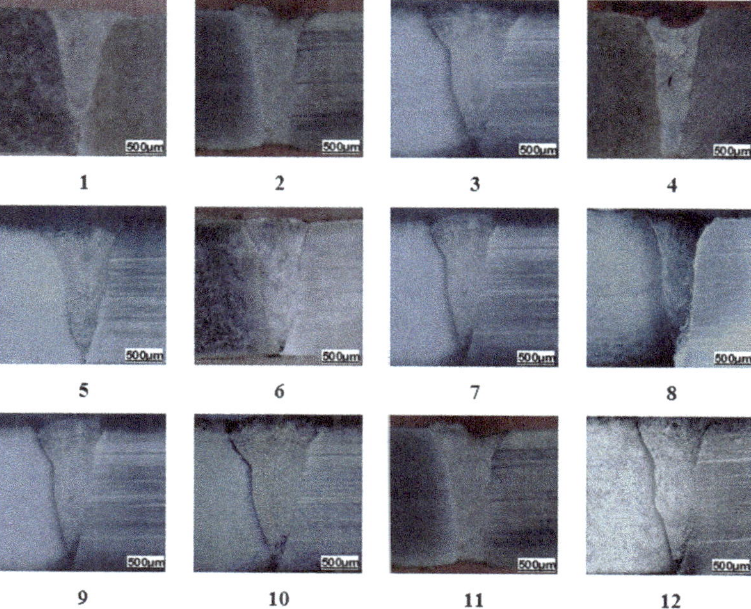

Figure 8. Macrograph of the weld geometry, depth, and width of appointed samples. The numbers shown are referred to the welding tests listed in Table 2.

A macro investigation showed narrow welded beads of 2205-304 stainless steel welded joints with complete fusion (Figure 8). A fast interaction of solid metals and molten liquids, as well as the formation of metallic vapors at high laser power intensity, results in instability in the weld pool during keyhole laser welding. On one side, it encourages deep laser penetration, and on the other hand, it controls the welded bead profile by governing the different patterns of heat transfer and fluid flow in the welded pool which can lead to various surface deformations [53].

During fusion welding, the deformation of the welded surface can assume a convex or concave profile (the convex weld bead profiles are profiles that protrude above the joining material, while those that form a dip are termed concave weld bead profiles). Few studies have been reported on the mechanisms that are responsible for the shape of weld bead profiles [54–56]. The effect of surface deformation and melt flow on weld bead formation was first investigated by Bradstreet [57] when correlated the bead profile of a convex-shaped profile to the surface tension, which controls edge wetting in the weld pool. Similarly, several reasons, such as the aspect ratio, heat transfer, melt flow, pressure balance, and welding speed account for the development of concave-shaped weld bead profiles [53,58–60].

The weld microstructure of various materials is highly complex because of different cooling rates in the laser welding process [61]. On the other hand, steadiness in the microstructures at different weld locations can be determined. The micrographic examination was performed on a cross-section of the 2205-304 dissimilar welded joints (Figure 9).

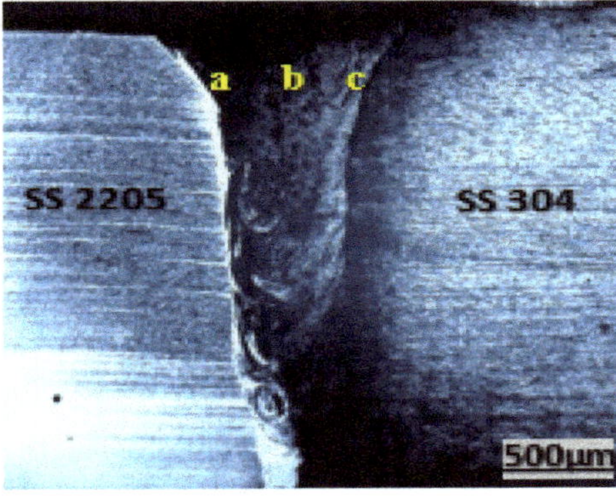

Figure 9. Micrograph of fiber-laser welding dissimilar 2205-304 stainless steels (SS).

The study was conducted to examine and analyse several points at the fusion zone and near the base metals. Figure 10a shows the microstructure (mainly a mixture of ferrite and austenite) present at the 2205 DSS base metals near the 2205 DSS (point "a" in Figure 9). Conversely, the microstructure at the position closer to the 2205 stainless steel was improved and additionally subtle than the initial dual phase assembly of the 2205 SS. A superior cooling rate is considered responsible for this behavior in laser welding [39,62,63]. Figure 10b showed the microstructures in the fusion zone (FZ), and at the boundary between the FZ and the base metals (point "b" in Figure 9). Larger grains were observed in the FZ, while fine grains were noted in the BM section. The fine grains in the FZ were bordered by the coarse grains of the two diverse metal sides due to dynamic recrystallization [64].

In Figure 10b, the longitudinal columnar grain structures were dispersed at the principal area of the FZ along the vertical direction. The area next to the BM was categorized by the equiaxed particles, as shown in Figure 10a,c (point "c" in Figure 9). Paralleled-lined structures were observed on the side of the ASS near the central equiaxed particle structures and the neighboring boundaries, are presented in Figure 10c.

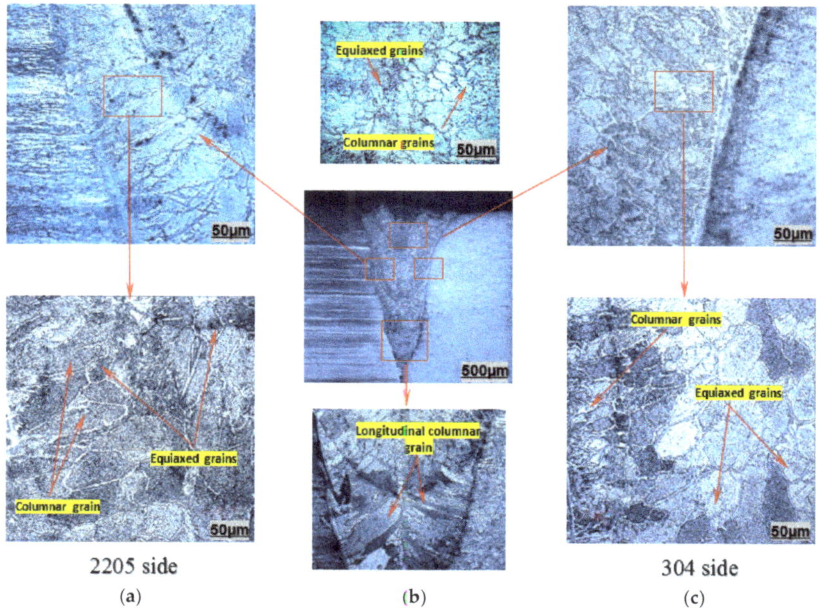

Figure 10. Microstructure of the fusion zone in 2205-304 dissimilar welding: Grain shape close to (**a**) 2205 side, (**b**) in the center line of the fusion zone, (**c**) close to 304 side.

A proper fusion of the base metals was achieved with the predetermined process parameters using fibre laser welding technique. The SEM observations of various weld metal microstructure at room temperature of some selected AISI 304 samples welded with duplex 2205 are listed in Figure 11. Generally, the final microstructure of all the weldments can be standardized by the chemical composition of the parent metals and the thermal cycle of the welding process, as the stainless steel welds had four potential solidification modes namely, fully ferritic, primary ferritic, primary austenitic, and fully austenitic [65,66].

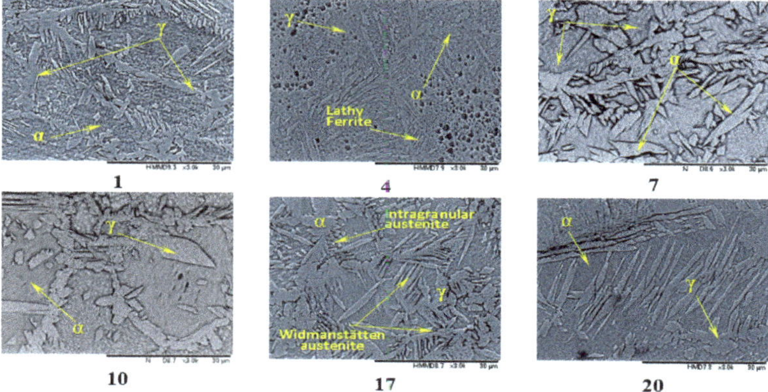

Figure 11. SEM of the weld zone of selected samples. The sample numbers displayed on the graphs refer to the welding tests listed in Table 2.

Type A (fully austenite) or AF (austenite to ferrite) principally refers to austenite solidification, which results in diverse austenite and ferrite morphologies. The FA (ferrite to austenite) mode signifies emaciated and/or stripped ferrite caused by ferrite to austenite transformation; F symbolizes acicular ferrite with a particle boundary of austenite, as illustrated in (Figure 9). A successive transformation of ferrite to austenite ensues in solid form within a temperature range of 1300–800 °C [33]. Additionally, an increase in the Cr_{eq}/Ni_{eq} ratio results in a higher ferrite propensity foundation and altering of the solidification approach from A and AF to FA or F [67].

SEM investigations (point b in Figure 9) revealed the formation of delta ferrite at the center line of the FZ due to the higher cooling level during laser welding, as shown in (Figure 11). On the 304 austenitic SS side (point c in Figure 9), a significant decrease in the austenite quantity was observed, and the microstructure largely comprised of ferrite materials (Figure 12). The chemical composition of the welded metals near 304 SS were studied using EDS; the line mapping of the weldment is shown in Figure 13, and the elements that were analyzed included both ferrite (Cr, Mo, Si, Fe) and austenite (Ni, C, Mn) stabilizing elements; while, the analyses and the corresponding values are presented in Table 3. Moreover, Table 3 also presents the chemical composition of the 304 stainless steel. Commencing the WRC-1992 diagram [68], it is expected that the weld microstructures would enclose a combination of austenite and ferrite, containing approximately 11–18% of ferrite. The Cr_{eq} value was equivalent to 19 and the Ni_{eq} value was close to 11, giving a Cr_{eq}/Ni_{eq} ratio of 1.73 and that means the solidification process led to ferritic formation, as previously reported [69,70].

Table 3. Chemical composition of 304SS base metal and the weld zone near 304 SS.

Position	Fe	Cr	Ni	Cu	Si	V	Mo	Mn
Base 304	54.46	14.8	5.96	1.48	0.53	0.09	0.18	1.49
WZ near 304	51.04	15.8	5.23	0.09	0.57	0.05	1.44	1.14

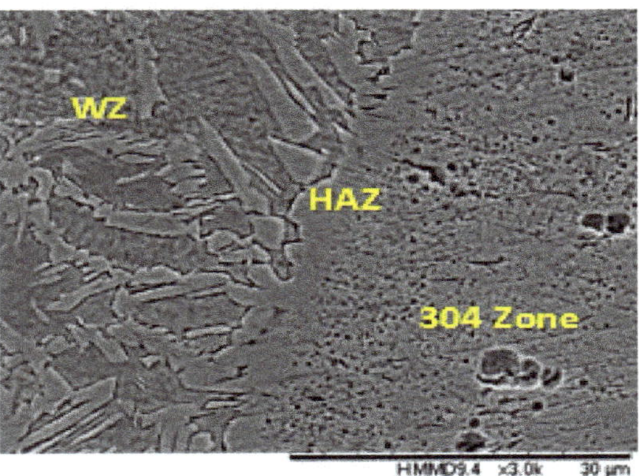

Figure 12. SEM of the weld zone microstructure at points near SS 304.

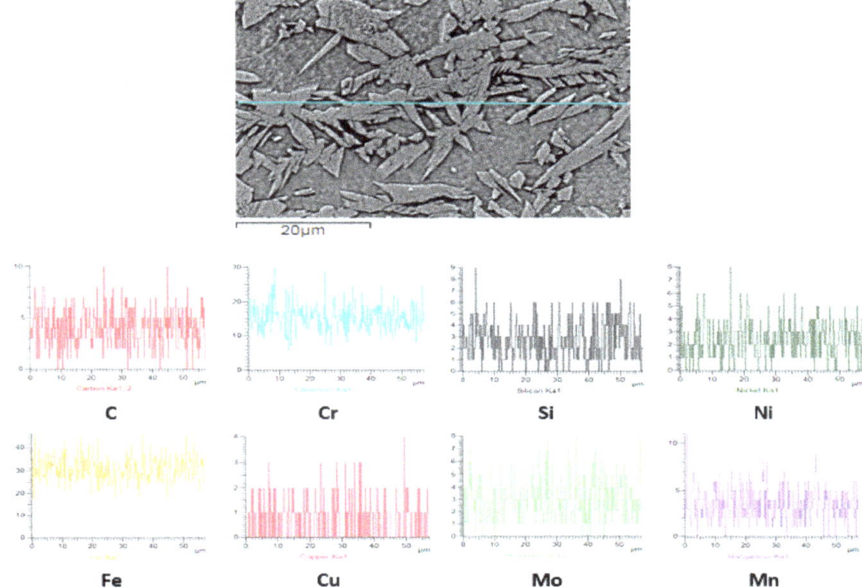

Figure 13. Energy dispersive spectrometer (EDS) line mapping analysis of duplex-austenitic stainless steel weldment.

At test number **5**, a solidification crack was detected on the side of 304 SS at a peak power of 1.8 kW due to the high peak power value. The crack was instigated from the center of the weld and it was directed towards the base metal, as illustrated in Figure 14. The diverse thermal and physical (heat capacity, thermal conductivity; and the relationship between hardness and temperature) behaviors of the materials, which were supposed to weld in DSS metal welding resulted in a general symmetrical distortion. In contrast to the duplex stainless steel, the tested specimens did not exhibit any extensive warp. This behavior of austenitic stainless steel is due to their less hardness and heat conductivity at elevated temperatures. This behavior has been recognized in the friction welding of different weldments, such as stainless steel to copper, steel to aluminum, and steel to titanium [26,71,72].

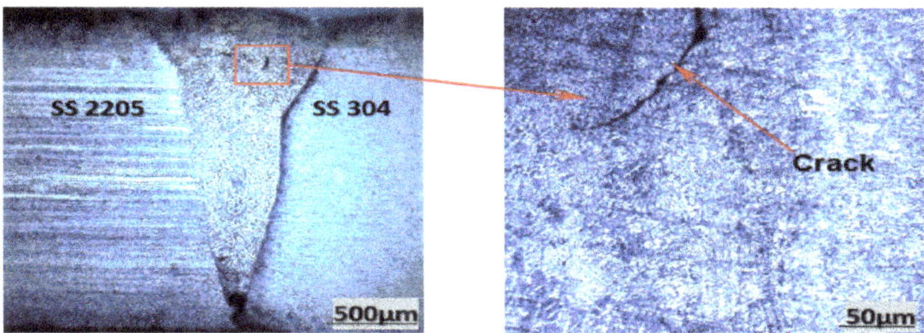

Figure 14. Solidification crack in the fusion zone at 304 side.

The basis of crack-susceptible microstructure in stainless steels is the evolution of impurity-developed and low-melting liquid films along the grain boundaries in the final phase of the solidification process [53,73]. The influence of ferrite content in welds on the solidification and cracking at room temperature has long been documented, but it has not been accurately linked with the cracking sensitivity. A greater focus has been on the solidification behavior of stainless steel welds [74–76]. For alloys that display a Cr_{eq}/Ni_{eq} ratio of less than 1.5, austenite is the key phase of solidification. With an increment in the Cr_{eq}/Ni_{eq} ratio (1.5 to 2.0), the primary solidification is promoted by ferrite. Duplex stainless steels are those that solidify completely to ferrite ($Cr_{eq}/Ni_{eq} > 2$) [39]. They exhibit cracking vulnerability to indeterminate alloys that experience complete austenitic and ferritic/austenitic solidifications [33]. In the welding process, most of the produced duplex weld metals are initially solidified in a fully ferritic mode. At the fusion line of the base metal, there is a higher rate of ferritic solidification within the weld metal. In the DSS weld, ferrite grains are encouraged to form a coarse columnar structure as a reasonable and epitaxial result of the growth of ferrite grains. Figure 15 showed that in the weld metal with excessive ferrite content, the austenite aggregate drops; the morphology modifies to granular form; and, fine precipitation appears in the ferrite. Austenite precipitates in the cooling process of a welded metal from a temperature of 1300 °C to 800 °C due to the higher free energy of the ferrite grain interfaces. The degree of precipitation is governed by the cooling rate and chemical composition of the compound [77,78]. However, the austenite morphology (Widmanstätten or intergranular) plates are largely based on the cooling rates and the size of the ferrite grains [33].

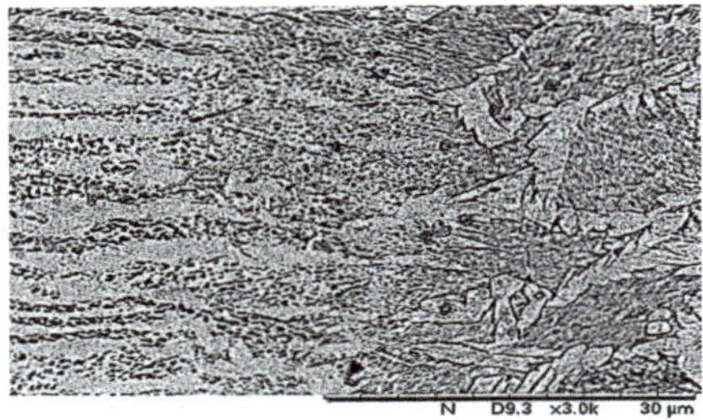

Figure 15. SEM of weld microstructures at a point near 2205 side.

3.3. Mechanical Tests

3.3.1. Microhardness

The microhardness profiles of the weld regions (304-AHAZ-FZ-DHAZ-2205) are presented in Figures 16 and 17. AHAZ refers to the austenitic heat-affected zone, while DHAZ refers to duplex heat-affected zone. Specimens were selected for the microhardness studies based on the peak power (Tests **2,3,4**) and pulse width (Tests **12,14,16**) of the welding process. All of the microhardness profiles in Figures 16 and 17 have the same tendency. The hardness increases in the weld zone and decreases towards the base metals.

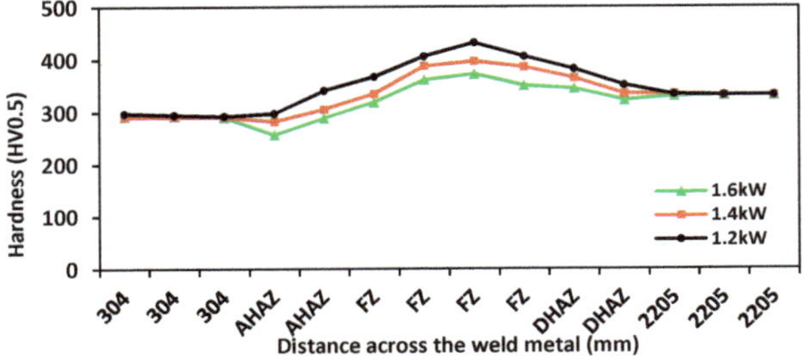

Figure 16. Microhardness profile along the welded line of dissimilar joints at different peak powers.

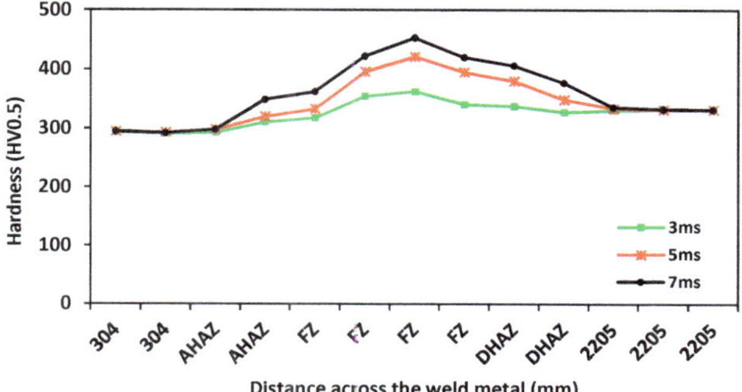

Figure 17. The cross-sectional Vickers microhardness profile of welds joint different pulse width.

For the peak power effect, the fusion zone presented a higher microhardness when compared to those of the base materials (AISI 304 and AISI 2205), likely being due to the rapid solidification effect. The microhardness and elemental (Cr, Fe, and Ni) redistribution gradients are positively correlated, which may be a special attribute of dissimilar joints [79]. The fusion zone of the laser keyhole welds had a rapid cooling rate of about 104 and 106 °C·s^{-1} [80] which not only increased the probabilities of under-cooling and nucleation with the attendant formation of fine structures, but also extended the solubility of the solutes, thereby preventing marked segregation and resulted in a supersaturated solid solution with new microstructures [81]. The microhardness of the weld HAZ was less than that of the weld metal, but greater than that of the base metals at both sides due to the earlier stated reasons. Figure 16 revealed the hardness dispersal across the welded regions, which indicates that the hardness was greater on the DSS side when compared to the ASS side. The BM regions showed an average hardness of 292 HV in the austenitic stainless steel 304 side and 332 HV for the duplex stainless steel 2205 side. The microhardness values increased in the FZ section, with a calculated maximum hardness value of 432 HV. This indicates that the FLW enhanced the hardness of the material to the level of about 32% and 23% for the duplex and austenitic stainless steel, respectively. The higher heating of laser welding is beneficial for the growth of ferrite, which is primed to immensely increase the hardness of the FZ region. Thus, the principal factor that is affecting the hardness in the FZ region is mainly ferrite formation.

For the pulse width effect, the operational parameters of sample 12, 14, and 16 differed in the pulse width (3, 5, and 7 ms, respectively). The welding parameters influenced the mode of solidification and cooling rate of the welding zone. Also, the microstructure of the weld zone was noted to change with the welding parameters as well. Moreover, the microhardness profile of the samples affected the variations in their cooling rate. In Figure 17, sample 16 (from Table 2) had the highest hardness value of 453 HV when compared to samples 12 and 14 with hardness values of 363 and 421 HV, respectively, due to the variation in the cooling rate. The cooling rate increases with the pulse duration.

3.3.2. Tensile Strength

The tensile tests performed on the fiber laser butt welded joints with complete penetration based on the impact of peak power (Tests **3**, **4**, **5** in Table 2). The tests were conducted following the relevant standards at room temperature. Three samples from each condition were subjected to the tensile strength test and the average of the triplicates was reported. Figure 18 presents the typical engineering stress-strain curve of the samples. The SEM analysis accomplished, to show the fracture manner of tensile test (brittle or ductile failures), as shown in (Figure 19a–c). Moreover, the EDS analysis was carried out to exhibit the redistribution of elements in the weld zone that form as a result to the fiber laser welding of dissimilar metals (2205-304).

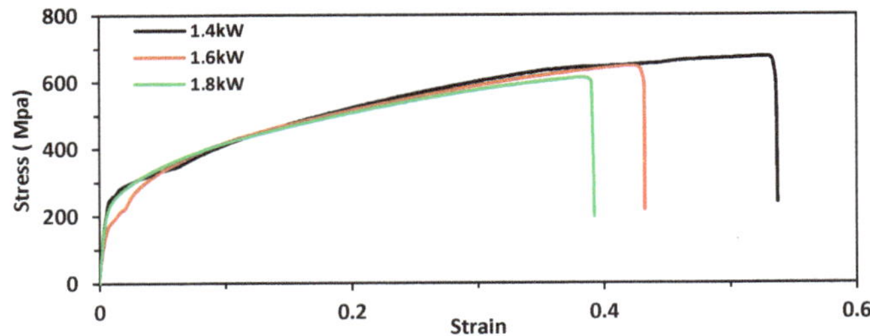

Figure 18. Typical trend of the 2205-304 in stress-strain diagram.

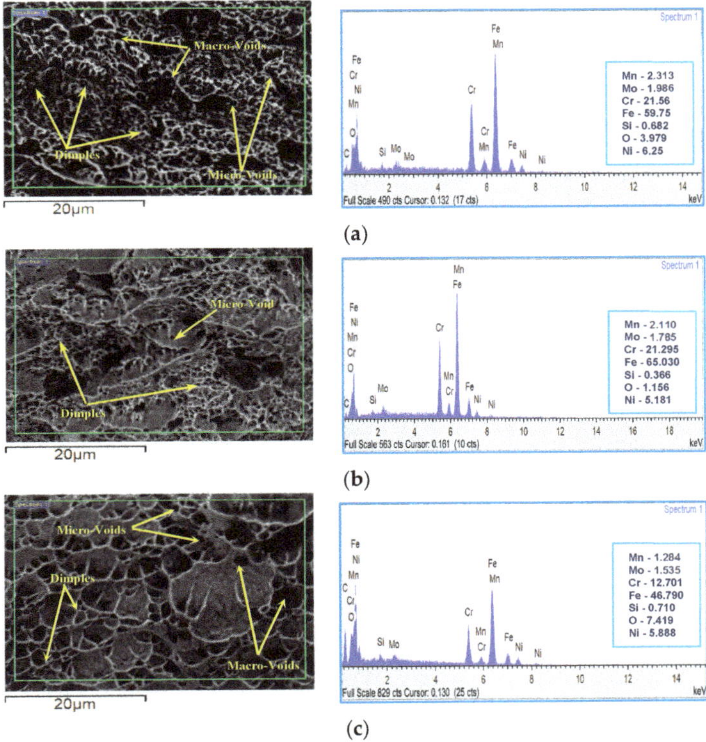

Figure 19. SEM image and EDS analysis of fiber-laser welded tensile-tested joints at (**a**) 1.4 kW; (**b**) 1.6 kW; and (**c**) 1.8 kW.

From the SEM fractographs of the tensile tested of peak power 1.4, 1.6, and 1.8 kW weldments using, the presence of dimpled facets with the macro/micro-voids coalesced in shiny, fibrous fringes were clearly shown, signifying the ductile mode of the fracture. Also, the 1.8 kW weldments showed shallow ridges with more voids and dimples, which contributed to the ductile fracture. The percentage elongation of the 1.4 kW weldments was also noted to be greater, which further confirmed the absence of any deleterious phase in the fusion zone.

The presence of Mn, Cr, Ni, Fe, Si, and Mo were confirmed from the EDS analysis. The fractography in Figure 19a showed a dimpled pattern, which indicates that a ductile fracture was shown, while in Figure 19b,c, deep dimples were depicted when compared to Figure 19a, which indicates an improved ductility. From Figure 19a, chromium, and manganese were observed to have the highest presence, indicating a higher tensile strength in the fractured joint when compared to the base material (AISI 2205 and AISI 304). The presence of oxygen, which suggests the oxidation of the material due to the presence of Fe, was noted in the three samples (Figure 19a–c). In Figure 19c, Mo notably had the lowest presence, pointing towards more ductility. A maximum percentage of Cr was observed in Figure 19a, which might be due to the increased tensile strength of the joint.

At a tensile stress range of 611, 646, and 674 MPa, respectively, the entire welded sample broke from the joint during the testing with a corresponding percentage elongation of about 39%, 43%, and 54%, respectively. During the tensile testing, failures at the weldment were due to the low strength of the welds due to their slightly increased hardness. The welded specimens and the base metal had similar tensile properties only that the ductility of the welded joint was somewhat lower when compared to that of the base metal because of the decreased amount of austenite in the welded joint as compared to the base metal. An increased Cr/Ni ratio resulted in the presence of ferrite in the solidified weld region and consequently increased the tensile strength of the weld material [67].

It is worthy to note that despite the higher volume of austenite in this study, an increase in the heat input did not significantly affect the tensile elongation of the laser welded joint which was likely due to the larger size of the fusion zone with a higher heat input. In other words, the decrease in the tensile elongation because of the fusion zone size was more than the austenite volume effect on increasing the tensile elongation.

4. Conclusions

Detailed experiments were carried out under diverse process environments, and based on the microstructural studies using OM, SEM, and EDS the following conclusions are made:

- Fiber-laser welding was effectively employed to produce austenitic–duplex stainless steels weldments.
- Weld bead profiles and hardness can be articulated in terms of the method parameters.
- The mechanical properties of the produced DSS-ASS joints were better when compared to that of the base metals due to the small HAZ that resulted from the laser welding and the advantageous effect of rapid solidification in the fusion zone.
- Using fiber-laser welding in the austenitic-duplex steel joint significantly increased the microhardness, largely because of the presence of ferrite structures. High heat inputs resulted in coarse grain size formation and a decrease in the microhardness values.
- In the 2205-304 joints, regions with a fully austenitic solidification mode were prone to solidification cracking.
- The welded joints had an average tensile strength of 611, 643, and 674 MPa, respectively, all satisfying the tensile strength requirements for engineering structures.
- From the SEM results, the presence of micro-voids and dimples confirmed the ductile mode of fracture.
- The small HAZ and the beneficial effects of rapid solidification in the fusion zone were demonstrated in the improved mechanical properties of the laser welds. The fusion zone microstructure demands further investigations using TEM. Also, there is a need to conduct a detailed study on the fractures behaviour and welding defects of dissimilar jointed materials.

Acknowledgments: The authors are grateful to Universiti Malaysia Pahang (UMP), Pekan, Malaysia, for the financial support given under grant no. RDU150378 and GRS 1503107.

Author Contributions: Ghusoon Ridha Mohammed and Mahadzir Ishak conceived the design, designed the experiments, analysed the data; performed the experiments and wrote the paper; Syarifah Nur Aqida Syed Ahmad and Hassan Abdulrssoul Abdulhadi provided the reagents, materials and analysis tools.

Conflicts of Interest: The authors declare no conflict of interest. The funding sponsors had no role in the design of the study, data collection, analyses of the data, writing of the manuscript, and the decision to publish the results.

References

1. Devendranath Ramkumar, K.; Singh, A.; Raghuvanshi, S.; Bajpai, A.; Solanki, T.; Arivarasu, M.; Arivazhagan, N.; Narayanan, S. Metallurgical and mechanical characterization of dissimilar welds of austenitic stainless steel and super-duplex stainless steel—A comparative study. *J. Manuf. Process.* **2015**, *19*, 212–232. [CrossRef]

2. Kumar, R.; Bhattacharya, A.; Bera, T.K. Mechanical and metallurgical studies in double shielded gmaw of dissimilar stainless steels. *Mater. Manuf. Process.* **2015**, *30*, 1146–1153. [CrossRef]

3. Martinsen, K.; Hu, S.J.; Carlson, B.E. Joining of dissimilar materials. *CIRP Ann. Manuf. Technol.* **2015**, *64*, 679–699. [CrossRef]

4. Wu, W.; Hu, S.; Shen, J. Microstructure, mechanical properties and corrosion behavior of laser welded dissimilar joints between ferritic stainless steel and carbon steel. *Mater. Des.* **2015**, *65*, 855–861. [CrossRef]

5. Li, C.L.; Fan, D.; Wang, B. Characteristics of TIG arc-assisted laser welding-brazing joint of aluminum to galvanized steel with preset filler powder. *Rare Met.* **2015**, *34*, 650–656. [CrossRef]

6. Zhang, Y.; Huang, J.; Chi, H.; Cheng, N.; Cheng, Z.; Chen, S. Study on welding–brazing of copper and stainless steel using tungsten/metal gas suspended arc welding. *Mater. Lett.* **2015**, *156*, 7–9. [CrossRef]

7. Filliard, G.; El Mansori, M.; Tirado, L.; Mezghani, S.; Bremont, C.; De Metz-Noblat, M. Industrial fluxless laser weld-brazing process of steel to aluminium at high brazing speed. *J. Manuf. Process.* **2017**, *25*, 104–115. [CrossRef]

8. Scotti, A. Mapping transfer modes for stainless steel gas metal arc welding. *Sci. Technol. Weld. Join.* **2000**, *5*, 227–234. [CrossRef]

9. Shao, L.; Shi, Y.; Huang, J.K.; Wu, S.J. Effect of joining parameters on microstructure of dissimilar metal joints between aluminum and galvanized steel. *Mater. Des.* **2015**, *66*, 453–458. [CrossRef]

10. Başyiğit, A.; Kurt, A. Investigation of the weld properties of dissimilar s32205 duplex stainless steel with aisi 304 steel joints produced by arc stud welding. *Metals* **2017**, *7*, 77. [CrossRef]

11. Handa, A.; Chawla, V. Investigation of mechanical properties of friction-welded AISI 304 with AISI 1021 dissimilar steels. *Int. J. Adv. Manuf. Technol.* **2014**, *75*, 1493–1500. [CrossRef]

12. Atapour, M.; Sarlak, H.; Esmailzadeh, M. Pitting corrosion susceptibility of friction stir welded lean duplex stainless steel joints. *Int. J. Adv. Manuf. Technol.* **2015**, *83*, 721–728. [CrossRef]

13. Lee, H.; Kim, C.; Song, J. An evaluation of global and local tensile properties of friction-stir welded DP980 dual-phase steel joints using a digital image correlation method. *Materials* **2015**, *8*, 5467. [CrossRef] [PubMed]

14. Lionetto, F.; Balle, F.; Maffezzoli, A. Hybrid ultrasonic spot welding of aluminum to carbon fiber reinforced epoxy composites. *J. Mater. Process. Technol.* **2017**, *247*, 289–295. [CrossRef]

15. Macwan, A.; Chen, D.L. Microstructure and mechanical properties of ultrasonic spot welded copper-to-magnesium alloy joints. *Mater. Des.* **2015**, *84*, 261–269. [CrossRef]

16. Kaçar, R.; Acarer, M. Microstructure–property relationship in explosively welded duplex stainless steel–steel. *Mater. Sci. Eng. A* **2003**, *363*, 290–296. [CrossRef]

17. Romoli, L.; Rashed, C.A.A. The influence of laser welding configuration on the properties of dissimilar stainless steel welds. *Int. J. Adv. Manuf. Technol.* **2015**, *81*, 563–576. [CrossRef]

18. Baghjari, S.H.; AkbariMousavi, S.A.A. Experimental investigation on dissimilar pulsed Nd:YAG laser welding of AISI 420 stainless steel to kovar alloy. *Mater. Des.* **2014**, *57*, 128–134. [CrossRef]

19. Khan, M.M.A.; Romoli, L.; Dini, G. Laser beam welding of dissimilar ferritic/martensitic stainless steels in a butt joint configuration. *Opt. Laser Technol.* **2013**, *49*, 125–136. [CrossRef]

20. Gao, M.; Chen, C.; Gu, Y.; Zeng, X. Microstructure and tensile behavior of laser Arc hybrid welded dissimilar Al and Ti alloys. *Materials* **2014**, *7*, 1590. [CrossRef] [PubMed]

21. Thomy, C.; Möller, F.; Sepold, G.; Vollertsen, F. Interaction between laser beam and arc in hybrid welding processes for dissimilar materials. *Weld. World* **2009**, *53*, 58–66. [CrossRef]

22. Ghorbani, S.; Ghasemi, R.; Ebrahimi-Kahrizsangi, R.; Hojjati-Najafabadi, A. Effect of post weld heat treatment (PWHT) on the microstructure, mechanical properties, and corrosion resistance of dissimilar stainless steels. *Mater. Sci. Eng. A* **2017**, *688*, 470–479. [CrossRef]

23. Phanikumar, G.; Chattopadhyay, K.; Dutta, P. Joining of dissimilar metals: Issues and modelling techniques. *Sci. Technol. Weld. Join.* **2011**, *16*, 313–317. [CrossRef]

24. Okano, S.; Tsuji, H.; Mochizuki, M. Temperature distribution effect on relation between welding heat input and angular distortion. *Sci. Technol. Weld. Join.* **2017**, *22*, 59–65. [CrossRef]

25. Ma, J.; Harooni, M.; Carlson, B.; Kovacevic, R. Dissimilar joining of galvanized high-strength steel to aluminum alloy in a zero-gap lap joint configuration by two-pass laser welding. *Mater. Des.* **2014**, *58*, 390–401. [CrossRef]

26. Kuryntsev, S.V.; Morushkin, A.E.; Gilmutdinov, A.K. Fiber laser welding of austenitic steel and commercially pure copper butt joint. *Opt. Lasers Eng.* **2017**, *90*, 101–109. [CrossRef]

27. Baldenebro-Lopez, F.; Gomez-Esparza, C.; Corral-Higuera, R.; Arredondo-Rea, S.; Pellegrini-Cervantes, M.; Ledezma-Sillas, J.; Martinez-Sanchez, R.; Herrera-Ramirez, J. Influence of size on the microstructure and mechanical properties of an aisi 304l stainless steel—A comparison between bulk and fibers. *Materials* **2015**, *8*, 451. [CrossRef] [PubMed]

28. Örnek, C.; Engelberg, D.L. Towards understanding the effect of deformation mode on stress corrosion cracking susceptibility of grade 2205 duplex stainless steel. *Mater. Sci. Eng. A* **2016**, *666*, 269–279. [CrossRef]

29. Lo, K.H.; Shek, C.H.; Lai, J. Recent developments in stainless steels. *Mater. Sci. Eng. R* **2009**, *65*, 39–104. [CrossRef]

30. Schulz, Z.J.; Reno, T. Duplex and super duplex stainless steels, and their applications in steam assisted gravity drainage (SAGD). In Proceedings of the Society of Petroleum Engineers Heavy Oil Conference, Calgary, AB, Canada, 12–14 June 2012.

31. Chai, G.; Kangas, P. Recent developments of advanced austenitic and duplex stainless steels for oil and gas industry. In *Energy Materials 2014*; John Wiley & Sons, Inc.: Hoboken, NJ, USA, 2015.

32. Rosales, J.; Cabrera, M.; Agrela, F. Effect of stainless steel slag waste as a replacement for cement in mortars. Mechanical and statistical study. *Constr. Build. Mater.* **2017**, *142*, 444–458. [CrossRef]

33. Bonollo, F.; Tiziani, A.; Ferro, P. Welding processes, microstructural evolution and final properties of duplex and superduplex stainless steels. In *Duplex Stainless Steels*; John Wiley & Sons, Inc.: Hoboken, NJ, USA, 2013; pp. 141–159.

34. Kocijan, A.; Merl, D.K.; Jenko, M. The corrosion behaviour of austenitic and duplex stainless steels in artificial saliva with the addition of fluoride. *Corros. Sci.* **2011**, *53*, 776–783. [CrossRef]

35. Hermawan, H.; Ramdan, D.; Djuansjah, J.R.P. Metals for biomedical applications. In *Biomedical Engineering—From Theory to Applications*; Fazel-Rezai, R., Ed.; InTech: Rijeka, Croatia, 2011; Chapter 17.

36. Talha, M.; Behera, C.K.; Sinha, O.P. Effect of nitrogen and cold working on structural and mechanical behavior of Ni-free nitrogen containing austenitic stainless steels for biomedical applications. *Mater. Sci. Eng. C* **2015**, *47*, 196–203. [CrossRef] [PubMed]

37. Mourad, A.H.I.; Khourshid, A.; Sharef, T. Gas tungsten arc and laser beam welding processes effects on duplex stainless steel 2205 properties. *Mater. Sci. Eng. A* **2012**, *549*, 105–113. [CrossRef]

38. Abuzeid, O.; Aljoboury, A.; Mourad, A.-H.; Alawar, A.; Zour, M.A. Characterization of Two Types of Stainless Steels Recommended for Manufacturing Brine Recirculation Pumps. In Proceedings of the ASME 2010 10th Biennial Conference on Engineering Systems Design and Analysis, Istanbul, Turkey, 12–14 July 2010; American Society of Mechanical Engineers: New York, NY, USA, 2010; pp. 603–607.

39. Mohammed, G.; Ishak, M.; Aqida, S.; Abdulhadi, H. Effects of heat input on microstructure, corrosion and mechanical characteristics of welded austenitic and duplex stainless steels: A review. *Metals* **2017**, *7*, 39. [CrossRef]

40. Karlsson, L.; Arcini, H. Low energy input welding of duplex stainless steels. *Weld. World* **2012**, *56*, 41–47. [CrossRef]

41. Kannengießer, T. Corrosion books: Welding metallurgy and weldability of stainless steel. By: JC Lippold, Dj Kotecki. *Mater. Corros.* **2006**, *57*, 94. [CrossRef]

42. Berretta, J.R.; de Rossi, W.; David Martins das Neves, M.; Alves de Almeida, I.; Dias Vieira Junior, N. Pulsed Nd:YAG laser welding of AISI 304 to AISI 420 stainless steels. *Opt. Lasers Eng.* **2007**, *45*, 960–966. [CrossRef]

43. Anawa, E.; Olabi, A.G. Optimization of tensile strength of ferritic/austenitic laser-welded components. *Opt. Lasers Eng.* **2008**, *46*, 571–577. [CrossRef]

44. Esfahani, M.N.; Coupland, J.; Marimuthu, S. Microstructure and mechanical properties of a laser welded low carbon-stainless steel joint. *J. Mater. Process. Technol.* **2014**, *214*, 2941–2948. [CrossRef]

45. Anawa, E.M.; Olabi, A.G. Using taguchi method to optimize welding pool of dissimilar laser-welded components. *Opt. Laser Technol.* **2008**, *40*, 379–388. [CrossRef]

46. Torkamany, M.J.; Sabbaghzadeh, J.; Hamedi, M.J. Effect of laser welding mode on the microstructure and mechanical performance of dissimilar laser spot welds between low carbon and austenitic stainless steels. *Mater. Des.* **2012**, *34*, 666–672. [CrossRef]

47. Hu, Y.; He, X.; Yu, G.; Ge, Z.; Zheng, C.; Ning, W. Heat and mass transfer in laser dissimilar welding of stainless steel and nickel. *Appl. Surf. Sci.* **2012**, *258*, 5914–5922. [CrossRef]

48. Yao, C.; Xu, B.; Zhang, X.; Huang, J.; Fu, J.; Wu, Y. Interface microstructure and mechanical properties of laser welding copper–steel dissimilar joint. *Opt. Lasers Eng.* **2009**, *47*, 807–814. [CrossRef]

49. Casalino, G.; Campanelli, S.L.; Ludovico, A.D. Laser-arc hybrid welding of wrought to selective laser molten stainless steel. *Int. J. Adv. Manuf. Technol.* **2013**, *68*, 209–216. [CrossRef]

50. Casalino, G.; Leo, P.; Mortello, M.; Perulli, P.; Varone, A. Effects of laser offset and hybrid welding on microstructure and imc in Fe-Al dissimilar welding. *Metals* **2017**, *7*, 282. [CrossRef]

51. Casalino, G.; Guglielmi, P.; Lorusso, V.D.; Mortello, M.; Peyre, P.; Sorgente, D. Laser offset welding of AZ31B magnesium alloy to 316 stainless steel. *J. Mater. Process. Technol.* **2017**, *242*, 49–59. [CrossRef]

52. Casalino, G.; Mortello, M. Modeling and experimental analysis of fiber laser offset welding of Al-Ti butt joints. *Int. J. Adv. Manuf. Technol.* **2016**, *83*, 89–98. [CrossRef]

53. Ghusoon, R.M.; Ishak, M.; Aqida, S.N.; Hassan, A.A. Weld bead profile of laser welding dissimilar joints stainless steel. *IOP Conf. Ser. Mater. Sci. Eng.* **2017**, *257*, 012072.

54. Jialie, P.Y.S.J.R. Study of humping tendency and affecting factors in high speed laser welding of stainless steel sheet. *Acta Metall. Sin.* **2012**, *48*, 1431–1436.

55. Rai, R.; Kelly, S.M.; Martukanitz, R.P.; DebRoy, T. A convective heat-transfer model for partial and full penetration keyhole mode laser welding of a structural steel. *Metall. Mater. Trans. A* **2008**, *39*, 98–112. [CrossRef]

56. Wei, P.S.; Yeh, J.S.; Ting, C.N.; DebRoy, T.; Chung, F.K.; Lin, C.L. The effects of prandtl number on wavy weld boundary. *Int. J. Heat Mass Transf.* **2009**, *52*, 3790–3798. [CrossRef]

57. Bradstreet, B. Effect of surface tension and metal flow on weld bead formation. *Weld. J.* **1968**, *47*, 314s–322s.

58. Eriksson, I.; Powell, J.; Kaplan, A.F.H. Melt behavior on the keyhole front during high speed laser welding. *Opt. Lasers Eng.* **2013**, *51*, 735–740. [CrossRef]

59. Krasnoperov, M.Y.; Pieters, R.R.G.M.; Richardson, I.M. Weld pool geometry during keyhole laser welding of thin steel sheets. *Sci. Technol. Weld. Join.* **2004**, *9*, 501–506. [CrossRef]

60. Seiler, M.; Patschger, A.; Tianis, L.; Rochholz, C.; Bliedtner, J. Experimental determination of influencing factors on the humping phenomenon during laser micro welding of thin metal sheets. *J. Laser Appl.* **2017**, *29*, 022413. [CrossRef]

61. Mohammed, G.R.; Ishak, M.; Aqida, S.N.; Abdulhadi, H.A. The effect of fiber laser parameters on microhardness and microstructure of duplex stainless steel. In Proceedings of the 2nd International Conference of Automotive Innovation & Green Energy Vehicle, Pahang, Malaysia, 2–3 August 2016.

62. El-Batahgy, A.M.; Khourshid, A.F.; Sharef, T. Effect of laser beam welding parameters on microstructure and properties of duplex stainless steel. *Mater. Sci. Appl.* **2011**, *2*, 1443–1451. [CrossRef]

63. Elmer, J.W.; Allen, S.M.; Eagar, T.W. Microstructural development during solidification of stainless steel alloys. *Metall. Trans. A* **1989**, *20*, 2117–2131. [CrossRef]

64. Fukumoto, S.; Tsubakino, H.; Aritoshi, M.; Tomita, T.; Okita, K. Dynamic recrystallisation phenomena of commercial purity aluminium during friction welding. *Mater. Sci. Technol.* **2002**, *18*, 219–225. [CrossRef]

65. Koseki, T. Solidification and solidification structure control of weld metals. *Weld. Int.* **2002**, *16*, 347–365. [CrossRef]

66. Suutala, N.; Takalo, T.; Moisio, T. The relationship between solidification and microstructure in austenitic and austenitic-ferritic stainless steel welds. *Metall. Trans. A* **1979**, *10*, 512–514. [CrossRef]

67. Bhattacharya, A.; Kumar, R. Dissimilar joining between austenitic and duplex stainless steel in double-shielded gmaw: A comparative study. *Mater. Manuf. Process.* **2016**, *31*, 300–310. [CrossRef]

68. Lippold, J.C. Welding metallurgy principles. In *Welding Metallurgy and Weldability*; John Wiley & Sons, Inc.: Hoboken, NJ, USA, 2015; pp. 9–83.

69. Suutala, N. Effect of solidification conditions on the solidification mode in austenitic stainless steels. *Metall. Trans. A* **1983**, *14*, 191–197. [CrossRef]

70. Ajith, P.M.; Sathiya, P.; Aravindan, S. Characterization of microstructure, toughness, and chemical composition of friction-welded joints of UNS S32205 duplex stainless steel. *Friction* **2014**, *2*, 82–91. [CrossRef]

71. Sammaiah, P.; Suresh, A.; Tagore, G.R.N. Mechanical properties of friction welded 6063 aluminum alloy and austenitic stainless steel. *J. Mater. Sci.* **2010**, *45*, 5512–5521. [CrossRef]
72. Tomashchuk, I.; Sallamand, P.; Belyavina, N.; Pilloz, M. Evolution of microstructures and mechanical properties during dissimilar electron beam welding of titanium alloy to stainless steel via copper interlayer. *Mater. Sci. Eng. A* **2013**, *585*, 114–122. [CrossRef]
73. Abdulhadi, H.; Ahmad, S.; Ismail, I.; Ishak, M.; Mohammed, G. Thermally-induced crack evaluation in h13 tool steel. *Metals* **2017**, *7*, 475. [CrossRef]
74. David, S.A.; Babu, S.S.; Vitek, J.M. Welding: Solidification and microstructure. *JOM* **2003**, *55*, 14–20. [CrossRef]
75. Kou, S. Weld metal solidification cracking. In *We'ding Metallurgy*; John Wiley & Sons, Inc.: Hoboken, NJ, USA, 2003; pp. 263–300.
76. Yoo, J.; Kim, B.; Park, Y.; Lee, C. Microstructural evolution and solidification cracking susceptibility of Fe–18mn–0.6C–xAl steel welds. *J. Mater. Sci.* **2015**, *50*, 279–286. [CrossRef]
77. Argandoña, G.; Palacio, J.; Berlanga, C.; Biezma, M.; Rivero, P.; Peña, J.; Rodriguez, R. Effect of the temperature in the mechanical properties of austenite, ferrite and sigma phases of duplex stainless steels using hardness, microhardness and nanoindentation techniques. *Metals* **2017**, *7*, 219. [CrossRef]
78. Corolleur, A.; Fanica, A.; Passot, G. Ferrite content in the heat affected zone of duplex stainless steels. *BHM Berg Hüttenmänn. Monatshefte* **2015**, *160*, 413–418. [CrossRef]
79. Šohaj, P.; Jan, V. Local changes of microhardness in dissimilar weld joints after high temperature exposure. *Key Eng. Mater.* **2014**, *586*, 249–252. [CrossRef]
80. Kathleen, M. *Metals Handbook: Welding, Brazing. and Soldering*, 9th ed.; American Society for Metals: Geauga County, OH, USA, 1983; Volume 6.
81. Nastac, L.; Stefanescu, D.M. An analytical model for solute redistribution during solidification of planar, columnar, or equiaxed morphology. *Metall. Trans. A* **1993**, *24*, 2107–2118. [CrossRef]

Article

Laser Welding of BTi-6431S High Temperature Titanium Alloy

Zhi Zeng [1,*], **J. P. Oliveira** [2], **Xianzheng Bu** [3], **Mao Yang** [1], **Ruoxi Li** [4] **and Zhimin Wang** [3,*]

[1] School of Mechatronics Engineering, University of Electronic Science and Technology of China, Chengdu 611731, China; 201521080159@std.uestc.edu.cn
[2] Department of Materials Science and Engineering, The Ohio State University, Columbus, OH 43221, USA; jp.oliveira@campus.fct.unl.pt
[3] Beijing Hangxing Machinery Manufacture Limited Corporation, Beijing 100013, China; nirubu@163.com
[4] College of Materials Science and Engineering, Beijing University of Technology, Beijing 100124, China; liruoxi_0927@126.com
* Correspondence: zhizeng@uestc.edu.cn (Z.Z.); d201677150@hust.edu.cn (Z.W.); Tel.: +86-028-6183-0229 (Z.Z.); +86-010-8810-3289 (Z.W.)

Received: 27 September 2017; Accepted: 13 November 2017; Published: 15 November 2017

Abstract: A new type of high temperature titanium alloy, BTi-6431S, has recently become the focus of attention as a potential material for aircraft engine applications, which could be used up to 700 °C. Pulsed laser welding was used to butt join the BTi-6431S titanium alloy in order to understand the feasibility of using fusion-based welding techniques on this material. The effect of laser energy on the microstructure and mechanical properties of the joints was investigated. The microstructural features of the joints were characterized by means of microscopy and X-ray diffraction. Tensile testing was conducted at both room temperature and high temperature to simulate potential service conditions. The results show that the microstructure of the laser welded joints consists of primary α phase and needle α′ phase, while the microstructure of the heat affected zone consists of α, β, and needle α′ phases. The tensile strength of the welded joints at room temperature was similar to that of the base material, despite a reduction in the maximum elongation was observed. This was related to the unfavorable microstructure in the welded joints. Nonetheless, based on these results, it is suggested that laser welding is a promising joining technique for the new BTi-6431S titanium alloy for aerospace applications.

Keywords: high temperature titanium alloy; laser welding; microstructure; mechanical properties; BTi-6431S

---

1. Introduction

Titanium alloys are widely used in the aerospace field to manufacture important structures which are used at high temperatures and/or under complex stress states, due to unarguable properties such as excellent corrosion resistance, high specific mechanical resistance, and excellent high temperature performance [1,2]. However, with the development of high performance flying technology, more strict requirements are put forward for the performance of titanium alloys in the aerospace field [3]. The need for improved fatigue resistance, mass reduction, and corrosion resistance in the aircraft industry has raised interest in using titanium alloys with higher specific strength to replace stainless steels and other high temperature alloys in some structural and mechanical parts [4,5].

Many studies have been performed to investigate the welding performance of titanium alloys. It is reported that tungsten inert gas (TIG) welding of Ti-6Al-4V alloy plates resulted in the highest impact strength as compared to laser and electron beam welding [6]. Gao et al. [7] found that for welding thin sheets of titanium alloys, TIG welding led to higher residual distortions and a coarse microstructure. Laser welding can be considerably adaptable for joining titanium alloys in large size and complex structures. Such a joining technique has recently been increasing in popularity as one of the preferred welding methods for most critical titanium components in the aerospace industry [8,9]. Nd:YAG pulsed laser beam welding is an advanced joining process in which melting and solidification of the weld zone takes place after each pulse by a more dense heat source [10]. Squillace et al. investigated the conduction and keyhole regime based on heat input range of Nd:YAG laser beam welding and concluded that weld morphology was highly dependent on the welding regime [11]. Campanelli et al. [12] studied the effect of laser welding parameters on Ti6Al4V alloy and found out that the tensile strength of the joints was of 80% of that of the base material, with fracture occurring in a brittle way, suggesting the need for a post-weld heat treatment to be applied on the welds. Shen et al. [13] applied laser beam welding to weld dissimilar titanium alloys, and the effect of laser beam offset on microstructural characterizations and mechanical properties of the joints were investigated. If proper welding procedures are found to effective join titanium alloys so that the joint achieves the desired properties, it can also result in the efficient utilization of materials and save on expenditure.

In general, welding of Ti alloys can be associated with some metallurgical problems. For example, in TIG welding Ti6246 alloys, the elongation of the weld metal is reported to be of only 0.4% [14]. This weld brittleness is related to the formation of brittle phases, precipitation of hard and brittle particles, and formation of hydrides in the weld metal [11,13,15]. Similarly, in laser welded Ti-6Al-4V, an $\alpha + \beta$ titanium alloy, it is reported that the hardness in the weld zone is 140 HV higher than that of the base material [16]. This increase in the weld hardness is related to the formation of secondary martensite due to the high cooling rates upon weld metal solidification [9,11]. Similar metallurgical problems are experienced when performing dissimilar welding of Ti-based alloys to other materials [17].

Recently, a new type of Ti-Al-Sn-Zr-Mo-Nb-W-Si series high temperature titanium alloy, named BTi-6431S, has become the focus of attention as a potential material for aircraft engine applications which could be used up to 700 °C [18]. It can meet the current high-speed aircraft demand for harsh working conditions, with broad application prospects. Until now, there has only been limited research on the microstructure, mechanical properties, and superplastic forming technology of the BTi-6431S alloy, with reports on the welding performance of this titanium alloy being scarce [13,18,19]. In this study, the effect of laser welding on the high temperature titanium alloy BTi-6431S was investigated. The microstructure and mechanical properties of the laser welded joint were analyzed and tested, providing a theoretical basis and experimental data for its application in future aircraft structures.

2. Materials and Experiments

The BTi-6431S titanium alloy sheets utilized for the laser welding were 200 mm × 100 mm × 2.2 mm. The chemical composition of the base material is listed in Table 1, in weight percentage. Prior to welding, the oxide film on the surface of each work piece was removed using a specific acid solution (4 mL hydrofluoric acid, 10 mL nitric acid, and 86 mL distilled water). The specimens were cleaned with acetone and ethanol, and subsequently dried.

Table 1. Chemical composition of BTi-6431S high temperature titanium alloy (wt %).

Ti	C	Si	Mo	N	O	Zr	Nb	Sn	W	Al	H	Fe
Bal.	0.0077	<0.1	1.26	0.0039	0.068	2.94	1.12	3.21	0.47	6.28	0.0052	0.018

Welding was performed using a Trump HL 4006D laser (TRUMPF Co. Ltd., Ditzingen, Germany) with a wavelength of 1.064 nm. The welding parameters use to perform similar butt joining of the BTi-6431S titanium alloy are depicted in Table 2.

Table 2. Laser welding parameters used to join the BTi-6431S titanium alloy.

Welding Parameters	
Base power (W)	850
Peak power (W)	1600
Pulse duration (ms)	20
Duty ratio (%)	50
Welding speed (mm/min)	1500
Focal length (mm)	120
Focal position (mm)	0
Spot diameter (μm)	400

After welding, cross sections of the joints were prepared for metallographic observation. The microstructural characterization was performed by optical microscopy using an Olympus PMG3 (Olympus Corporation, Tokyo, Japan). X-ray diffraction (XRD) was performed to determine the phases present in the base material and weld regions using a Philips XRD system (Royal Dutch Philips Electronics Ltd., Amsterdam, The Netherlands) operating at 40 kV and 30 mA equipped with a Cu source, which is used as line 0.05° of step size in 2θ and 2 s/step. To examine the variation in the chemical composition across the weld, electron microprobe analysis (EPMA) was performed on the as-polished samples using a JEOL 733 SuperProbe (JEOL Ltd., Tokyo, Japan) operating at 10 kV and 50 nA. A micro-hardness profile across the weld beads of the welded coupons was obtained using a Leco® Microhardness Tester LM248AT (LECO Corporation, Saint Joseph, MO, USA). All the hardness readings were obtained using a load of 500 g with a dwell time of 15 s.

Wire electrodischarge machining (WEDM), abrasive waterjet (AWJ), or laser cutting could be alternative processes to conventional milling to produce tensile test specimens. These processes have the advantage of being extensively used in the sheet metal forming industry, whose effect on tensile strength were lower than 5% [20]. For mechanical characterization of the welded joints, dog-bone shaped specimens were prepared after WEDM (6 mm/min, 0.18 mm Mo wire, V_d = 80 V, I = 3 A) with dimensions depicted in Figure 1. Tensile tests were performed at room temperature and at 650 °C following the ASTM E8-13a standard. A 30 kN Instron 5582 universal tensile testing machine (Instron Corporation, Norwood, MA, USA) using a travel speed of 2 mm/min was used. The tensile testing direction was perpendicular to the weld interface and the welded region was located in the center of the dog-bone specimens. The maximum stress and strain of the samples were recorded. The base material was also tested at both room and high temperature for comparison of the effect of laser welding on the mechanical properties. Following the tensile experiments, the fracture surfaces of the specimens were investigated by scanning electron microscope (SEM) using a Zeiss Supra 55 SEM (Zeiss Group, Heidenheim, Germany).

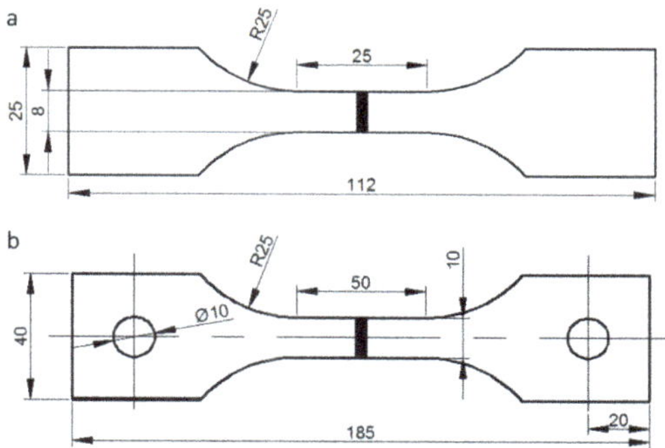

Figure 1. Dimensions of the dog-bone specimens used for tensile testing: (**a**) sample design for room temperature testing; (**b**) sample design for high temperature testing. (Unit: mm).

3. Appearance of Welded Joints

3.1. Appearance of Welded Joints

Figure 2a,b exhibits the appearance of the face and root of the welded joints. It can be observed that the bead formations were uniform with a bright silver color, which indicates that the surfaces of the weld were not contaminated by oxygen, nitrogen, or carbon. However, the surfaces of the joints were slightly concave and misaligned as observed in Figure 2c.

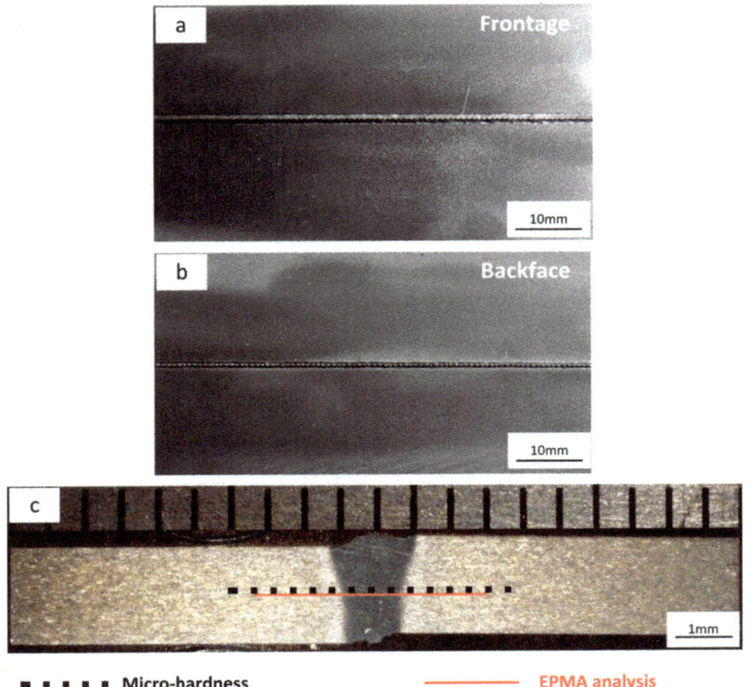

Figure 2. Appearance of BTi-6431S laser welding joint: (**a**) face of the welded joint; (**b**) root of the welded joint; (**c**) macro view of the surface of weld with indication of the location of the microhardness and EPMA testing.

3.2. Microstructural Characterization

The microstructure of the BTi-6431S titanium alloy base material consisted of the β phase which was distributed at the elongated α grain boundaries (shown in Figure 3). This microstructure resulted in the dislocation plug product in the phase boundaries, which could improve the materials toughness, creep resistance, and creep rupture strength at high temperatures, especially for aircraft engine application [21].

The overall morphology of the laser welded joints can be depicted in Figure 4a. The laser welded joint consisted of four zones, which were weld zone (WZ), high temperature heat effect zone (H-HAZ), low temperature heat effect (L-HAZ), and base material (BM).

Some pores were identified in the fusion zone of the welded joint. Several factors may explain the presence of pores: keyhole instability, pre-existence of pores in the base material and poor shielding conditions [22,23]. The weld zone was a typical basket weave pattern (Figure 4b). A large amount of primary acicular structures was uniformly distributed in the α-phase substrate. It has been illustrated that the main phase present in the weld zone was the α phase, as identified by XRD analysis (Figure 4c).

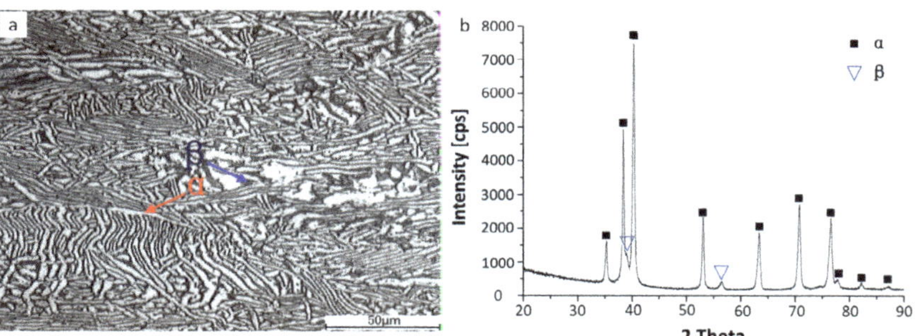

Figure 3. Microstructure and phase composition of base metal: (**a**) optical microscopy; (**b**) X-ray diffraction pattern.

The microstructural transformations of the welded joint primarily depends on the initial microstructure of the base metal and the temperature cycle experienced during the welding process. The key aspects of the temperature cycle include the heating rate, maximum heating temperature, the dwelling time at high temperature, and the cooling rate [24]. During the welding process, the liquid metal within the weld initially crystallized to form the β phase. During the subsequent cooling process, owing to the rapid cooling rate, part of the β phase could not transform into the α phase through atomic diffusion, and it mainly transformed into α′ martensite via shear transformation. The shear transformation mode depends on the regular short-range migration of the atoms in the β phase [25]. For titanium-based alloys [26–28], the starting and finish temperatures of the martensite transformation depends on the chemical composition of the alloy. In general, as the amount of the β stabilizer elements increases, there is a greater resistance for the phase transition required for lattice reconfiguration. Therefore, a higher degree of supercooling was required for the phase transformation, and the martensite start transition temperature reduced. Additionally, for Ti-based alloys, diffusion-controlled transformations are favored for slow cooling rates, whereas high cooling rates as those found in laser welding favor the shear transformation [29]. As such, part of the β phase transformed into the α′ phase. Zhang et al. [30] demonstrated that the β phase stabilizer elements did not only enrich the β phase but also reduced the amount of α′ martensite. As shown in Figure 4c, the XRD results demonstrated that only the diffraction peaks corresponding to the α phase could be observed. As both α′ martensite and acicular α exhibited a hexagonal structure and their lattice constants were similar, there was a superimposition of the diffraction peaks of both the α and α′ phases, thus preventing identification of both phases through this technique.

As the distance from the boundary of the fusion zone increased, the maximum temperature and cooling rate decreased. The distinct HAZ microstructural features are mainly due to the different weld thermal cycles. The optical micrograph in Figure 4a demonstrates that the grain size increased in the high temperature heat affected zone (H-HAZ) towards the low temperature heat affected zone (L-HAZ) and that the microstructure mainly consisted of acicular α and martensite α′ in the H-HAZ. This was mainly because the peak heating temperature is assumed to be higher than the β transition temperature and that the α phase completely transformed into the β phase during the welding process [13]. During the subsequent cooling process, the β phase transformed into acicular α and martensite α′ as a result of the rapid cooling rate. However, the cooling rate was still slower than that of the WZ and the amount of martensite α′ in the H-HAZ was less than that in the WZ.

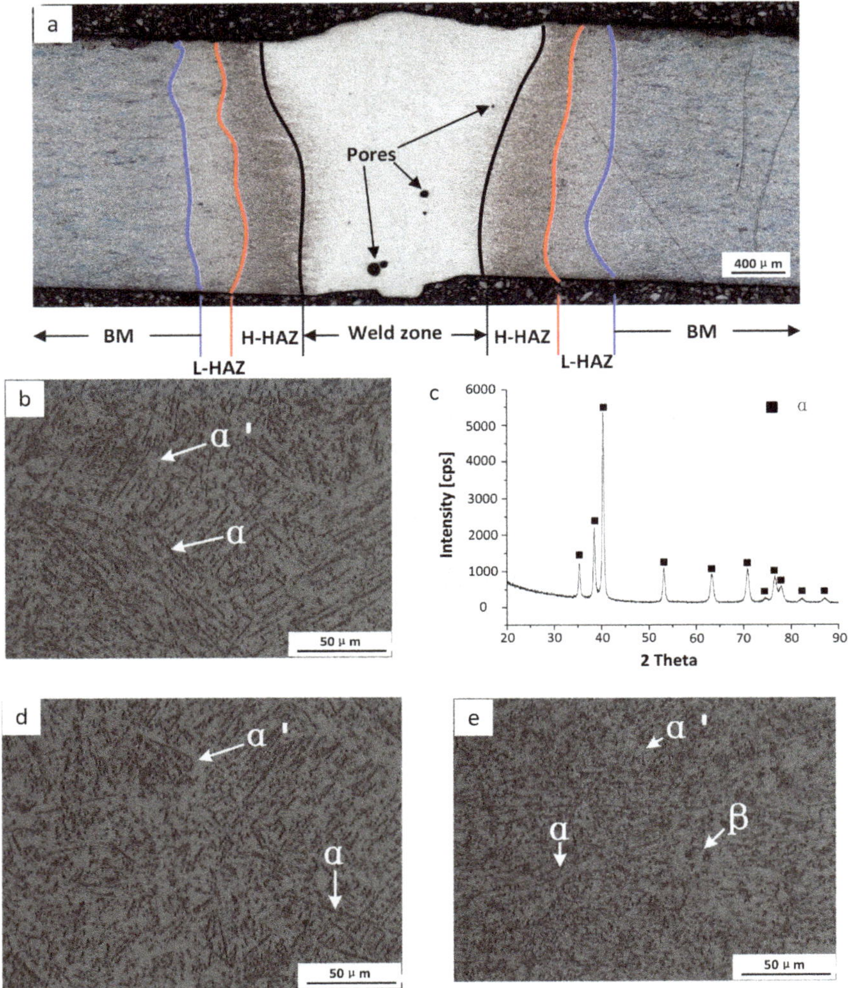

Figure 4. Macro and microstructure of joint and X-ray diffraction pattern: (**a**) macrostructure; (**b**) microstructure of the welded region; (**c**) X-ray diffraction pattern of the welded region; (**d**) microstructure of high temperature heat affected zone; (**e**) microstructure of low temperature heat affected zone.

The L-HAZ extends from a position where the weld thermal cycle has no observable effect on the original microstructure to a position where the temperature is lower than the β transition temperature. The microstructure of the L-HAZ differed from the microstructure adjacent to the fusion line where the temperature was above the β phase transition temperature (H-HAZ). Figure 4e showed that the microstructure in the L-HAZ mainly consisted of a mixture of secondary α′, primary α, and prior β phase. In the L-HAZ, as the distance between the base metal and the fusion line decreased, the prior β phase gradually transformed into the secondary α′ phase. This was mainly because of the peak heating temperature, which is assumed to be lower than the β transition temperature and that the temperature of the L-HAZ reached the zone of the α and β phase transition temperature [31,32]. As the temperature in the L-HAZ was not high enough and the retention time was not long enough, the microstructure consisted of α and β phase under a high temperature and there was still a significant amount of primary α phase retained. In the subsequent cooling process, transformation of the α phase did not occur, and the secondary α phase formed at the grain boundaries and grain interiors of the β phase. Thus, the microstructure of the L-HAZ exhibited a mixture of secondary α′, primary α, and prior β phase. The temperature increases as the distance from the fusion line decreases. Therefore, the closer to the fusion line, the more prior β phase would transfer into the secondary α′ phase during the subsequent cooling process.

The chemical composition variation along the welded joint of major alloying elements in the fusion zone of BTi-6431S, determined by EPMA, are presented in Figure 5. It can be observed that the distribution of Al, Nb, Sn, Zr, and Mo from the fusion zone to the base material does not change significantly. This is a sign of a sound weld since losing alloying elements during welding in the fusion zone due to evaporation is not desirable due to possible detrimental effects on the mechanical properties of the joint. For Al, Nb, Zr, and Mo there are slight fluctuations (about 8%) in the distribution of these alloying elements in the L-HAZ and BM regions. This can be explained in terms of the presence of α and β phases in both the L-HAZ and BM. Indeed, due to a small testing zone (about 5 μm) and coarser α and β sizes, depending on whether the measurement location was on α or β phase, a different chemical composition was measured as these alloying elements have different affinities for α and β phases [33].

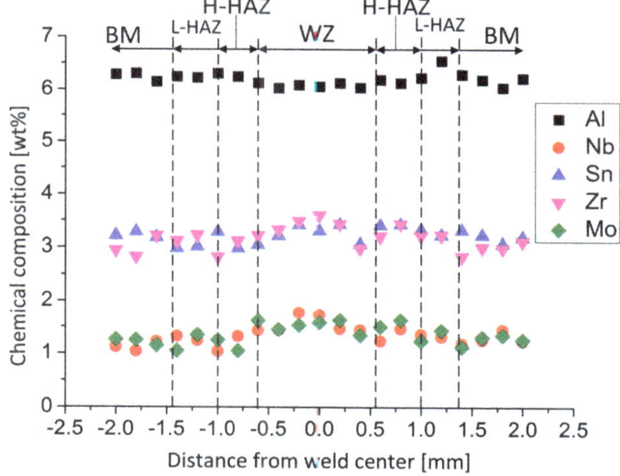

Figure 5. Chemical composition of BTi-6431S laser weld joint.

3.3. Mechanical Properties of the Joints

In order to map the variation in the mechanical properties across the weld, microhardness measurements were performed. The variation in the microhardness across the material is plotted in Figure 6. It can be seen that by moving from the BM toward the L-HAZ and H-HAZ the hardness increases in the welded specimen. For the BM, the hardness varies from 351 to 364 HV. This small variation can be related to the indent size and α/β size. The average indent diameter was 24 μm while α and β had the respective size of 32.5 ± 15.3 μm and 12.7 ± 5.3 μm as determined by digital image analysis taking as reference the microstructure of the base material depicted in Figure 3a. Therefore, hardness can be measured either mostly on the β phase, or partly on α and partly on β. Since β is softer and more ductile than α, there will be some variation in the measured hardness of the base material. In contrast, in the weld zone the microstructure was uniform which increased the hardness values to 412 ± 4 HV with α and secondary α′ phases as previously presented. Additionally, an increased amount of harder acicular martensite α resulted in a significantly higher hardness than the BM. In the

WZ, it is noted that a small kink in the hardness peak leading two micro-hardness peaks was observed. Also, the variation in the hardness in along the WZ is not significant. In the H-HAZ, closer to WZ, more martensite α' is formed due to a higher peak temperature. The increasing amount of α' from the base material/heat affected zone interface towards the fusion boundary justifies the increasing hardness observed in Figure 6. Almost a symmetric hardness profile was observed for the two sides of the weld centerline. The observed symmetry in the microhardness profiles with respect to the weld centerline indicates that the two weld parts are affected thermally almost in the same way.

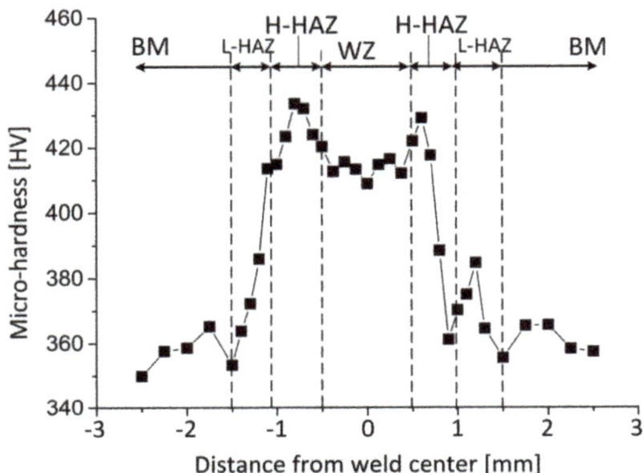

Figure 6. Microhardness profile along the cross section of the laser welded joint.

A summary of the results of the tensile tests conducted at room temperature for welded specimens and base material are depicted in Table 3. The tensile strength for the welds was determined from three dog-bone tensile samples extracted from the laser welded sheet. At room temperature, the average ultimate tensile strength (UTS) was 1050 MPa, the average elongation was 4.9% and the yield strength (YS) is 993 MPa. One must be aware that the tensile properties of a given sample specimen may be dependent on the methodology used to obtain it, as described in the literature [20,34,35]. However, since the same process was used for all test specimens any potential detrimental effect is diluted within the test itself.

Fracture of the welded joint occurred in the heat affected zone. Compared to the base material, the tensile strength is nearly the same, while the elongation is decreased significantly. According to [36], the strength of the existing phases demonstrates the following order: $\alpha' > \alpha > \beta$. Additionally, α' possesses very high hardness and low ductility and toughness [37]. The microstructure of the fusion zone consisted of martensite α' and a slender acicular α phase, whereas the microstructure of the heat affected zone consisted of acicular α phase, α' martensite, and β phase. The combination of coarse grain size, presence of α' martensite, and sharp hardness variation promoted fracture in the heat affected zone of the joints. The equiaxed morphology of the grains in the base material in terms of its mechanical properties is better than the microstructure which composes the heat affected and fusion zones. The present of coarse grain structures and phases with very distinct mechanical properties, may cause localized deformation during tensile testing and therefore premature rupture in one of these regions may occur. When the sample is tensile loaded, the dislocations are able to pile up between the α and β phases, which could prevent the micro-cracks initiation-extension and present better toughness than the welds [38]. The tensile results were consistent with the microstructural distribution of the welded joints, and it suggested that the post-weld heat treatment (PWHT) can improve the mechanical properties in brittle Ti alloy welds [18]. Additionally, it was reported that when welding Ti-Al alloys as the one used in this investigation, careful control of the welding parameters is necessary to obtain a sound joint [39,40].

Table 3. Summary of the mechanical testing performed on the welded specimens and base material at room temperature (RT) and 650 °C. For the base material the average results presented are based on a total of three samples tested at both temperatures.

Mechanical Testing Summary	Sample Reference	Temperature (°C)	UTS (MPa)	YS (MPa)	Elongation (%)	Fracture Location
Welded specimens	W1	RT	1072	988	5.1	HAZ
	W2	RT	1040	1002	4.8	HAZ
	W3	RT	1045	991	5.0	HAZ
	W4	650	776	683	4.4	HAZ
	W5	650	755	672	4.2	HAZ
	W6	650	765	681	4.5	HAZ
Base material	BM1	RT	1060 ± 15	982 ± 13	16.5 ± 0.3	-
	BM2	650	890 ± 17	824 ± 16	13.2 ± 0.5	-

When the tensile tests were performed at 650 °C, the yield strength and tensile strength drastically decreases to approximately 679 MPa and 765 MPa (about 82% and 86% of that of the base material at the same testing temperature), respectively. However, no significant reduction in the elongation of the welded joints tested at room and high temperature was observed. It should be noted, however, that the BTi-6431S laser weld could still support significant loads at high temperature, which is of significant importance for aircraft engine design using this new material. Similar to what was observed on the room temperature tensile tests, all the welded specimens fractured in the heat affected zone with evidence of necking and it could be observed that the surfaces were oriented at 45° from the tensile axis.

Figure 7 depicts the fracture surfaces of the welded specimens in the tensile testing at room temperature and 650 °C. The fracture surfaces demonstrated that the failure was mixed mode combining brittle and ductile features for the samples tested at room temperature. It was observed from the SEM analysis that the fracture surface tends to be mainly intergranular, and there are still some small and shallow dimples (ranging between 5 and 10 μm in size) which were shown in Figure 7c. Some of the pores that were observed in the microstructural observation performed using optical microscopy can also be observed in the fracture surface of the welded joint.

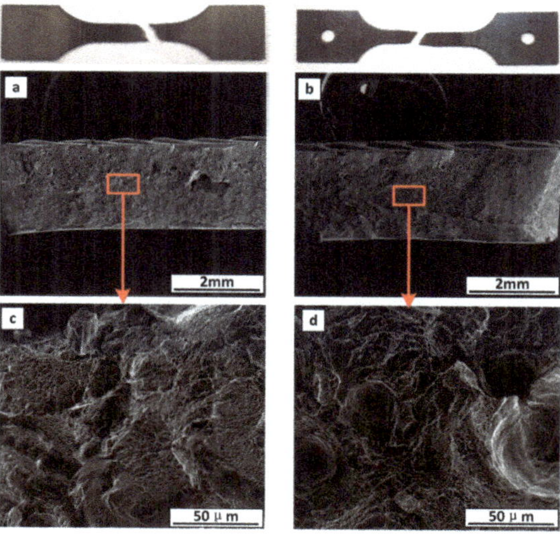

Figure 7. Fracture surface of welded joints: (**a**) fracture surface after tensile test performed at room temperature; (**b**) fracture surface after tensile test performed at 650 °C; (**c**) high magnification of the red box in (**a**); (**d**) high magnification of the red box in (**b**).

4. Conclusions

The quality of the laser welded BTi-6431S titanium alloy sheet in terms of tensile properties, microhardness, chemical composition distribution, and microstructure in the as-welded condition was investigated. The following conclusion can be drawn:

1. The base material microstructure of the BTi-6431S high temperature titanium alloy consisted of long strip and equiaxed α phases, with a small amount of β phase distributed between the α phase. The welded region was composed of primary α phase and a significant amount of acicular α' phase, showing basket weave pattern which resulted from the high degree of supercooling during the laser welding process. The microstructure in heat affected zone contained primary α phase, a small amount of β phase, and acicular α' phase.

2. The microhardness values of the weld zone and heat affected zone were higher than that of base metal. This was due to the fast cooling rate during the laser welding process, resulting in a large amount of α' martensitic phase distributed in the weld zone and the heat affected zone near the fusion line.

3. In this experiment, the tensile strength of the welded joint at room temperature was basically equivalent to the base metal due to the high strength phase distribution and nearly unchanged chemical composition. The tensile strength at high temperature decreased to about 765 MPa which tends to be mainly cleavage fracture with some dimples. The plasticity of the joint was obviously lower than that of the parent metal, and the elongation was about 5%. This was related to the formation of α' martensite in both the heat affected and fusion zone. All the specimens fractured in the heat affected zone with evidence of necking and it could be observed that the surfaces were oriented at 45° with the tensile axis.

Acknowledgments: Zhi Zeng acknowledge Science and Technology Planning Project of Guangdong Province (Grant No. 2016A010102002) and the Fundamental Research Funds for the Central Universities (Grant No. ZYGX2015J149).

Author Contributions: Zhi Zeng and Zhimin Wang conceived and designed the experiments; Xianzheng Bu performed the experiments; Ruoxi Li analyzed the data; Mao Yang contributed materials/analysis tools; Zhi Zeng and J. P. Oliveira wrote the paper.

References

1. Schwab, H.; Palm, F.; Kuhn, U.; Eckert, J. Microstructure and mechanical properties of the near-beta titanium alloy Ti-5553 processed by selective laser melting. *Mater. Des.* **2016**, *105*, 75–80. [CrossRef]

2. Zeng, Z.; Wu, X.; Yang, M.; Peng, B. Welding distortion prediction in 5A06 aluminum alloy complex structure via inherent strain method. *Metals* **2016**, *6*, 214. [CrossRef]

3. Babu, N.K.; Sundara Raman, S.G.; Murthy, C.V.S.; Reddy, G.M. Effect of beam oscillation on fatigue life of Ti-6Al-4V electron beam weldments. *Mater. Sci. Eng. A* **2017**, *471*, 113–119. [CrossRef]

4. Peters, M.; Kumpfert, J.; Ward, C.H.; Leyens, C. Titanium alloys for aerospace applications. *Adv. Eng. Mater.* **2010**, *5*, 419–427. [CrossRef]

5. Lu, K. The future of metals. *Science* **2010**, *328*, 319–320. [CrossRef] [PubMed]

6. Balasubramanian, T.S.; Balakrishnan, M.; Balasubramanian, V.; Manickam, M.A.M. Influence of welding processes on microstructure, tensile and impact properties of Ti-6Al-4V alloy joints. *Chin. J. Nonferrous Met.* **2011**, *21*, 1253–1262. [CrossRef]

7. Gao, X.L.; Zhang, L.J.; Liu, J.; Zhang, J.X. A comparative study of pulsed Nd:YAG laser welding and TIG welding of thin Ti6Al4V titanium alloy plate. *Mater. Sci. Eng. A* **2013**, *559*, 14–21. [CrossRef]

8. Lei, Z.L.; Dong, Z.J.; Chen, Y.B.; Huang, L.; Zhu, R.C. Microstructure and mechanical properties of laser welded Ti-22Al-27Nb/TC4 dissimilar alloys. *Mater. Sci. Eng. A* **2013**, *559*, 909–916. [CrossRef]

9. Winco, K.C.; Yung, B.R.; Leec, W.B.; Fenn, R. An investigation into welding parameters affecting the tensile properties of titanium welds. *J. Mater. Proc. Technol.* **1997**, *63*, 759–764.

10. Ghaini, F.M.; Hamedi, M.J.; Torkamany, M.J.; Sabbaghzadeh, J. Weld metal microstructural characteristics in pulsed Nd:YAG laser welding. *Scr. Mater.* **2007**, *56*, 955–958.

11. Squillace, A.; Prisco, U.; Ciliberto, S.; Astarita, A. Effect of welding parameters on morphology and mechanical properties of Ti-6Al-4V laser beam welded butt joints. *J. Mater. Proc. Technol.* **2012**, *212*, 427–436. [CrossRef]

12. Campanelli, S.L.; Casalino, G.; Mortello, M.; Angelastro, A.; Ludovico, A.D. Microstructural characteristics and mechanical properties of Ti6Al4V alloy fiber laser welds. *Procedia CIRP* **2015**, *33*, 429–434. [CrossRef]

13. Zhang, H.; Hu, S.; Shen, J.; Li, D.; Bu, X. Effect of laser beam offset on microstructure and mechanical properties of pulsed laser welded BTi-6431S/TA15 dissimilar titanium alloys. *Opt. Laser Technol.* **2015**, *74*, 158–166. [CrossRef]

14. Mullins, F.D.; Becker, D.W. Property-microstructure relationships in metastable-beta titanium alloy Weldments. *Weld. Res. Suppl.* **1980**, *6*, 177–182.

15. Sabol, J.C.; Marvel, C.J.; Watanabe, M.; Pasang, T.; Misiolek, W.Z. Confirmation of the ω-phase in electron beam welded Ti-5Al-5V-5Mo-3Cr by high-resolution scanning transmission electron microscopy: An initial investigation into its effects on embrittlement. *Scr. Mater.* **2014**, *92*, 15–18. [CrossRef]

16. Akman, E.; Demir, A.; Canel, T.; Sinmazcelik, T. Laser welding of Ti6Al4V titanium alloys. *J. Mater. Proc. Technol.* **2009**, *209*, 3705–3713. [CrossRef]

17. Oliveira, J.P.; Miranda, R.M.; Fernandes, F.M.B. Welding and joining of NiTi shape memory alloys: A review. *Prog. Mater. Sci.* **2017**, *88*, 412–466. [CrossRef]

18. Zhang, W.J.; Song, X.Y.; Hui, S.X.; Ye, W.J.; Wang, Y.L. Effect of single annealing on microstructure and mechanical properties of BTi-6431S titanium alloy. *Chin. J. Nonferrous Met.* **2013**, *23*, 1530–1535.

19. Wang, X.X.; Wang, W.Q.; Ma, H.H. Microstructure and mechanical properties of high temperature and high strength BTi-6431S alloy at 700 °C. *Chin. J. Nonferrous Met.* **2010**, *20*, 792–795.

20. Krahmer, D.M.; Polvorosa, R.; de Lacalle, L.N.L.; Alonso-Pinillos, U.; Abate, G.; Riu, F. Alternatives for specimen manufacturing in tensile testing of steel plates. *Exp. Tech.* **2016**, *40*, 1555–1565. [CrossRef]

21. Casavola, C.; Pappalettere, C.; Pluvinage, G. Fatigue resistance of titanium laser and hybrid welded joints. *Mater. Des.* **2011**, *32*, 3127–3135. [CrossRef]

22. Zhao, H.; Debroy, T. Pore formation during laser beam welding of die-cast magnesium alloy AM60B-mechanism and remedy. *Weld. J.* **2001**, *80*, 204–210.

23. Zeng, Z.; Panton, B.; Oliveira, J.P.; Han, A.; Zhou, Y.N. Dissimilar laser welding of NiTi shape memory alloy and copper. *Smart Mater. Struct.* **2015**, *24*, 125036. [CrossRef]

24. Lee, J.Y.; Ko, S.H.; Farson, D.F.; Yoo, C.C. Mechanism of keyhole formation and stability in stationary laser welding. *J. Phys. D Appl. Phys.* **2002**, *35*, 1570–1576 [CrossRef]

25. Zhang, B.G.; Shi, M.X.; Chen, G.Q.; Feng, J.C. Microstructure and defect of titanium alloy electron beam deep penetration welded joint. *Chin. J. Nonferrous Met.* **2012**, *22*, 2633–2637. [CrossRef]

26. Deng, A.H. Martensitic transformation of titanium alloys. *Chin. J. Nonferrous Met.* **1999**, *20*, 193–199.

27. Thompson, S.A. An overview of nickel-titanium alloys used in dentistry. *Int. Endod. J.* **2000**, *33*, 297–310. [CrossRef] [PubMed]

28. Donachie, M.J. *Titanium: A Technical Guide*, 2nd ed.; ASM International: Novelty, OH, USA, 2000.

29. Threadgill, P.L. The prospects for joining titanium aluminides. *Mater. Sci. Eng. A* **1995**, *192*, 640–646. [CrossRef]

30. Zhang, H.T.; He, P.; Feng, J.C.; Wu, H.Q. Interficial microstructure and strength of the dissimilar joint Ti 3 Al/TC4 welded by the electron beam process. *Mater. Sci. Eng. A* **2006**, *425*, 255–259. [CrossRef]

31. Liu, L.; Du, X.; Zhu, M.; Chen, G. Research on the microstructure and properties of weld repairs in TA15 titanium alloy. *Mater. Sci. Eng. A* **2007**, *445*, 691–696. [CrossRef]

32. Hsieh, C.T.; Chu, C.Y.; Shiue, R.K.; Tsay, L.W. The effect of post-weld heat treatment on the notched tensile fracture of Ti-6Al-4V to Ti-6Al-6V-2Sn dissimilar laser welds. *Mater. Des.* **2014**, *59*, 227–232. [CrossRef]

33. Junaid, M.; Baig, M.N.; Shamir, M.; Khan, F.N.; Rehman, K.; Haider, J. A comparative study of pulsed laser and pulsed TIG welding of Ti-5Al-2.5Sn titanium alloy sheet. *J. Mater. Proc. Technol.* **2017**, *242*, 24–38. [CrossRef]

34. Silva, C.M.; Rosa, P.A.; Martins, P.A. Innovative testing machines and methodologies for the mechanical characterization of materials. *Exp. Tech.* **2016**, *40*, 569–581. [CrossRef]

35. Han, Z.; Weatherley, D.; Puscasu, R. A relationship between tensile strength and loading stress governing the onset of mode I crack propagation obtained via numerical investigations using a bonded particle model. *Int. J. Num. Anal. Meth. Geomech.* **2017**, 1–13. [CrossRef]
36. Lu, W.; Shi, Y.; Lei, Y.; Li, X. Effect of electron beam welding on the micro-structures and mechanical properties of thick TC4-DT alloy. *Mater. Des.* **2012**, *34*, 509–515. [CrossRef]
37. Sun, Z.; Annergren, I.; Pan, D.; Mai, T.A. Effect of laser surface remelting on the corrosion behavior of commercially pure titanium sheet. *Mater. Sci. Eng. A* **2003**, *345*, 293–300. [CrossRef]
38. Wang, Z.M.; Yao, W.; Shi, K. Research on laser welding and superplastic process for superalloy BTI-62421S. *Aeronaut. Manuf. Technol.* **2011**, *16*, 28–31.
39. Arenas, M.F.; Acoff, V.L. Analysis of Gamma titanium aluminide welds produced by gas tungsten arc welding. *Weld. J.* **2003**, *82*, 110–115.
40. Cao, J.; Qi, J.; Song, X.; Feng, J. Welding and joining of titanium aluminides. *Materials* **2014**, *7*, 4930–4962. [CrossRef] [PubMed]

Article

Microstructures and Mechanical Properties of Dissimilar Al/Steel Butt Joints Produced by Autogenous Laser Keyhole Welding

Li Cui [1,*], Boxu Chen [1], Wei Qian [1], Dingyong He [1] and Li Chen [2]

[1] College of Materials Science and Engineering, Beijing University of Technology, Beijing 100124, China; thecbx92@emails.bjut.edu.cn (B.C.); qwn@emails.bjut.edu.cn (W.Q.); dyhe@bjut.edu.cn (D.H.)

[2] High Energy Density Beam Processing Technology Department, Aeronautical Manufacturing Technology Research Institute, Beijing 100024, China; ouchenxi@163.com

* Correspondence: cuili@bjut.edu.cn; Tel.: +86-10-6739-2523

Received: 30 September 2017; Accepted: 8 November 2017; Published: 10 November 2017

Abstract: Dissimilar Al/steel butt joints of 6.0 mm thick plates have been achieved using fiber laser keyhole welding autogenously. The cross sections, interface microstructures, hardness and tensile properties of Al/steel butt joints obtained under different travel speeds and laser beam offsets were investigated. The phase morphology and thickness of the intermetallic compound (IMC) layers at the interface were analyzed by scanning electronic microscopes (SEM) using the energy-dispersive spectrometry (EDS) and electron back-scattered diffraction (EBSD) techniques. The results show that travel speeds and laser beam offsets are of considerable importance for the weld shape, morphology and thickness of IMC layers, and ultimate tensile strength (UTS) of Al/steel butt joints. This proves that the IMC layers consist of Fe_2Al_5 phases and Fe_4Al_{13} phases by EBSD phase mapping. Increasing laser beam offsets from 0.3 mm to 0.7 mm significantly decreases the quantity of Fe_4Al_{13} phases and the thickness of Fe_2Al_5 layers at the interface. During tensile processing, the Fe_2Al_5 layer with the weakest bonding strength is the most brittle region at the interface. However, an intergranular fracture that occurred at Fe_2Al_5 layers leads to a relatively high UTS of Al/steel butt joints.

Keywords: Al/steel joints; laser keyhole welding; IMC layers; tensile properties; EBSD phase mapping

---

1. Introduction

Dissimilar joining of Al/steel joints has become increasingly significant in industrial applications with a particular weight-saving interest [1,2]. Although research on joining dissimilar steel/Al alloy joints with fusion-based welding processes is extensive, most previous studies have been focused on lapped joints or overlap joints of less than 2 mm thick sheets for automotive applications [2,3]. In recent years, the development of fast-speed vessels requires improved solutions to the dissimilar joining of Al alloy superstructures to the steel hull for achieving weight reduction in the shipbuilding industry [4,5]. Thus, the dissimilar joining of thick Al/steel joints has received considerable attention in shipbuilding applications [6]. In general, an explosive Al/steel transition joint is used in the shipyard for the dissimilar joining of Al/steel joints. However, the cost of the transition joints bonded by explosive welding is high due to the complex manufacturing process [4,5,7]. Therefore, it is highly important to develop new welding process for dissimilar joining Al/steel joints directly.

It is well known that dissimilar joining Al/steel joints is extremely challenging due to the huge disparity in thermal–physical properties between steels and Al alloys [4,8,9]. One of the main issues associated with welding Al/steel joints is the formation of thick intermetallic compound (IMC) layers at the interface [10,11]. While the IMC layer is necessary for achieving a reliable Al/steel joint, the brittleness of Al/steel joints is increased and the mechanical property is reduced if the IMC layer thickness is beyond a certain range [12,13]. The key to improving the mechanical properties of the joints is to control the formation of IMCs and limit IMCs thickness within 10 μm [14]. Therefore, considerable effort has been put into inhibiting the growth of IMC layers at the interface of Al/steel joints. The results show that welding processes allowing precise control to potentially minimize heat input (by an optimization of process parameters) are required to obtain thin IMC layers [15,16]. In view of this, a laser is regarded as a desirable heating source for joining dissimilar steel to Al alloys because of its characteristics of high power density, low welding heat input, and precise control of the location of laser focus, etc., which allow better control of the IMC layer thickness at the interface [7,17,18].

Some work has been reported on the laser welding-brazing process for dissimilar steel/Al alloy joining. During the welding, metals with a low melting point (e.g., Al alloy) and the filler wire are melted by laser or a laser-arc hybrid heating source, while metals with a high melting point (e.g., steel) is brazed by the molten metal in the solid state [1,16,17]. Thus, the resulted joint has dual characteristics: it is a welded joint on the low melting point metal side and a brazed joint on the high melting point metal side [19]. Compared to the fusion-welding processes, this process is effective for limiting the mixture of the dissimilar molten metal and inhibiting the formation of brittle IMCs at the interface. However, it is inefficient when the laser is irradiated at the low melting point Al alloy side, because Al has a high reflectivity for lasers. In addition, using laser welding–brazing can achieve an acceptable steel/Al alloy joint, but the wettability is not good when liquid Al alloy meets the solid steel. So the addition of flux, coating or other methods is usually applied to improve wettability [14,18,20,21].

Laser keyhole welding is one of the most widely used of the various laser-welding processes in dissimilar joints. With the laser keyhole-welding technique, rather good mechanical properties were obtained by irradiating steels placed on Al sheets, and limiting Al to steels mixed with a reduced steel penetration in Al alloys [8]. This indicates that the laser keyhole welding process was mainly used to lapped or overlapped joints [12,13,15,22–24]. It can be, but rarely is, used for welding Al/steel butt joints requiring accurate groove preparations, flux and filler wires [24–27]. Although the use of filler wires or flux leads to some improvement in the joint properties, laser welding with filler materials has usually been considered too difficult for industrial applications because there are too many parameters and the requirements for wire positioning are too stringent [28], thereby reducing efficiency and productivity. Thus, the possibility to work in autogenous laser keyhole welding is of important technological interest, as it allows a reduction of the time taken in joint preparation and has fewer parameters to control, making it easier to obtain reproducible and stable results [29].

In the present study, an autogenous laser keyhole welding process was introduced to join 6 mm thick steel to Al alloy plates in a butt configuration. The primary objective of this work was to understand the influences of travel speeds and laser beam offsets on the weld shapes, the IMC layers and ultimate tensile strength (UTS) of Al/steel butt joints. Using electron back-scattered diffraction (EBSD) phase mapping, the phase identification and the fracture behavior of IMC layers at the interface were investigated. Based on microstructual and tensile testing, correlations between the processing parameters, IMC layers, and the UTS of Al/steel butt joints were established.

2. Materials and Methods

The materials studied in the present work were 6 mm thick 5083 alloy and Q235 low carbon steel plates arranged in a butt configuration, and their nominal compositions are shown in Table 1. Before welding, the oxides on the surface of the aluminum alloy sheet were removed with abrasive paper, and then the surface of the aluminum alloy and steel plates were cleaned with acetone to remove grease and residue. With particular attention to the butt joint, the Al and steel plate edges of the butt joint were also ground, brushed using a stainless steel wire brush, and cleaned thoroughly with ethanol in order to remove the oxide layer in the welding gap.

Table 1. Nominal composition of Q235steel and AA5083 alloy (wt %).

Materials	Si	Mn	S	P	Cr	Ni	Cu	C	Fe	Mg	Zn	Ti	Other	Al
Q235	≤0.35	0.3~0.7	≤0.045	≤0.045	≤0.30	≤0.30	≤0.30	≤0.2	Base	-	-	-	-	-
5083	≤0.40	0.4~1.0			0.05~0.25		≤0.10		0.4	4.0~4.9	0.25	≤0.15	0.15	Bal.

The autogenous laser keyhole welding was carried out by a 6 kW YLR-6000 fiber laser (IPG Photonics, Oxford, MA, USA) with an emission wavelength of 1.07 μm delivering in continuous wave mode through an output fiber core diameter of 100 μm. The laser beam was transmitted through a processing fiber with a diameter of 300 μm, collimated by a lens of 150 mm focal length, and then focused on the materials by a focusing lens with 200 mm focal length. Accordingly, the spot size of focused laser beam was approximately 0.3 mm when the beam-defocused distance was 0 mm. Throughout the experiments, the primary and back-shielding gas, using ultrahigh purity argon gas, was supplied at flow rates of 15 L/min.

The schematic illustration of the welding setup and the principle of laser keyhole-welded Al/steel butt joints is presented in Figure 1. During welding, the focused laser beam was directed onto the steel plate to generate a keyhole, and the solidification of the welding pool formed a weld penetration. Important process parameters for laser keyhole welding affected the weld shapes, including laser power (P), travel speed (v), defocused distance (f), and laser beam offset (Δd). P and v are generally important process variables affecting the heat input, as in regular arc welding. Δd refers to the distance from the laser focus to the interface illustrated in Figure 1b, and is an important parameter only in laser welding, which also controls the welding thermal cycles and cooling rates at the interface [24,30,31]. As a result, the process parameters of Δd and v were considered for studying the process parameter influence upon the laser keyhole welded Al/steel butt joints. Thus, $\Delta d = 0.3$–0.7 mm and $v = 0.6$–1.2 m/min were varied discretely under the constant conditions of $P = 3.25$ kW and $f = 0$ mm.

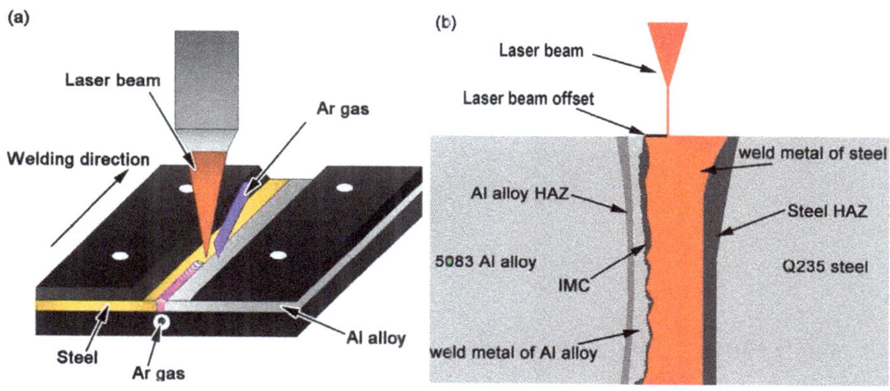

Figure 1. Schematic illustration of laser keyhole welding of Al/steel butt joints: (**a**) welding setup; (**b**) principle of laser keyhole welding.

The quality of welded joints is significantly affected by the defects generated in the process of welding dissimilar materials. Thus, after welding, the welding quality was evaluated according to the Chinese welding procedure qualification code DL/T868-2004 and the international standard EN ISO 6520-1 for dissimilar welded joints. An acceptable joint is one in which the weld ripples, the weld width is uniform, and the weld surface should have no visible porosity or hot cracking. Also, there should be no excessive asymmetry or top and root concavity. After the weld surface was checked, X-ray inspection was performed, and the determined quality should be less than grade

II. Then, the acceptable Al/steel butt joints were cut by electro-discharge machining, and this was followed by grinding, polishing and etching. Keller's reagent and Nital acid were used for etching the Al alloys and steels, respectively. The weld appearance and cross-sections were examined using a BX51M optical microscope (Olympus, Tokyo, Japan). The top width (W_1) and bottom width (W_2) of the weld formed in the steel side, and the area (S) of the heat-affected zone (HAZ) formed in the Al alloy side, were used to evaluate the weld shapes of the Al/steel butt joints, as illustrated in Figure 2.

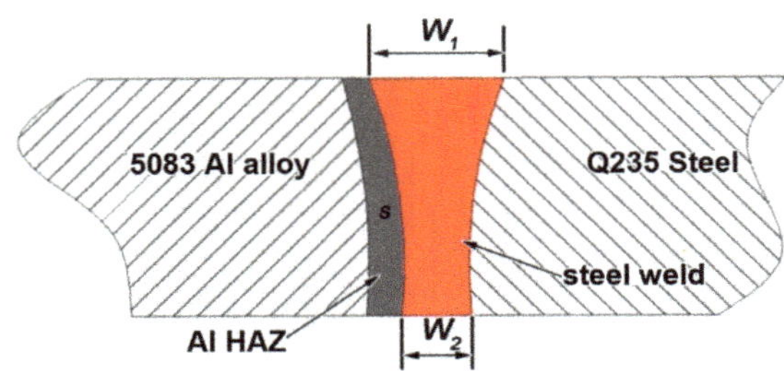

Figure 2. Diagram of dimensions of the weld cross section in the Al/steel butt joints.

Vickers microhardness was measured using a HXD-10007 digital and intelligent micro-hardness tester (Caikon, Shanghai, China) with a load of 100 N and 10 s holding time. For each value, the average of five measurements was taken. Uniaxial transverse tensile tests of the Al/steel butt joints were carried out in accordance with the GB/T 228-2002 and GB/T 2651-2001 standards at room temperature using a MTS810 testing machine (MTS, Eden Prairie, MN, USA) operating with a load rate of 1 mm/min. The average UTS value was determined by calculating three tensile specimens.

The IMC layers at the interface of the Al/steel butt joints were preliminarily examined using scanning electron microscopy (SEM, FEI, Hillsboro, OR, USA) with a QUANTA FEG 650 equipped with an energy-dispersive X-ray spectrometer (EDS). Then, EBSD phase mapping was examined in a rectangular zone using a step size of 0.3 μm between two measurements in order to further identify the IMC phases. The EBSD analysis was conducted by a Quanta FEI 650 SEM field emission gun operated at 30 kV and 1.0 nm. The phases and thickness of the IMC layers were analyzed with the HKL-Channel software package (Oxford Instruments, Witney, Oxon, UK). After tensile testing, the failed samples were observed using the SEM and EBSD techniques to study the facture behavior of the IMC layers.

3. Results

3.1. Weld Appearance of Al/Steel Butt Joints

Figure 3 shows a representative weld appearance of an Al/steel butt joint under the conditions of $\Delta d = 0.6$ mm, $v = 1.0$ m/min, $P = 3250$ W and $f = 0$ mm. It is obvious that the top and bottom appearance of the weld are uniform, with regular ripples, except for a little irregularity in proximity to the Al alloys. This means that a relatively acceptable weld appearance of Al/steel butt joints has been achieved by the autogenous laser keyhole-welding process under the optimum welding conditions.

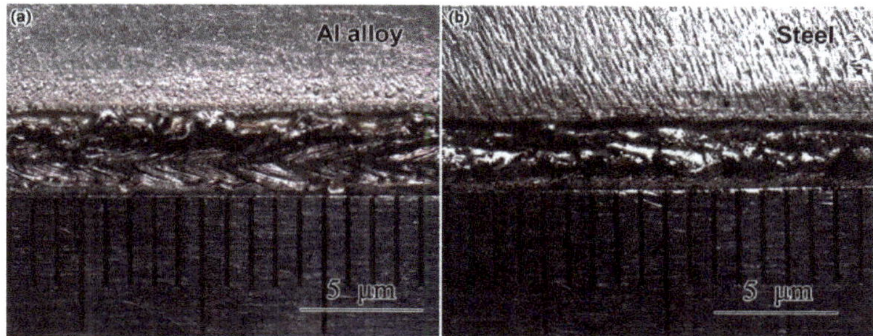

Figure 3. Weld appearance of Al/steel butt joints produced by autogenous laser keyhole welding: (**a**) top surface; (**b**) bottom surface.

3.2. Cross Sections of Al/Steel Butt Joints

Figure 4 shows the cross sections of the Al/steel butt joints obtained at different travel speeds under the condition of Δd = 0.6 mm, P = 3.25 kW and f = 0 mm. For all the travel speeds, each weld is fully penetrated in keyhole mode. The cross sections of the welds at a travel speed of 0.6 m/min and 0.8 m/min have a "champagne glass" shape, while the weld at 1.0 m/min has a "nail" tape. The welds at 1.1 m/min and 1.2 m/min fall between the two types, with a hybrid of both "nail" and "champagne glass" shapes. Thus, the travel speed plays a major role in forming the weld shape of Al/steel butt joints. In addition, it is obvious that the weld width and weld area decrease rapidly as travel speed increases from 0.6 m/min to 1.0 m/min. At slower travel speeds of 0.6 and 0.8 m/min, a relatively large keyhole weld is generated due to the relatively high welding heat input compared with that of 1.0 m/min. As travel speeds increase from 1.0 to 1.2 m/min, although the amount of time that the laser beam is over a particular zone along the joint is less [32], the resulting level of the top part of the welds increases to some extent, whereas the middle part and bottom part of the welds have a relatively regular interface, resulting in a welding–brazing Al/steel butt joint. Thus, the optimum travel speeds for the best cross section obtained by laser keyhole welding is 1.0 and 1.1 m/min.

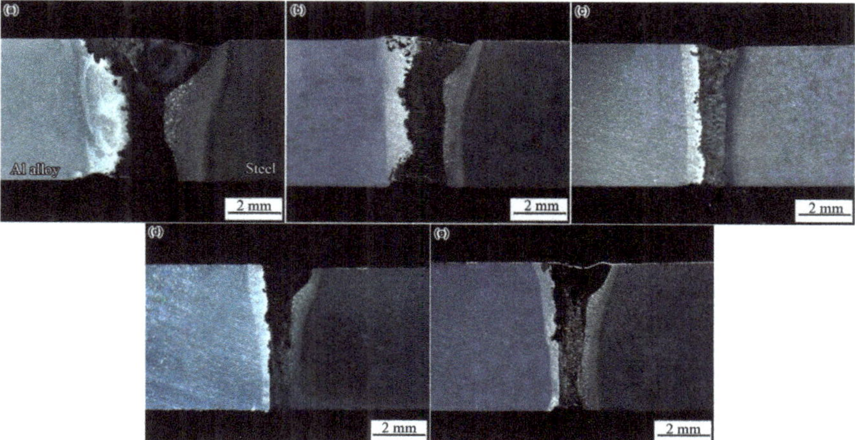

Figure 4. Effect of travel speeds on the cross-sections of Al/steel butt joints: (**a**) 0.6 m/min; (**b**) 0.8 m/min; (**c**) 1.0 m/min; (**d**) 1.1 m/min; (**e**) 1.2 m/min.

Figure 5 shows the cross sections of Al/steel butt joints obtained at different laser beam offsets under the condition of v = 1.0 m/min, P = 3.25 kW and f = 0 mm. All the welds possess a similar "nail" type penetration, meaning that the laser beam offsets have no obvious influence on the weld type of Al/steel butt joints. Regarding the weld quality, at relatively low laser beam offsets of 0.3–0.5 mm, the top surfaces of the welds are slightly concave, which is usually caused by excessive metallurgical reactions resulting from the high welding heat input [31]. As the laser beam offset increases to 0.6 and 0.7 mm, the concavity of the welds is considerably reduced due to the decrease of the heat input. In addition, at relatively low laser beam offsets of 0.3–0.4 mm, the interface boundary is seriously uneven along the joint thickness because of the excessive metallurgical reactions happening at the interface. When the laser beam offsets continuously increase to 0.6 and 0.7 mm, the excessive metallurgical reaction of the interface is lessened. As a result, the acceptable laser beam offsets for laser keyhole welding of Al/steel butt joints is 0.6 and 0.7 mm.

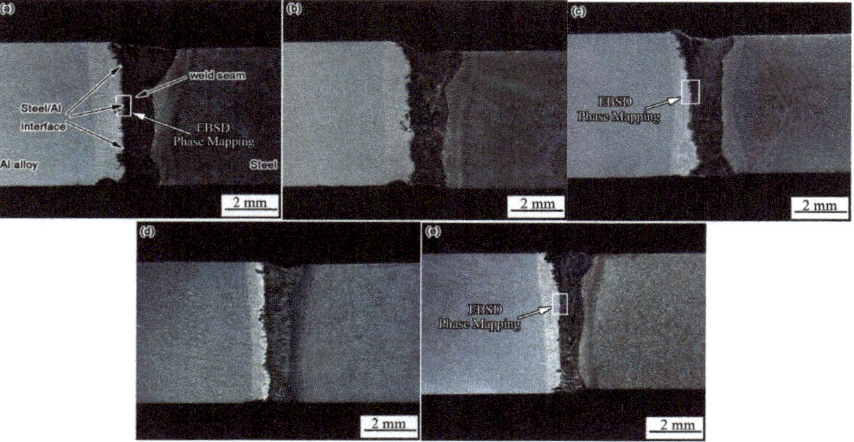

Figure 5. Effect of laser beam offsets on the cross-sections of Al/steel butt joints: (**a**) 0.3 mm; (**b**) 0.4 mm; (**c**) 0.5 mm; (**d**) 0.6 mm; (**e**) 0.7 mm.

However, the thermal cycle suffered at the interface from the top to the bottom of the joints is different, because laser welding has a high temperature gradient of through-thickness, which induces uneven distribution of reaction layer morphology, leading to different reaction modes. At the top part of the steel/Al alloy joints, the interface generated an excessive metallurgical reaction showing fusion joint characteristics as a result of the relatively high heat input. However, in some regions of the interface, such as in the middle part, or the one close to the bottom, a relatively regular interface line due to the reduced heat input is revealed. It is found that a narrow zone of solid steel HAZ is formed between the fusion weld and the Al alloy, and the interfacial reaction mode is transformed from fusion-melting to brazing, which proves that a welding–brazing butt joint has been obtained by the keyhole laser welding process.

The effects of travel speeds and laser offsets on W_1, W_2, S are illustrated in Figure 6. As shown in Figure 6a, S decreases sharply from 6.84 to 3.42 mm^2 with the increase of travel speeds from 0.6 to 1.0 m/min, and decreases slowly from 2.76 to 2.10 mm^2 with increasing travel speeds from 1.1 to 1.2 m/min; indicating that S is decreased continuously with increasing travel speeds from 0.6 to 1.2 m/min. The decrease of S is due to the fact that the laser beam interacts with the steels for a shorter period of time as travel speed increases [33], and thus less molten weld metal is produced, causing S to decrease continuously. As shown in Figure 6b, the S decreases sharply from 4.56 to 2.64 mm^2, and then decreases slowly from 2.22 to 1.92 mm^2 with increasing laser offsets from 0.3 to 0.7 mm. The decrease of S with increasing laser beam offsets is due to the fact that the heat input at the interface is decreased continuously with increasing laser beam offsets [9]. This indicates that the travel speeds and laser beam offsets directly impact the molten zone of the Al alloy side, which contributes to the final weld quality of the interface.

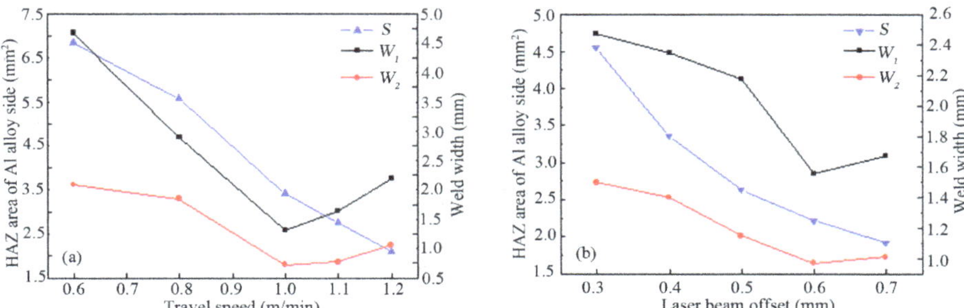

Figure 6. Relationship between process parameters and weld width or the heat-affected zone (HAZ) of the Al side: (**a**) travel speeds; (**b**) laser beam offsets. Blue color corresponds to HAZ; Black and red color correspond to top width (W_1) and bottom width (W_2) of the welds.

Concerning the weld width, the W_1 decreases sharply from 4.92 to 1.56 mm, and then increases slowly from 1.81 to 2.44 mm, as travel speed increases from 0.6 to 1.2 m/min. The W_2 decreases gradually from 2.34 to 0.97 mm with the increase of travel speeds from 0.6 to 1.0 m/min, and then increases a little from 1.02 to 1.31 mm with the increase of travel speeds from 1.0 to 1.2 m/min. This indicates that travel speed has a significant effect on W_1, but has less influence on W_2. In the case of laser beam offsets, W_1 decreases gradually from 2.47 to 1.67 mm and W_2 decreases a little from 1.5 to 1.0 mm, as the laser beam offsets increase from 0.3 to 0.7 mm, indicating that a varying laser beam offset influences W_1 and W_2 somewhat. Consequently, varying travel speeds and laser beam offsets have a more pronounced effect on S than on weld width.

3.3. Microstructures of Al/Steel Joint

The typical cross section of an Al/steel joint with a laser beam offset 0.3 mm was observed and analyzed. Several different zones shown in Figure 7a and the characteristics of each are discussed in detail. Figure 7b,d,f presents an enlarged view of zones A, B and C as shown in Figure 7a. Figure 7b shows the microstructure of the steel-fusion zone, steel HAZ and steel base metal, showing the variation of grain morphology along with the decline of heat input. Figure 7c,e shows high-magnification microstructures of steel fusion and steel HAZ, corresponding to zones marked by white dashes and red dashes shown in Figure 7b, respectively. It can be noticed that the steel-fusion zone has a coarser lath structure, which contains martensite microstructures after rapid cooling, while the steel HAZ consists of granular bainite. Figure 7d shows an enlarged view of zone B shown in Figure 7a, and displays the typical microstructure of a steel seam which is mainly composed of columnar crystal. Grain size is from 100 to 200 μm. Moreover, columnar crystal was nearly perpendicular to the fusion boundary, which is induced by the higher cooling rate in this direction and the characteristic of the preferred growth of crystallized grains [24].

Figure 7f displays an enlarged view of zone C shown in Figure 7a, revealing the microstructure of the Al side of the Al/steel joint. It can be seen that the welding joint is formed between the molten aluminum weld metal and local molten steel. IMC layers were produced along the Al/steel interface, with broken IMC phases in the neighboring Al alloy weld metal. In this area, the changes of grain size can be observed, which may confirm the fact that heat input affects the process of crystallization and growth. Figure 7g shows the Al fusion zone marked by white dashes in Figure 7f. The microstructure of the base metal is very different to that of the fusion zone. The external Al fusion zones close to the base metal (shown in Figure 7g), are characterized by dendritic growth, which corresponds to the zones with higher solidification rates [34].

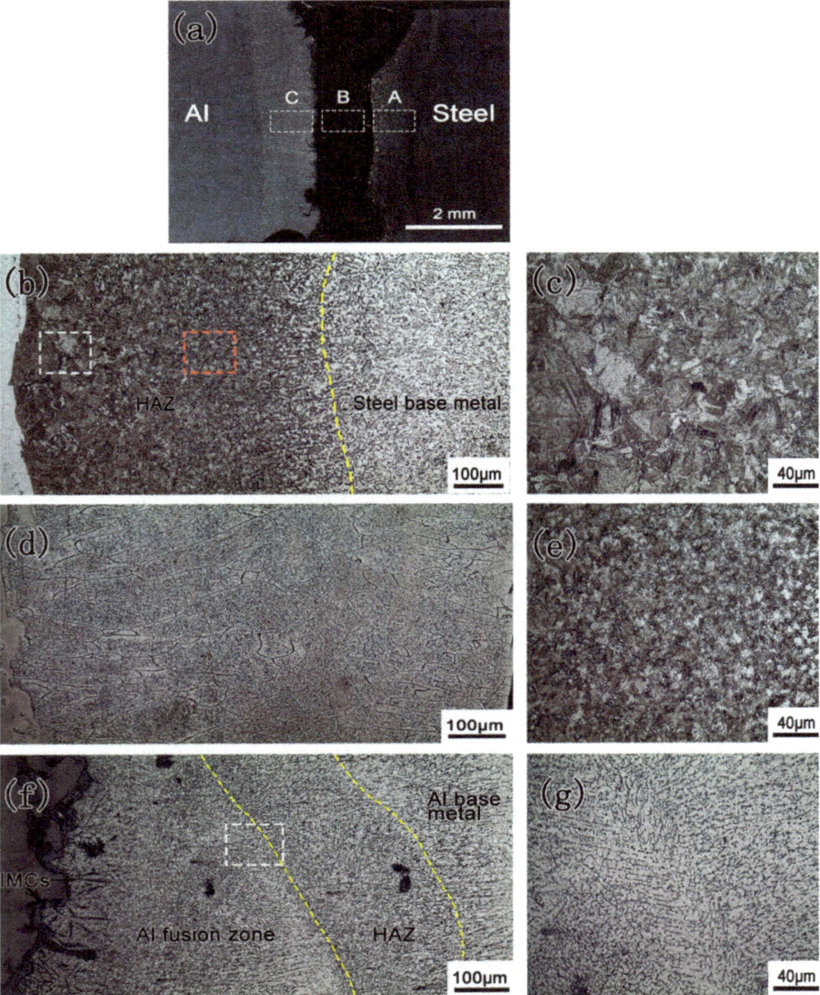

Figure 7. Cross-section microstructure of laser welding–brazing joint of aluminum and steel: (**a**) cross section of joint; (**b**) enlarged zone A in (**a**); (**c**) enlarged zone marked by white dashes in (**b**); (**d**) enlarged zone B in (**a**); (**e**) enlarged zone marked by red dashes in (**b**); (**f**) enlarged zone C in (**a**); (**g**) enlarged zone marked by white dashes in (**f**).

3.4. Phase Identification of Intermetallic Compound (IMC) Layers

Figure 8 shows a back-scattered scanning electron (BSE) micrograph of a thick IMC layer at the interface of Al/steel butt joints produced under the same conditions of $\Delta d = 0.4$ mm, $P = 3.25$ kW, $f = 0$ mm and $v = 1.0$ m/min. It was found that the IMC layers were composed of a large needle-like phase ("1") dispersed in the Al alloy side, a compact lath structure ("2") having a relatively smooth boundary adjacent to the steel side, and a fine serrated phase ("3") grown from the "2" layer. An elemental composition analysis from EDS was conducted on the location "1", "2" and "3", and the results are shown in Figure 8. The composition of location "1" is 77.09 at % Al and 22.91 at % Fe, which is similar to that of location "3", and the one for location "2" is 72.81 at % Al and 27.19 at % Fe. Compared to the measured composition with the Fe–Al phase diagram, the compositions of location "1" and "3" are in accordance with $FeAl_3$, and the location "2" can be the Fe_2Al_5 phase. Therefore, EDS analysis shows that the IMC layers at the interface consists of $FeAl_3$ and Fe_2Al_5 phases. Concerning the formation sequence and phase growth, some references have shown that the Fe_2Al_5 would be formed first. An indication that the Fe_2Al_5 is the phase that is formed first is its wavy interface towards the $FeAl_3$, which indicates in a binary system that it was a former interface between a liquid and a solid, while the straight interface of the Fe_2Al_5 on the steel side indicates solid-state diffusion [35]. A further indication of the prior formation of the Fe_2Al_5 is the fact that each Fe_2Al_5 grain is a direct neighbor of several $FeAl_3$ grains, thus it appears likely that the $FeAl_3$ nucleated at the interface of the already formed Fe_2Al_5 [36].

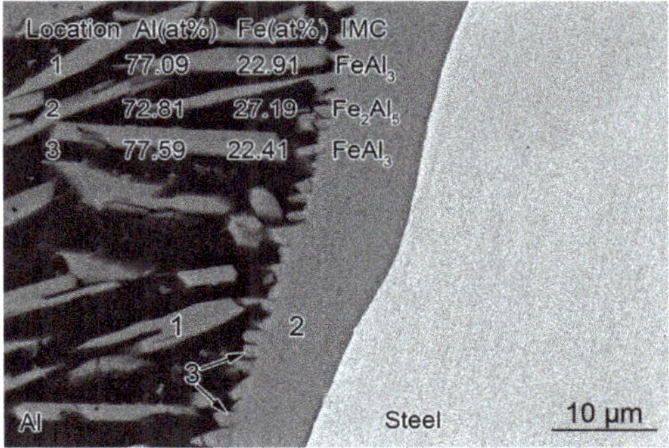

Figure 8. Back-scattered scanning electron (BSE) micrograph and energy dispersive spectrometry (EDS) analysis result of intermetallic compound (IMC) layers at the interface.

An interface region including the Al alloy, IMC layers and steel base metal was measured using EBSD phase mapping to obtain statistically significant phase information formed at the steel/Al alloy butt joints, as shown in Figure 9. The phase type of the IMC layers was accurately determined by Kikuchi patterns and the lattice constants using EBSD phase mapping performed at a selected rectangular zone near the middle part (plotted in Figure 5) of the Al/steel butt joints. The colors in phase mapping is based on a color-coded legend, in which the red, yellow, blue and green correspond to the α-Fe, α-Al, Fe_2Al_5, and Fe_4Al_{13} phases, respectively. As shown in Table 2, the Fe_4Al_{13} phase had a monoclinic unit cell with a = 15.49 Å, b = 8.08 Å, c = 12.48 Å, and β = 107.72°, and the Fe_2Al_5 phase had an orthorhombic crystal with lattice constants of a = 7.66 Å, b = 6.42 Å, c = 4.22 Å, and β = 90°, which is consistent with that of the previous studies [25,37]. Note that the Fe_4Al_{13} phase here was the same as the equilibrium $FeAl_3$ phase identified by preliminary EDS analysis. Figure 8 displays EBSD phase mapping of the IMC layers at the interface for the Al/steel butt joints obtained at a laser beam offset of 0.3 mm. According to Table 2 and Figure 8, the needle-like, serrated Fe_4Al_{13} phase and lath Fe_2Al_5 phases were formed at the interface, which agrees with the early results obtained by Dybkov [38] and Shi et al. [37]. They have proposed that the reaction products are a solid solution based upon the Fe_4Al_{13} and Fe_2Al_5 [15]. Therefore, by EDS and EBSD analysis, it can be confirmed that the IMC layers at the interface were composed of Fe_4Al_{13} and Fe_2Al_5 phases.

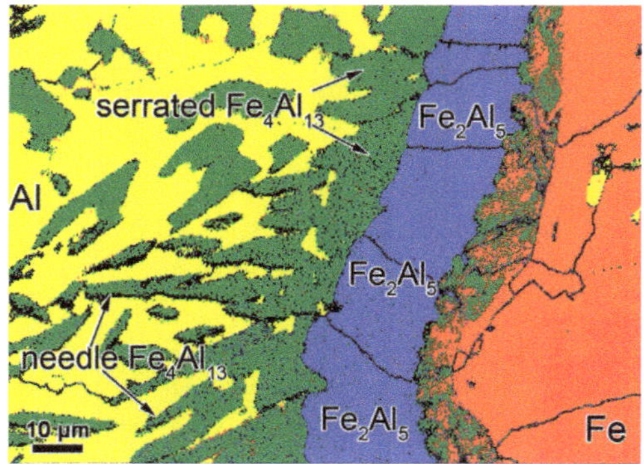

Figure 9. EBSD phase mappings of the interface of Al/steel joints at a laser beam offset of 0.3 mm.

Table 2. Intermetallic compound (IMC) phases identified using electron back-scattered diffraction (EBSD) phase mapping.

Symbol	Phase	*a*	*b*	*c*	Alpha	Beta	Gamma	Space Group
θ	Fe$_4$Al$_{13}$	15.49 Å	8.08 Å	12.48 Å	90.00°	107.72°	90.00°	12
η	Fe$_2$Al$_5$	7.66 Å	6.42 Å	4.22 Å	90.00°	90.00°	90.00°	63

3.5. Morphology and Thickness of IMC Layers

Figure 10 shows the BSE micrographs of the IMC layers at the interface of the Al/steel butt joints with different travel speeds. At a travel speed of 0.6 m/min, the IMC morphology formed at the interface exhibits a number of large needle-like FeAl$_3$ phases and thick Fe$_2$Al$_5$ layers owing to the high heat input induced by the slow travel speeds. With the increase of travel speeds from 0.8 to 1.2 m/min, the quantity of large needle-like FeAl$_3$ phases was significantly decreased and the thickness of the Fe$_2$Al$_5$ layers was also reduced. In particular, the large needle-like FeAl$_3$ phases dispersed in the Al alloy completely disappeared when the travel speed was increased from 1.0 m/min to 1.2 m/min. The effect of the travel speeds on the quantity of the needle-like FeAl$_3$ may be attributed to the decrease of the cooling rate of the interface [11]. Figure 10f displays the thickness of the IMC layers as a function of travel speeds. Here, the thickness of Fe$_2$Al$_5$ layers is used to evaluate the IMC layer because of its regularity. The thickness of Fe$_2$Al$_5$ layers decreased sharply from 47.7 to 15.1 µm from 0.6 to 0.8 m/min, and then decreased slowly from 7.9 to 4.7 µm with the increase of travel speeds from 1.0 to 1.2 m/min. The decrease of Fe$_2$Al$_5$ layer thickness at the interface can be explained by the reduction of linear energy input with the increase of travel speeds [33].

The IMC layers at the interface of the Al/steel butt joints with different laser beam offsets are shown in Figure 11. At relatively low laser beam offsets of 0.3–0.5 mm, it was obvious that a great number of large needle-like FeAl$_3$ phases and thick Fe$_2$Al$_5$ layers were formed at the interface. With increasing laser beam offsets from 0.6 to 0.7 mm, the quantity of large needle-like FeAl$_3$ phases dispersed in the Al alloy significantly decreased, and only compact lath Fe$_2$Al$_5$ layer and fine serrated FeAl$_3$ phases attached to Fe$_2$Al$_5$ layer were observed at the interface. Therefore, that laser beam offset decreases greatly the quantity of large needle-like FeAl$_3$ phases at the interface. Concerning the effect of laser beam offsets on the IMC layer thickness, it was found that the thickness of Fe$_2$Al$_5$ layers decreased quickly from 21.0 to 9.6 µm with the increase of laser beam offsets from 0.3 to 0.4 mm, and then decreased gradually from 7.8 to 4.1 µm as laser beam offsets increase from 0.5 to 0.7 mm, as shown in Figure 11f. The decrease of the Fe$_2$Al$_5$ layer thickness with the increase of laser beam offsets was mainly caused by the decrease in the maximum temperature and the reaction time at the interface [30,39]. With the increase of the laser offsets, the maximum temperature at the interface tends to be lower and the time for diffusion and dissolution of Fe atoms decreases, giving a thinner IMC layer.

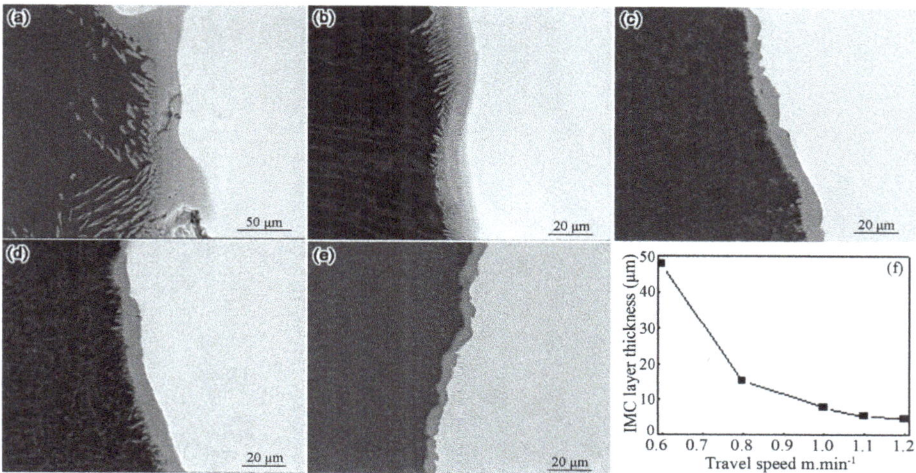

Figure 10. IMC layer microstructures at the interface of the Al/steel butt joints with different travel speeds: (**a**) 0.6 m/min; (**b**) 0.8 m/min; (**c**) 1.0 m/min; (**d**) 1.1 m/min; (**e**) 1.2 m/min; (**f**) thickness of the IMC layers (contained in the second panel).

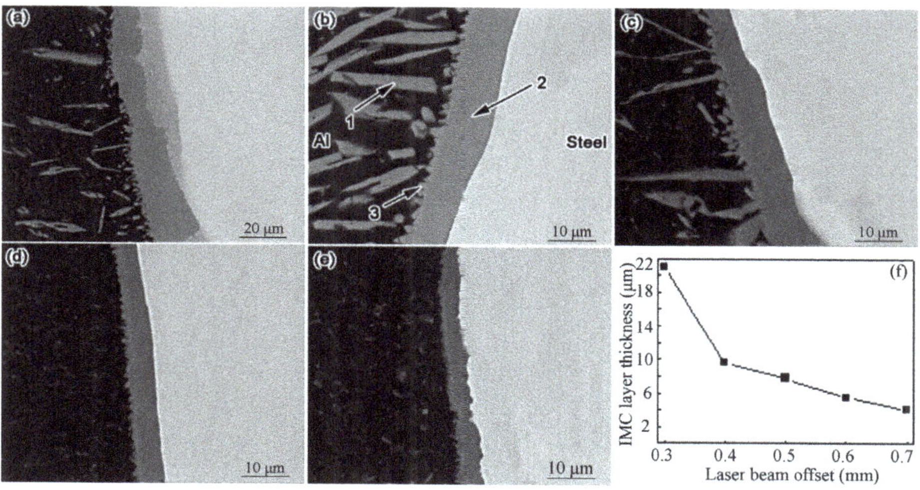

Figure 11. IMC layer microstructures at the interface of the Al/steel butt joints with different laser beam offsets: (**a**) 0.3 mm; (**b**) 0.4 mm; (**c**) 0.5 mm; (**d**) 0.6 mm; (**e**) 0.7 mm; (**f**) thickness of IMC layers.

3.6. Microhardness Profile of Al/Steel Butt Joints

Figure 12 shows the microhardness distribution of the Al/steel butt joints with different laser beam offsets. It was found that the average microhardness of 5083 Al alloys and Q235 steels is 74.1 and 142.3 HV, while the weld and HAZ microhardness at the steel side reach 258.4 and 185.5 HV, respectively. The higher hardness in the weld is mainly caused by the steel hardening because of the quenching effect, and the high hardness in the HAZ is related to the high heating and cooling rate experienced in the HAZ resulting in a fine microstructure [40].

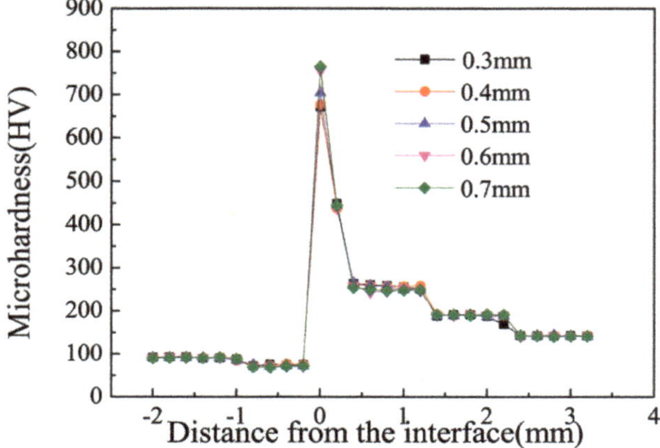

Figure 12. Microhardness distribution in the Al/steel butt joints.

The hardness at the interface of the Al/steel butt joints increased abruptly in the range from 600 to 800 HV from the laser beam offset of 0.3 to 0.7 mm, as seen in Figure 11a. The higher hardness at the interface was due to the presence of the brittle Fe_2Al_5 and Fe_4Al_{13} phases [13]. In addition, the lowest hardness of 0.3 mm was 671.3 HV, and the highest hardness was 765.4 HV for the laser beam offset of 0.7 mm. The difference in hardness at the interface between the laser beam offsets of 0.3 mm and 0.7 mm was closely associated with the different morphology of the IMC layers generated at the interface. As mentioned above, at a laser beam offset of 0.3 mm, the IMC layers at the interface were composed of a thick lath Fe_2Al_5 layer, large needle-like and fine serrated Fe_4Al_{13} phases, due to the excessive reactions at the interface as a result of the high heat input caused by a small laser beam offset of 0.3 mm. However, at a relatively higher laser beam offset of 0.7 mm, the IMC layers at the interface consisted of only thin lath Fe_2Al_5 layer and fine serrated Fe_4Al_{13} phases as a result of the lesser heat input. Thus, at the laser beam offset of 0.7 mm, the hardness at the interface mainly depended on Fe_2Al_5 layers, which have a higher hardness than Fe_4Al_{13} layers [41]. This explained why the joint with a laser beam offset of 0.7 mm had the higher hardness at the interface.

3.7. Tensile Properties of the Al/Steel Butt Joints

The UTS of the Al/steel butt joints was influenced significantly by travel speeds and laser beam offsets, as shown in Figure 13. Figure 13a shows the UTS of the Al/steel butt joints increased rapidly from 64.0 to 110.4 MPa with the increase of travel speeds from 0.6 to 1.0 m/min, and then decreased to 77.3 MPa at the travel speed of 1.2 m/min, which indicates that the UTS of the Al/steel butt joints reached its maximum at the travel speed of 1.0 m/min. It can be seen from Figure 13b that the UTS of the Al/steel butt joints was increased from 42.0 to 107.0 MPa by increasing laser beam offsets from 0.3 to 0.6 mm, while the UTS of the Al/steel butt joints decreased to 96.0 MPa as laser beam offset continuously increased to 0.7 mm. Thus, the maximum UTS of 107.0 MPa was obtained at a laser beam offset of 0.6 mm. Consequently, the UTS of the Al/steel butt joints reached its maximum in the conditions of a laser beam offset of 0.6 mm or a travel speed of 1.0 m/min.

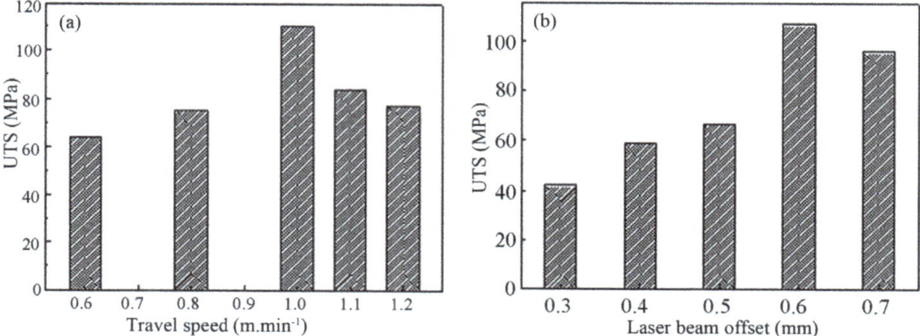

Figure 13. Effect of laser-welding parameters on the ultimate tensile strength (UTS) of the Al/steel butt joints: (**a**) travel speeds; (**b**) laser beam offsets.

The relatively low UTS of Al/steel butt joints is primarily due to the formation of brittle IMC phases at the interface, which is the main cause of embrittlement [42]. All the Al/steel butt joints failed at the IMC layers at the interface, as shown in Figure 14. During the tensile test, the crack firstly initiated at the top part ("I" zone in Figure 14a) or the bottom part ("III" zone in Figure 14c) of the IMC layers. Finally, it propagated into the middle part ("II" zone in Figure 14b) of the IMC layer, which led to the failure of the Al/steel butt joint. The fracture occurring at the IMC layer was related to the brittleness and morphology of the IMC layers at the interface. At the top part of the IMC layer, a thick Fe_2Al_5 layer and a number of the large needle-like Fe_4Al_{13} phases were observed owing to the high heat input. For the bottom part of the IMC layer, some welding undercuts were observed at the root of the joint in the Al alloy side close to the interface, which was more inclined to initiate microcracks at high loads during the tensile test. Therefore, it can be supposed that the crack starts from the bottom or top part, and propagates into the middle part of the IMC layer at the interface.

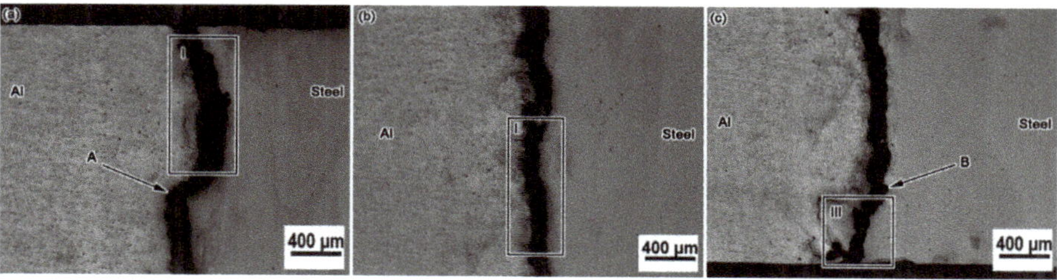

Figure 14. Fracture path of the interface at the steel/Al alloy butt joint: (**a**) top position (**b**) middle position (**c**) bottom position.

Figure 15 shows typical results from SEM investigations of fracture surfaces at the steel side and Al alloy side of the position "II" in the Al/steel butt joints, as this region is the final failure location of the joint during tensile tests. The fracture surface at the Al alloy side can be seen to be mostly flat, and the microcrack grows through the grains. Also, there was evidence of some broken Fe_4Al_{13} phases and the needle-like Fe_2Al_5, as shown in Figure 15a, which depicts brittle transgranular fracture. The factographic surface in the steel side shown in Figure 15b presented the typical cleavage fracture mode, with river pattern strips of particular orientation on the fracture surface, which occupies the largest fraction area of the sample surface exhibiting interfacial failure. Even though the failure of all the samples was always located at the brazing interface during the tensile test, the specimens endured some plastic deformation, as shown in Figure 15c; some regions of the factographic surface in the steel side appear to show some dimples accompanied by some local plasticity, and the crack grew along the grain boundary, showing some intergranular fracture characteristics. Although these fractures were brittle, the latter had a certain resistance to the crack propagation of the IMC layers at the steel side, which enhanced its mechanical resistance at the steel side. As a result, the IMC layer region close to the Al base material was the most brittle region having the weakest bonding strength at the interface.

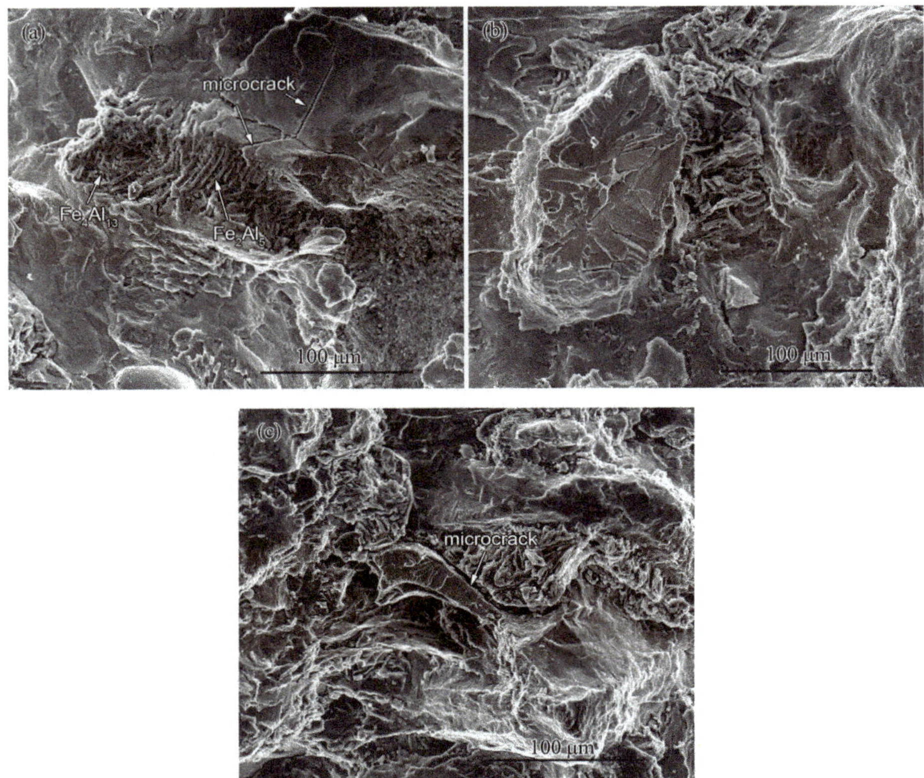

Figure 15. Scanning electronic microscope (SEM) micrographs of the middle position at the interface of the steel/Al alloy butt joint: (**a**) Al alloy side; (**b**) steel side; (**c**) steel side showing some dimples.

4. Discussion

4.1. Effect of IMC Layer Thickness on Ultimate Tensile Strength (UTS)

The relationship between the IMC layer thickness and the UTS of Al/steel joints has been proposed by Borrisutthekul [12], showing that the UTS was increased with decreasing thickness of the reaction layer. In the present study, increasing travel speeds from 0.6 to 1.0 m/min improved the UTS of Al/steel joints, as shown in Figure 13a, due to the decreased thickness of the IMC layer continuously from 47.7 to 7.9 µm (Figure 10f). In Figure 13b, with the increase of laser beam offsets from 0.3 to 0.6 mm, the UTS of Al/steel joints was increased as the IMC layer thickness decreased from 21.0 to 5.6 µm (Figure 11f). In these cases, the increased tendency of the UTS with decreasing IMC layer thickness was clearly observed, which is in good agreement with the results reported by Borrisutthekul [12].

However, with continuously increasing travel speeds to 1.2 m/min or increasing laser beam offset to 0.7 mm, the UTS of the Al/steel joints was not increased continuously with further reductions in IMC layer thickness. This indicates that the UTS of Al/steel joints does not always increase with a decrease in IMC layer thickness [3]. Gao et al. [17] has proposed that there is an optimal range of IMC layer thickness for improving the UTS of Al/steel joints, because an IMC layer that is too thin or too thick results in low UTS. The reasons are due to the fact that an IMC layer that is too thin means an insufficient interface reaction leads to the lack of fusion, and the IMC layer that is too thick resulting from an excessive growth of the IMC phases increased the interface brittleness [17]. Therefore, it is concluded that there is a relationship between optimum IMC layer thickness and the UTS of Al/steel butt joints. However, this is not the sole condition. When comparing the thickness of the IMC layer of the joints welded with a welding speed of 0.6 m/min and those welded with a laser offset of 0.5 mm, it is found that the thickness for a welding speed of 0.6 m/min is about ~48 µm, and the joint has a low UTS of 64.0 MPa. For a laser beam offset of 0.5 mm, the IMC layers have thin thickness of 7.8 µm, whereas the joints also exhibit low UTS of 66.5 MPa. Therefore, for thickness of more than 10 µm, the increased tendency of UTS with decreasing IMC layer thickness is clearly observed, which is in good agreement with the results reported by Borrisutthekul [12]. However, with continuously decreasing thickness to less than 10 µm, the UTS of Al/steel joints not only largely depends on the IMC layer thickness, but also on the IMC phase type and morphology at the interface.

4.2. Fracture Behavior of IMC Layers

It is well accepted that the UTS of Al/steel joints largely depends on the IMC layers (thickness and phase type) at the interface [43]. However, regarding the fracture behavior of IMC layers during tensile processing, there are some contradictory results. Springer et al. pointed out that the crack propagated in the brittle IMC phases and cut through the remainder of the steel at the irregular, wavy steel/Fe_2Al_5 interface [44]. However, Chen et al. claimed that the Fe_2Al_5 layer was the primary and most brittle phase at the interface [43]. These observations are evidence that the fracture of IMC layers associated with tensile testing are not well understood. Thus, in the present study, an EBSD phase mapping of the fractured surfaces of Al/steel butt joints under three laser beam offsets was conducted in order to understand the fracture behavior of the IMC layers at the interface. At a laser beam offset of 0.3 mm, it is evident that the crack travels through Fe_2Al_5 layers and the fracture path is smooth, as shown in Figure 16a, while the Fe_4Al_{13}/Fe_2Al_5 and Fe_2Al_5/steel phase interface bond well. This indicates that the Fe_2Al_5 layer is the weakest in terms of bonding strength of the IMC layers at the interface, and a transgranular fracture that occurred at the Fe_2Al_5 layers is evident due to the higher hardness of the Fe_2Al_5 layers [41] (Figure 12). Figure 16b shows a brittle transgranular fracture still occurring at Fe_2Al_5 layer at a laser beam offset of 0.5 mm, but some second cracks can be observed in the Al base material, showing that some deformation has happened at the interface. This explains why the Al/steel butt joint at a laser beam offset of 0.5 mm has relatively high UTS compared to that of a laser beam offset of 0.3 mm. When the laser beam offset is 0.7 mm (Figure 16c), it clearly reveals that the crack travels along the boundary of the Fe_2Al_5 grains and the fracture path shows the obvious irregularity, which demonstrates evidence that an intergranular fracture occurred at the Fe_2Al_5 layers. Thus, the intergranular fracture occurring at the Fe_2Al_5 layer is thought to be attributable to the relatively high UTS of the Al/steel butt joints with a laser beam offset of 0.7 mm. Consequently, the present study of EBSD phase mapping shows that the Fe_2Al_5 layer is the most brittle region having the weakest bonding strength, which is entirely consistent with a previous study reported by Chen et al. [43]. Moreover, an intergranular fracture that occurred at the Fe_2Al_5 layers led to a relatively high UTS of the Al/steel butt joints.

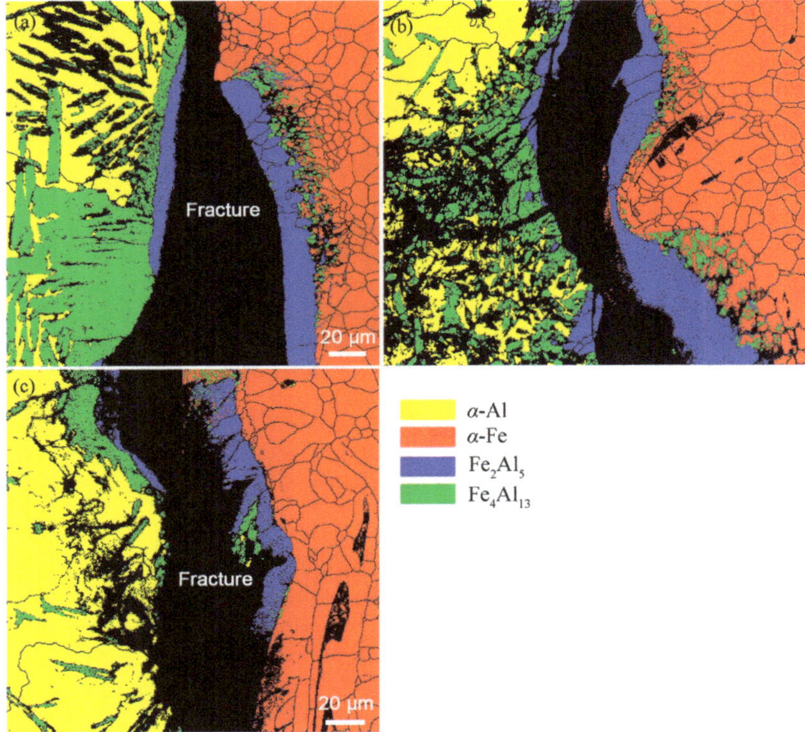

Figure 16. EBSD phase mapping of the failed tensile samples for the Al/steel joint with different laser beam offsets: (**a**) 0.3 mm; (**b**) 0.5 mm; (**c**) 0.7 mm.

5. Conclusions

Based on the above results and discussion, butt joints of 6.0 mm thick 5083 Al alloys to Q235 steel plates have been achieved by using autogenous laser keyhole welding. The following conclusions can be drawn from this work:

(1) Varying travel speeds and laser beam offsets have a more pronounced effect on the area of HAZ S than on weld width. By increasing the laser beam offsets and welding speed, the S in the Al alloy side is decreased continuously. However, travel speed has a significant effect on W_1, but has less influence on W_2, whereas varying laser beam offset influences W_1 and W_2 a little.

(2) The thickness of Fe_2Al_5 layers decreases from 47.7 to 4.7 µm with increasing travel speeds from 0.6 to 1.2 m/min. As laser beam offset increases from 0.3 to 0.7 mm, the thickness of Fe_2Al_5 layers decreases from 21.0 to 4.1 µm. Thus, increasing travel speeds or laser beam offsets obviously decreases the quantity of long needle-like $FeAl_3$ phases and the thickness of the Fe_2Al_5 layer.

(3) The UTS of Al/steel butt joints is influenced significantly by the tested travel speeds and laser beam offsets. The UTS reaches its maximum under the conditions of a travel speed of 1.0 m/min or a laser beam offset of 0.6 mm. There should be a matching relationship between the IMC layer thickness and UTS of Al/steel butt joints.

(4) EBSD phase mapping proves that the IMC layers consist of Fe_2Al_5 phases and Fe_4Al_{13} phases. Increasing laser beam offsets from 0.3 to 0.7 mm significantly decreases the quantity of Fe_4Al_{13} phases and the thickness of Fe_2Al_5 layers at the interface of Al/steel butt joints.

(5)   EBSD phase mapping proves that the Fe_2Al_5 layer is the most brittle region having the weakest bonding strength. An intergranular fracture occurring at the Fe_2Al_5 layers leads to the relatively high UTS of Al/steel butt joints.

Acknowledgments: This work was supported by the National Natural Science Foundation of China (grant number 51475006) and the Key Program of Science and Technology Projects of Beijing Municipal Commission of Education (grant number KZ201610005004). The authors are grateful to Steve Shi from TWI Ltd. (Cambridge, UK) for providing language help and editing this manuscript.

Author Contributions: Li Cui and Dingyong He conceived and designed the experiments; Wei Qian performed the welding experiments; Boxu Chen analyzed the data; Li Chen contributed analysis tools; Li Cui wrote the paper.

Conflicts of Interest: The authors declare no conflict of interest.

References

1. Thomy, C.; Vollertsen, F. Laser-MIG hybrid welding of aluminium to steel—Effect of process parameters on joint properties. *Weld. World* **2012**, *56*, 124–132. [CrossRef]
2. Qin, G.; Lei, Z.; Su, Y.; Fu, B.; Meng, X.; Lin, S. Large spot laser assisted gma brazing–fusion welding of aluminum alloy to galvanized steel. *J. Mater. Process. Technol.* **2014**, *214*, 2684–2692. [CrossRef]
3. Mathieu, A.; Shabadi, R.; Deschamps, A.; Suery, M.; Matteï, S.; Grevey, D.; Cicala, E. Dissimilar material joining using laser (aluminum to steel using zinc-based filler wire). *Opt. Laser Technol.* **2007**, *39*, 652–661. [CrossRef]
4. Meco, S.; Pardal, G.; Ganguly, S.; Williams, S.; Mcpherson, N. Application of laser in seam welding of dissimilar steel to aluminium joints for thick structural components. *Opt. Laser Eng.* **2015**, *67*, 22–30. [CrossRef]
5. Tricarico, L.; Spina, R. Experimental investigation of laser beam welding of explosion-welded steel/aluminum structural transition joints. *Mater. Des.* **2010**, *31*, 1981–1992. [CrossRef]
6. Lahdo, R.; Springer, A.; Pfeifer, R.; Stefan, K.; Ludger, O. High-power laser welding of thick steel-aluminum dissimilar joints. *Phys. Procedia* **2016**, *83*, 396–405. [CrossRef]
7. Atabaki, M.M.; Ma, J.; Liu, W.; Kovacevic, R. Hybrid laser/arc welding of advanced high strength steel to aluminum alloy by using structural transition insert. *Mater. Des.* **2015**, *75*, 120–135. [CrossRef]
8. Song, J.L.; Lin, S.B.; Yang, C.L.; Ma, G.C.; Liu, H. Spreading behavior and microstructure characteristics of dissimilar metals tig welding–brazing of aluminum alloy to stainless steel. *Mater. Sci. Eng. A* **2009**, *509*, 31–40. [CrossRef]
9. Kreimeyer, M.; Wagner, F.; Vollertsen, F. Laser processing of aluminum–titanium-tailored blanks. *Opt. Laser Eng.* **2005**, *43*, 1021–1035. [CrossRef]
10. Shah, L.H.; Ishak, M. Review of research progress on aluminum–steel dissimilar welding. *Mater. Manuf. Process.* **2014**, *29*, 928–933. [CrossRef]
11. Shao, L.; Shi, Y.; Huang, J.K.; Wu, S.J. Effect of joining parameters on microstructure of dissimilar metal joints between aluminum and galvanized steel. *Mater. Des.* **2015**, *66*, 453–458. [CrossRef]
12. Borrisutthekul, R.; Yachi, T.; Miyashita, Y.; Mutoh, Y. Suppression of intermetallic reaction layer formation by controlling heat flow in dissimilar joining of steel and aluminum alloy. *Mater. Sci. Eng. A* **2007**, *467*, 108–113. [CrossRef]
13. Chen, S.H.; Huang, J.H.; Ma, K.; Zhang, H.; Zhao, X. Influence of a Ni-foil interlayer on Fe/Al dissimilar joint by laser penetration welding. *Mater. Lett.* **2012**, *79*, 296–299. [CrossRef]
14. Huang, J.K.; He, J.; Yu, X.Q.; Li, C.L.; Fan, D. The study of mechanical strength for fusion-brazed butt joint betweenaluminum alloy and galvanized steel by arc-assisted laser welding. *J. Manuf. Process.* **2017**, *25*, 126–133. [CrossRef]
15. Shi, Y.; Zhang, H.; Takehiro, W.; Tang, J.G. CW/PW dual-beam YAG laser welding of steel/aluminum alloy sheets. *Chin. J. Lasers* **2010**, *48*, 732–736.
16. Mei, S.W.; Gao, M.; Yan, J.; Zhang, C.; Li, G.; Zeng, X.Y. Interface properties and thermodynamic analysis of laser–arc hybrid welded Al/steel joint. *Sci. Technol. Weld. Join.* **2013**, *18*, 293–300. [CrossRef]
17. Gao, M.; Chen, C.; Mei, S.; Wang, L.; Zeng, X. Parameter optimization and mechanism of laser–arc hybrid welding of dissimilar al alloy and stainless steel. *Int. J. Adv. Manuf. Technol.* **2014**, *74*, 199–208. [CrossRef]

18. Dharmendra, C.; Rao, K.P.; Wilden, J.; Reich, S. Study on laser welding–brazing of zinc coated steel to aluminum alloy with a zinc based filler. *Mater. Sci. Eng. A* **2011**, *528*, 1497–1503. [CrossRef]
19. Laukant, H.; Wallmann, C.; Korte, M.; Glatzel, U. Flux-less joining technique of aluminium with zinc-coated steel sheets by a dual-spot-laser beam. *Adv. Mater. Res.* **2005**, *6–8*, 163–170. [CrossRef]
20. Zhou, D.W.; Liu, J.S.; Lu, Y.Z.; Xu, S.H. Effect of adding powder on joint properties of laser penetration welding for dual phase steel and aluminum alloy. *Opt. Laser Technol.* **2017**, *94*, 171–179. [CrossRef]
21. Chen, R.; Wang, C.L.; Jiang, P.; Shao, X.; Zhao, Z.; Gao, Z.; Yue, C. Effect of axial magnetic field in the laser beam welding of stainless steel to aluminum alloy. *Mater. Des.* **2016**, *109*, 146–152. [CrossRef]
22. Seffer, O.; Springer, A.; Kaierle, S. Investigations on remote laser beam welding of dissimilar joints of aluminum alloys and steel with varying sheet thicknesses for car body construction. *J. Laser Appl.* **2017**, *29*, 022414. [CrossRef]
23. Kouadri-David, A. Study of metallurgic and mechanical properties of laser welded heterogeneous joints between dp600 galvanised steel and aluminium 6082. *Mater. Des.* **2014**, *54*, 184–195. [CrossRef]
24. Sun, J.H.; Qi, Y.; Gao, W.; Huang, J. Investigation of laser welding on butt joints of al/steel dissimilar materials. *Mater. Des.* **2015**, *83*, 120–128.
25. Zhang, M.J.; Chen, G.Y.; Zhang, Y.; Wu, K.R. Research on microstructure and mechanical properties of laser keyhole welding–brazing of automotive galvanized steel to aluminum alloy. *Mater. Des.* **2013**, *45*, 24–30. [CrossRef]
26. Sun, J.H.; Huang, J.; Yan, Q.; Li, Z.G. Fiber laser butt joining of aluminum to steel using welding-brazing method. *Int. J. Adv. Manuf. Technol.* **2016**, *85*, 2639–2650. [CrossRef]
27. Zhu, Z.T.; Wan, Z.D.; Li, Y.X.; Chen, H. Intermediate layer, microstructure and mechanical properties of aluminum alloy/stainless steel butt joint using laser-MIG hybrid welding-brazing method. *Int. J. Mod. Phys. B* **2017**, *31*, 1744035. [CrossRef]
28. Salminen, A. The filler wire—Laser beam interaction during laser welding with low alloyed steel filler wire. *Mechanika* **2010**, *37*, 67–74.
29. Tomashchuk, I.; Sallamand, P.; Cicala, E.; Peyre, P.; Grevey, D. Direct keyhole laser welding of aluminum alloy AA5754 to titanium alloy Ti6Al4V. *J. Mater. Process. Technol.* **2015**, *217*, 96–104. [CrossRef]
30. Casalino, G.; Mortello, M.; Peyre, P. Yb–yag laser offset welding of aa5754 and t40 butt joint. *J. Mater. Process. Technol.* **2015**, *223*, 139–149. [CrossRef]
31. Zhou, L.; Li, Z.Y.; Song, X.G.; Tan, C.W.; He, Z.Z.; Huang, Y.X.; Feng, J.C. Influence of laser offset on laser welding-brazing of al/brass dissimilar alloys. *J. Alloys Compd.* **2017**, *717*, 78–92. [CrossRef]
32. Majumdar, B.; Galun, R.; Weisheit, A.; Mordike, B.L. Formation of a crack-free joint between ti alloyand al alloy by using a high-power CO_2 laser. *J. Mater. Sci.* **1997**, *32*, 6191–6200. [CrossRef]
33. Cui, L.; He, D.Y.; Fu, G.; Li, X.Y.; Jiang, J.M. Effect of fiber laser–mig hybrid process parameters on weld bead shape and tensile properties of commercially pure titanium. *Mater. Manuf. Process.* **2010**, *25*, 1309–1316.
34. Sánchez-Amaya, J.M.; Delgado, T.; De Damborenea, J.J.; Lopez, V.; Botana, F.J. Laser welding of AA 5083 samples by high power diode laser. *Sci. Technol. Weld. Join.* **2013**, *14*, 78–86.
35. Bouche, K.; Barbier, F.; Coulet, A. Intermetallic compound layer growth between solid iron and molten aluminium. *Mater. Sci. Eng. A* **1998**, *249*, 167–175. [CrossRef]
36. Leonardo, A.; Dominique, E.; Christian, H.S.; Arenholz, E.; Jank, N.; Bruckner, J.; Pyzalla, A.R. Intermetallic FexAly-phases in a steel/Al-alloy fusion weld. *J. Mater. Sci.* **2007**, *42*, 4205–4214.
37. Shi, H.X.; Qiao, S.; Qiu, R.F.; Zhang, X.J.; Yu, H. Effect of welding time on the joining phenomena of diffusion welded joint between aluminum alloy and stainless steel. *Mater. Manuf. Process.* **2012**, *27*, 1366–1369. [CrossRef]
38. Dybkov, V.I. Interaction of 18Cr-10Ni stainless steel with liquid aluminium. *J. Mater. Sci.* **1990**, *25*, 3615–3633. [CrossRef]
39. Miao, Y.G.; Han, D.F.; Yao, J.Z.; Li, F. Microstructure and interface characteristics of laser penetration brazed magnesium alloy and steel. *Sci. Technol. Weld. Join.* **2013**, *15*, 97–103. [CrossRef]
40. Sierra, G.; Peyre, P.; Deschaux-Beaume, F.; Stuart, D.; Fras, G. Steel to aluminium key-hole laser welding. *Mater. Sci. Eng. A* **2007**, *447*, 197–208. [CrossRef]
41. Bouayad, A.; Gerometta, C.; Belkebir, A.; Ambari, A. Kinetic interactions between solid iron and molten aluminium. *Mater. Sci. Eng. A* **2003**, *363*, 53–61. [CrossRef]

42. Nascimento, R.M.D.; Martinelli, A.E.; Buschinelli, A.J.A. Review article: Recent advances in metal-ceramic brazing artigo revisão: Avanços recentes em brasagem metal-cerâmica. *Cerâmica* **2003**, *49*, 178–198. [CrossRef]
43. Chen, J.; Li, J.; Amirkhiz, B.S.; Liang, J.; Zhang, R. Formation of nanometer scale intermetallic phase at interface of aluminum-to-steel spot joint by welding–brazing process. *Mater. Lett.* **2014**, *137*, 120–123. [CrossRef]
44. Springer, H.; Kostka, A.; Santos, J.F.D.; Raabe, D. Influence of intermetallic phases and kirkendall-porosity on the mechanical properties of joints between steel and aluminium alloys. *Mater. Sci. Eng. A* **2011**, *528*, 4630–4642. [CrossRef]

 *metals*

Article

Experimental Investigation on Electric Current-Aided Laser Stake Welding of Aluminum Alloy T-Joints

Xinge Zhang [1,2,*], Liqun Li [2], Yanbin Chen [2], Zhaojun Yang [1] and Xiaocui Zhu [1]

[1] School of Mechanical Science and Engineering, Jilin University, Changchun 130025, China; yangzj@jlu.edu.cn (Z.Y.); zxc@jlu.edu.cn (X.Z.)

[2] State Key Laboratory of Advanced Welding and Joining, Harbin Institute of Technology, Harbin 150001, China; lilqhit@163.com (L.L.); chenyanbinhit@163.com (Y.C.)

* Correspondence: zhangxinge@jlu.edu.cn; Tel.: +86-431-8509-5839

Received: 30 July 2017; Accepted: 30 October 2017; Published: 1 November 2017

Abstract: In the present study, aluminum alloy T-joints were welded using the laser stake-welding process. In order to improve the welding quality of the T-joints, an external electric current was used to aid the laser stake-welding process. The effects of the process parameters on the weld morphology, mechanical properties, and microstructure of the welded joints were analyzed and discussed in detail. The results indicate that the aided electric current should be no greater than a certain maximum value. Upon increasing the aided electric current, the weld width at the skin and stringer faying surface obviously increased, but there was an insignificant change in the penetration depth. Furthermore, the electric current and pressing force should be chosen to produce an expected weld width at the faying surface, whereas the laser power and welding speed should be primarily considered to obtain an optimal penetration depth. The tensile shear specimens failed across the faying surface or failed in the weld zone of the skin. The specimens that failed in the weld of the skin could resist a higher tensile shear load compared with specimens that failed across the faying surface. The microstructural observations and microhardness results demonstrated that the tensile shear load capacity of the aluminum alloy welded T-joint was mainly determined by the weld width at the faying surface.

Keywords: laser stake welding; aluminum alloy T-joint; aided electric current; weld morphology; mechanical properties

---

1. Introduction

In industrial fields, T-joint structures have been extensively designed and used to decrease weight and improve the structure strength [1–4]. For T-joint structures, adhesive bonding or the welding process is usually employed to join the skin and stringer [5–8].

The laser welding process is considered a competitive technology for stake-welded T-joints because of its ability to form high-quality welds with a high aspect ratio (depth-to-width ratio), a small heat-affected zone (HAZ) and a high welding speed [9–11]. Romanoff et al. [12] evaluated the rotational stiffness of a steel-welded T-joint using laser stake welding, and their results indicated that the welded T-joint stiffness was mainly influenced by the weld thickness when the skin plate was thicker. Meng et al. [13] studied the weld geometry, microstructure and microhardness of typical laser-welded T-joints. A desired bead geometry of laser lap-welded steel T-joints can be produced by adopting a low welding speed, a thin face plate and a small gap. Li et al. [14] provided a mathematical model to study the porosity induced by the keyhole during the laser stake welding of steel T-joints, which indicated that the porosity formation was primarily attributable to keyhole collapse. In order to combine both advantages, laser-arc hybrid welding has been attempted as a new method to stake weld T-joints. Wang et al. [15] investigated the weld characteristics of laser-TIG (Tungsten Inert Gas) hybrid stake welding of a titanium alloy T-joint. It was found that the gap tolerance between the skin and stringer significantly improved using laser-arc hybrid welding, and it was especially suitable to stake weld T-joints with a thick skin component under high-precision assembly conditions. Hou et al. [16,17] performed a series of comparative investigations on magnesium alloy T-joints welded using different welding methods, including TIG welding, laser welding, laser-TIG welding, and laser-TIG welding with a filler wire. However, the skin of the T-joint was always excessively melted as the laser-arc hybrid stake welded the T-joint.

Friction stir welding is considered one of the most promising welding processes; this is a solid-state joining method for joining aluminum alloys. Fratini et al. [18] welded T-joints of two types of aluminum alloys using friction stir welding, and their results showed that high-quality AA6082 alloy T-joints were welded; however, there were poor results for the AA2024 alloy. Cui et al. [19] fabricated AA6061 T-joints via friction stir welding using three different combinations of skins and stringers. The results showed that increasing the welding speed leads to more welding defects. Zhao et al. [20] focused on the relations between the defects and welding parameters and the tensile properties of the T-joints. Some T-joints without tunnel defects could only be obtained with a traverse speed of 100 mm/min in their experiments, and the welding parameters influenced the features and sizes of the kissing-bond defects. Urbikain et al. [21] put forward a new method that was a combination of two processes, friction drilling and form tapping, for the rapid and economical welding of nutless joints. The combined method achieves a quick way for the production of threaded holes on couples of dissimilar metal alloys, as it is the case of steels and aluminum alloys.

In previous studies, steel T-joints and titanium alloy T-joints stake welding were always carried out by laser welding and laser-arc hybrid welding, respectively, and the combination of friction drilling and a form tapping process was used to join dissimilar metal alloy T-joints. Aluminum alloy T-joints were mainly accomplished via friction stir stake welding. Compared with friction stir welding, laser welding has advantages such as a high welding speed, low distortion, and ease of automation [22–25]. However, reports on the laser stake welding of aluminum alloy T-joints are rare, although this welding technology has great potential in aviation, aerospace, marine and automobile industries. Therefore, to investigate the feasibility and improve the weld quality of T-joints, in this study, aluminum alloy T-joints were welded via electric current-aided laser stake welding. After welding, the weld morphology, mechanical properties and microstructure of the welded joints were analyzed and discussed in detail.

2. Materials and Methods

The base metal used in this investigation was a 5A06 aluminum alloy plate. Plates that were 150 mm × 80 mm × 2 mm were used as the skin components, and plates that were 150 mm × 30 mm × 10 mm were used as the stringer components.

The chemical composition of the base metal is presented in Table 1. Before welding, the aluminum alloy plates were mechanically cleaned followed by acetone solution cleaning.

Table 1. Chemical composition of the 5A06 aluminum alloy (wt %).

Material	Mg	Mn	Si	Cu	Zn	Fe	Ti	Al
5A06	5.8–6.8	0.5–0.8	0.4	0.1	0.2	0.4	0.02	Balance

A CO_2 laser (DC030; Rofin-Sinar, GmbH, Hamburg, Germany) source operating in continuous wave (CW) mode was used. A combination device (PSG6130 and PSI6200; Bosch Rexroth Company China, Shanghai, China) with an adjustable output electric current was used to supply the aided electric current, which could vary from 2.0 to 13.0 kA. As shown in Figure 1, the 5A06 aluminum alloy T-joints were welded via electric current-aided laser stake welding, as a rolling electrode and stringer component were connected to the positive and negative terminals of the power supply, respectively.

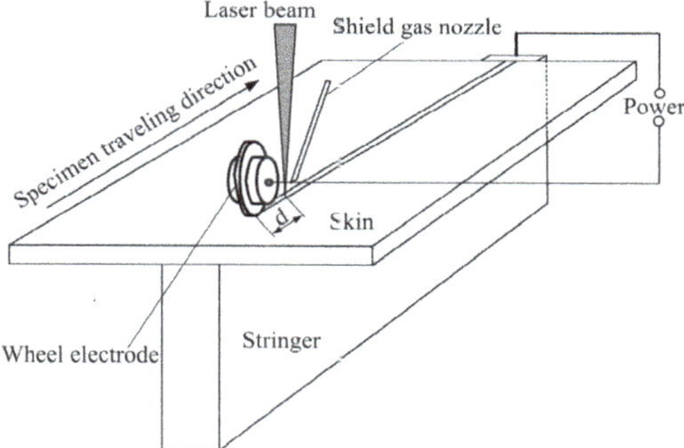

Figure 1. Schematic diagram of the set-up of electric current-aided laser stake welding of a T-joint.

During the electric current-aided laser stake-welding process, the laser beam was focused on the top of the skin surface using a 190 mm focal length lens, and the focused spot diameter was 0.2 mm. The wheel electrode touched the top of the skin surface. While the T-joint specimens moved with the traveling device, the wheel electrode rotated about a fixed axis. The distance, *d* (illustrated in Figure 1), between the position of the bottom of the wheel electrode and the position of the laser beam focus was fixed at 25 mm. The flow rate of the argon shielding gas was 25 L/min.

Following the welding process, the welded T-joints were carried out to create specimens for metallographic analysis, microhardness and tensile shear tests. Considering the material and shape of the welded joint, the wire electro-discharge machining (WEDM) method was used to manufacture the test specimens [26]. The Instron5500R material tensile machine was used to perform the tensile shear test at a strain rate of 2 mm/min. The self-design clamping was used for the tensile shear test. The schematic diagram of the specimen assembly is shown in Figure 2; the compensation plates were used to avoid the bending moment. To perform the metallographic examination, the welded samples were sectioned, mounted, polished, and etched using a solution of 1 vol % HCl, 1.5 vol % HF, 2.5 vol % HNO_3 and 95 vol % H_2O for 5 s. The appearance of the weld cross-section and microstructure were observed using an optical microscope (Eclipse E200, Nikon Instruments (Shanghai) Co., Ltd., Shanghai, China). The weld depth and width measurements were obtained from the resulting optical graphs. The microhardness test was performed using a microhardness testing machine (HXD-1000TM; Zhuhai Precision Instrument Co., Ltd., Zhuhai, China) with a 150 g test load. The joint fracture characteristics were examined via scanning electron microscopy (SEM; S-4700; Hitachi High-Technologies in China, Suzhou, China) with an energy-dispersive X-ray spectrometer (EDS), and the SEM images were gained in the secondary electron imaging mode using a 12.5 mm working distance, a 20 kV accelerating voltage and a 40 µA emission current. The experimental equipment used in this study is listed in Table 2.

Table 2. The experimental equipment used in this study.

Equipment	Model, Supplier
CO_2 laser source	DC030; Rofin-Sinar, GmbH
Electric current supply device	PSG6130 and PSI6200; Bosch Rexroth Company
Tensile test machine	Instron5500R, Adivision of Illinois Tool Works Inc.
Optical microscope	Eclipse E200, Nikon Instruments, Ltd.
Microhardness test machine	HXD-1000TM; Zhuhai Precision Instrument Co., Ltd.
Scanning electron microscopy	S-4700; Hitachi, Ltd.

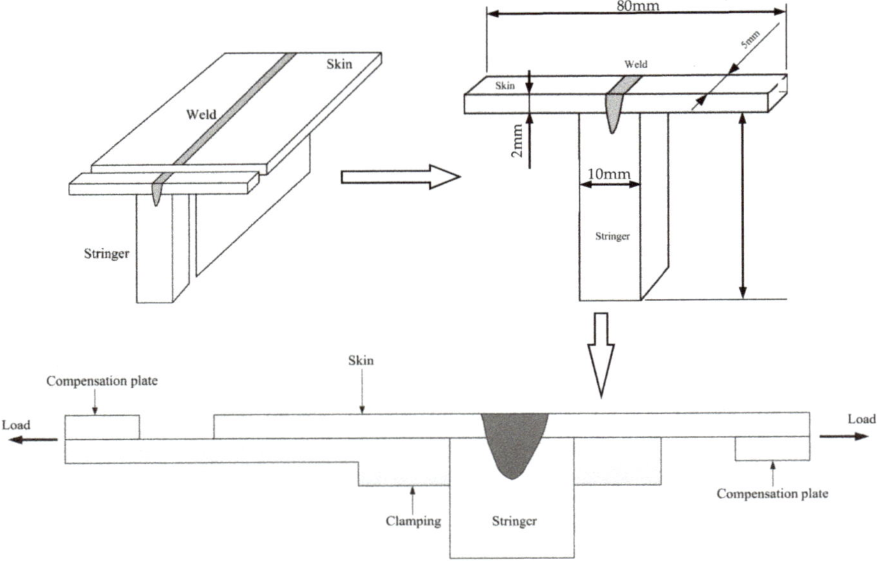

Figure 2. Schematic diagram of the specimen preparation and assembly for the tensile shear tests.

3. Experimental Results and Discussion

3.1. Weld Morphology of the T-Joint

Figure 3 shows typical macrographs of the weld morphology of T-joints produced using different electric current-aided laser stake-welding processes. There was almost no root gap between the skin and stringer owing to the pressing by the wheel electrode, as shown in Figure 3. The weld morphology of the laser welding is a "wedge" profile, whereas the weld morphology noticeably changed in the electric current-aided laser-welded joints. The weld morphology was nearly a "cylindrical" profile, whereas the aided electric current was 3.5 kA, as shown in Figure 3b. The weld morphology exhibited an "amphoral" profile when the aided electric current increased to 5.0 kA, as shown in Figure 3c. However, when the electric current increased to 6.0 kA, the stability of the process clearly reduced, which caused the production of many spatters and serious concavity in the weld center, as shown in Figure 3d. When the electric current was 7.0 kA, the fluid metal in the molten pool flowed in a turbulent manner, and the weld quality was so bad that there was no effective joint between the skin and stringer, as shown in Figure 3e. Thus, the aided electric current in the systematic experimental investigation was varied from 0 to 5.0 kA in the following study.

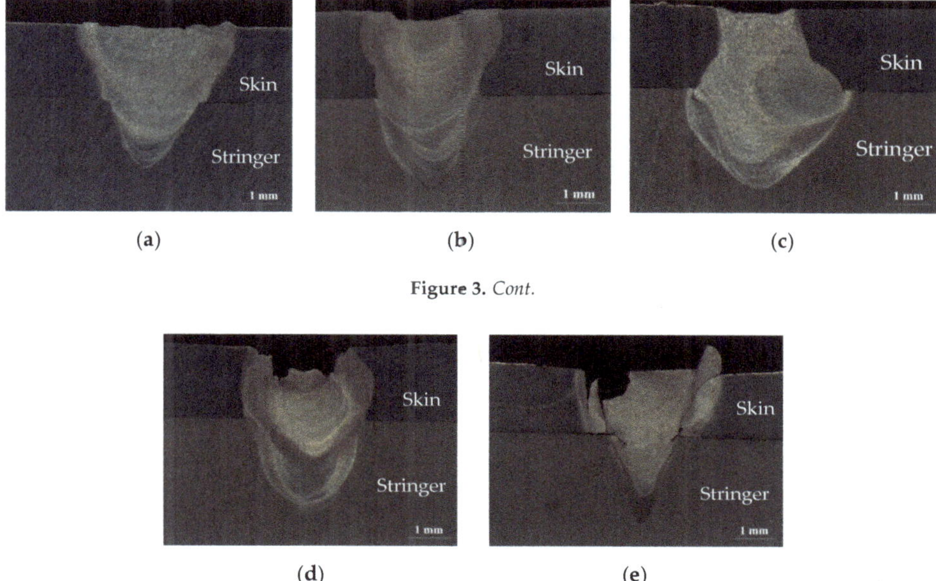

Figure 3. *Cont.*

Figure 3. Typical macrographs of the weld morphology produced with different aided electric currents (fixed process parameters: laser power, P = 1.6 kW; welding speed, v = 0.72 m/min; pressing force, F = 1.1 kN): (**a**) 0 kA; (**b**) 3.5 kA; (**c**) 5.0 kA; (**d**) 6.0 kA; (**e**) 7.0 kA.

3.2. Effects of Process Parameters on the Weld Morphology

For the stake welded T-joints, the total depth of the weld demonstrates that the penetration capability and the weld width at the faying surface may significantly determine the loading capacity of T-joints. To determine the influence of the process parameters on the weld morphology, two geometric dimensions were extracted, namely, the total depth of the weld (H) and the weld width at the faying surface (W), as shown in Figure 4.

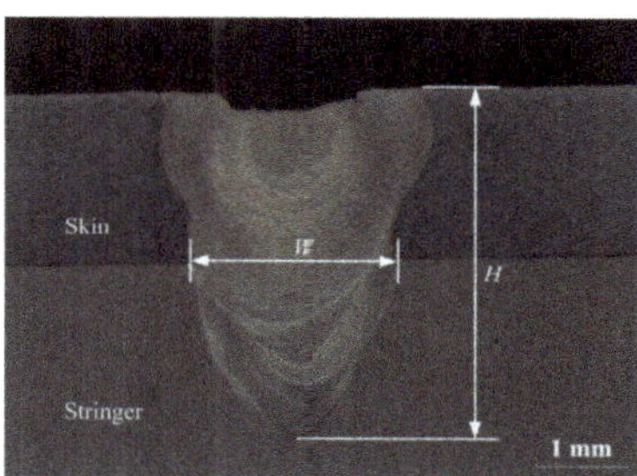

Figure 4. Definitions of the geometric dimensions of the weld morphology.

During the electric current-aided laser stake welding of the T-joints, the process parameters mainly included the electric current (I), pressing force (F; applied on the T-joint surface through the wheel electrode), laser power (P), and welding speed (v). Figure 5 indicates the effects of the single process parameters on the geometric dimensions (H and W) of the weld while the other three process parameters were held constant. Using a higher electric current, the weld width at the faying surface

(*W*) and penetration depth (*H*) both became larger, but there was little change in the penetration depth (*H*), as shown in Figure 5a. As the aided electric current increased from 0 to 5.0 kA, the weld width at the faying surface varied from 1.3 to 3.2 mm, and the penetration depth ranged from 3.3 to 4.2 mm. It is effective to increase the weld width at the faying surface for T-joints using an aided electric current. Figure 5b displays the effect of pressing force on the weld morphology (*H*, *W*) using the same laser power, electric current and welding speed. Upon increasing the pressing force, the penetration depth increased, whereas the weld width at the faying surface decreased. The results showed that the weld width at the faying surface visibly decreased, and the penetration depth exhibited a slower increase. During the electric current-aided laser stake welding of the T-joint, when the electric current is passed through the skin to the stringer, the resistance heat should provide preheating of the skin. The higher the aluminum alloy temperature is, the greater the absorptivity of the aluminum alloy [10]. Therefore, it is beneficial to improve the absorption of the laser on the top of the skin surface, as a result of increasing the penetration depth (*H*) and weld width at the faying surface (*W*). In addition, the aided electric current flowed through the faying surface between the skin and stringer, which contributed to additional heat generation ($Q = I^2Rt$, where Q is heat energy, I is current, R is contact resistance, and t is time) resulting from contact electrical resistance at the faying surface [27]. Therefore, the weld width at the faying surface (*W*) clearly increased upon using the aided electric current, and a higher aided electric current led to a larger weld width at the faying surface (*W*). However, the contact electrical resistance decreased with increasing pressing force; this reduced additional heat generation at the faying surface, and consequently, the weld width at the faying surface (*W*) decreased by increasing the pressing force.

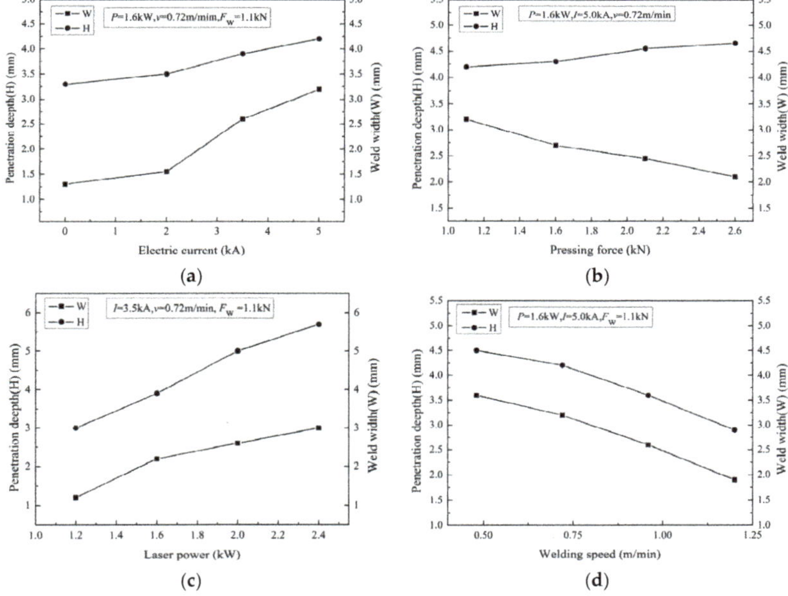

Figure 5. Effects of the different process parameters on the weld morphology (*W*, *H*): (**a**) electric current; (**b**) pressing force; (**c**) laser power; (**d**) welding speed.

Figure 5c displays the effect of the laser power on the weld morphology (*H*, *W*) with the same electric current, welding speed and pressing force. As shown in Figure 5c, the penetration depth increased linearly with increasing laser power, whereas the penetration depth increased sharply from 3.0 to 5.7 mm with an increasing laser power from 1.2 to 2.4 kW. The weld width at the faying surface also increased with increasing laser power; however, the growth declined slowly. Figure 5d shows the effect of welding speed on the weld morphology (H, W) with the same electric current, laser power and pressing force. The weld width at the faying surface and penetration depth both clearly decreased when the welding speed increased. The curves for the weld width and penetration depth similarly changed with increasing welding speed. During the laser welding process, a keyhole is created in a liquid molten pool via evaporation of the base metal. The keyhole develops a deep penetration weld and a high depth-to-width ratio weld [10]. With increasing laser power, the weld width at the faying surface (W) and penetration depth (H) increase; however, the variation in the penetration depth (H) is sharper. The welding speed and laser power determine the rate of energy input to the specimen, and a faster welding speed brings about lower energy input, identified by a lower laser power. The weld width and penetration depth (H) at the faying surface (W) both decrease with an increase in welding speed.

As a consequence of the above results, the electric current and pressing force should be chosen to produce an expected weld width at the faying surface, whereas the laser power and welding speed should be primarily considered to obtain an optimal penetration depth for the electric current-aided laser stake welding of aluminum alloy T-joints.

3.3. Tensile Shear Test

The experimental results indicated that the weld width at the faying surface (W) has a significant effect on the tensile shear loading capacity of welded T-joints. The relationships between the weld width at the faying surface (W) and tensile shear load are provided in Figure 6. The loads sharply increase at first and then decrease slowly with an increase of the weld width at the faying surface (W). Figure 6 shows that the welded T-joint tensile shear load rises from 1280 to 2910 N and then decreases to 2850 N, when the weld width at the faying surface (W) increases from 1.2 to 3.6 mm. The results suggest that the joint tensile shear loading capacity improvement is mainly associated with increasing the weld width at the faying surface (W).

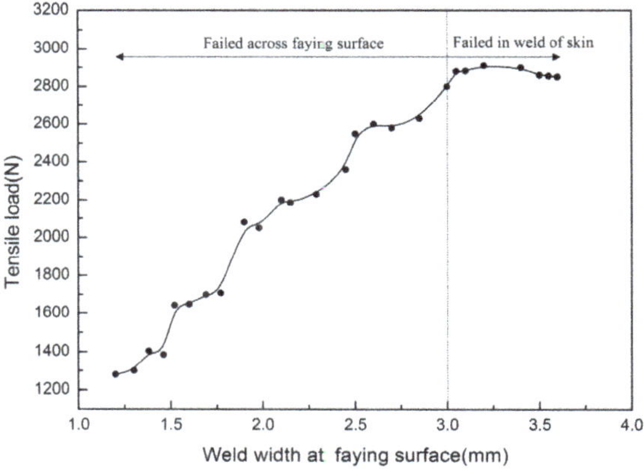

Figure 6. Relationship between weld width at the faying surface (W) and tensile shear load.

During the tensile shear tests, some tensile shear specimens failed across the faying surface and the others failed in the weld zone of the skin, as shown in Figure 7. As shown in Figure 6, the weld width at the faying surface (W) also has an effect on the failure mode of the welded T-joint. For a relatively small weld width at the faying surface (W; less than 3.0 mm), the tensile shear specimens failed across the faying surface, as shown in Figure 7a. For a relatively large weld width at the faying surface (W; not less than 3.0 mm), the tensile shear specimens failed in the weld of the skin, as shown in

Figure 7b. The specimens that failed in the weld of the skin could resist higher tensile shear loads compared with the specimens that failed across the faying surface. Figure 8 shows the load-strain curves of the test specimens of different failed modes. While the test specimen failed across the faying surface, the load reached its maximum and then decreased rapidly, as shown in Figure 8a. While the test specimen failed in the weld of the skin, the load reached the maximum, and then slowly decreased until the joint broke, as shown in Figure 8b. The failure surfaces were observed using SEM. For two types of failed specimens, different fracture surface morphologies were observed at a higher magnification. As shown in Figure 9a, as the tensile shear test specimens failed across the faying surface, the fracture surfaces were flat, and shear fracture features with a plastic deformation along the shear direction were observed. When the tensile shear test specimens failed in the weld of the skin, the fracture surfaces were typical normal fracture features with a ductile failure mode, as shown in Figure 9b.

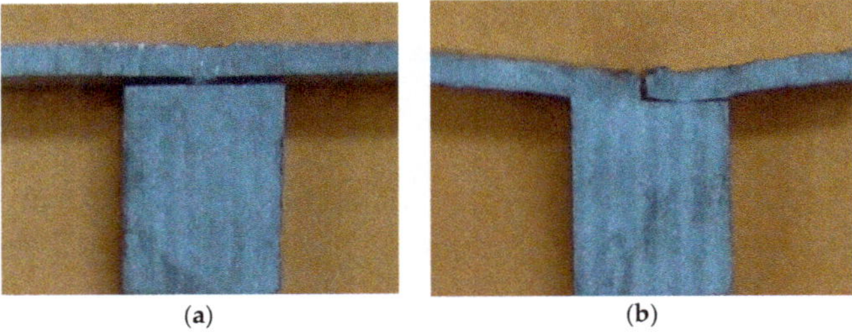

Figure 7. Macrographs of the failed tensile shear test specimens: (**a**) failed across the faying surface; (**b**) failed in the weld of the skin.

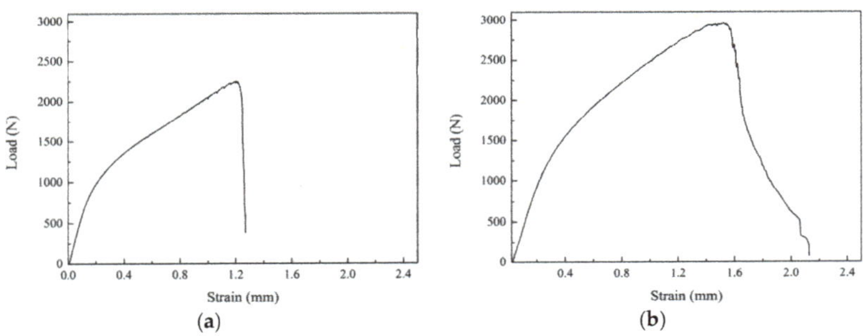

Figure 8. Load-strain curves of the failed tensile shear test specimens: (**a**) failed across the faying surface; (**b**) failed in the weld of the skin.

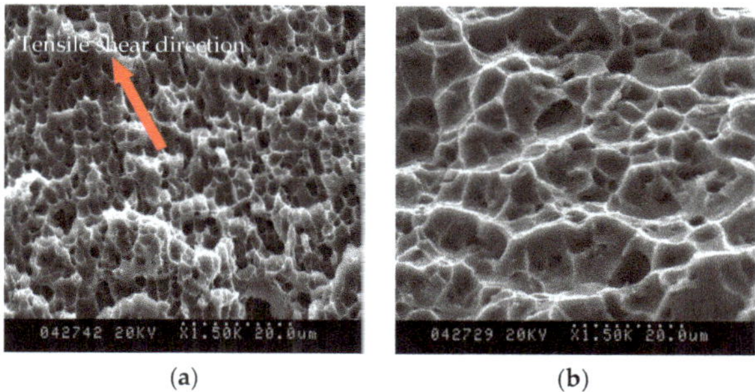

(a)   (b)

Figure 9. Scanning electron microscopy (SEM) images showing the fracture morphology of the failed specimens: (**a**) failed across the faying surface; (**b**) failed in the weld of the skin.

3.4. Microstructure

Optical micrographs of the laser-welded and different electric current-aided laser-welded aluminum alloy T-joints (failed across the faying surface and in the weld of the skin, respectively) are shown in Figure 10. The welded joints consisted mainly of the weld zone, the HAZ and the base metal. The experimental results indicate that the fusion line is distinct for a rapid heating/cooling rate during the laser welding process and the HAZ is narrow, as shown in Figure 10a. As shown in Figure 10b,c, the HAZ width is related to the aided electric current, and a higher electric current produces a larger HAZ width during electric current-aided laser welding.

Figures 11a and 12 show the microstructure of the base metal and weld center, respectively. Gao [28] and Wang [29] have investigated the phase components of the 5A06 alloy base metal and its weld metal using X-ray diffraction (XRD) analysis Their results showed that the 5A06 alloy base metal and its weld metal all mainly consisted of α-Al phase and β phase (Al_3Mg_2); meanwhile, the other phases were few in quantity and thus there were no obvious phase peaks in the XRD patterns. On the basis of the XRD analysis results in the references [28,29], the phase particles were executed using EDS analysis in this study, and the microstructure observation and EDS analysis results show that the base metal had a distributed second strengthening phase β (Al_3Mg_2) in the aluminum matrix (α-Al) together with relatively few insoluble particles of Mg_2Si and Al_6(Fe, Mn), as illustrated in Figure 11. The weld zones for the laser welding and the electric current-aided laser welding consisted of α-Al and phase β with a dispersed distribution, as shown in Figure 12a. The second strengthening phase β in the weld zone of the electric current-aided laser-welded joint was longer than that of the laser weld because of the lower cooling rate. From Figure 12b,c, it is shown that the phase β slightly increased in size and decreased in quantity, with a higher electric current due to the higher heat input.

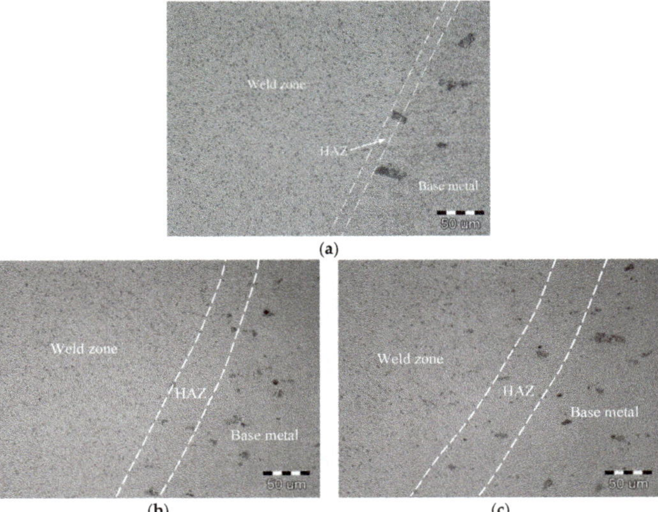

Figure 10. Microstructure of the welded joint: (**a**) laser welding; (**b**) 3.5 kA electric current-aided laser welding; (**c**) 5.0 kA electric current-aided laser welding.

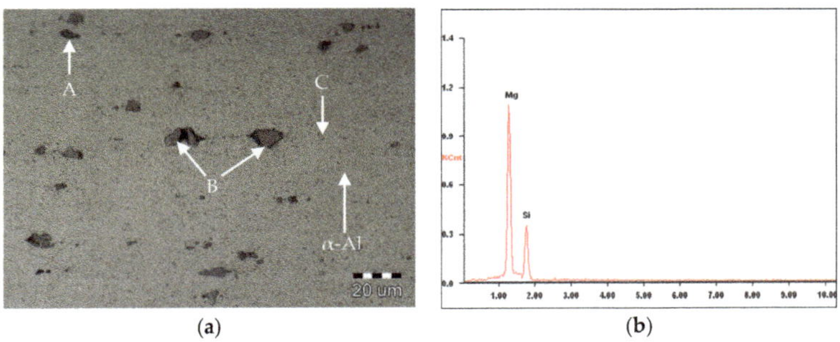

Figure 11. *Cont.*

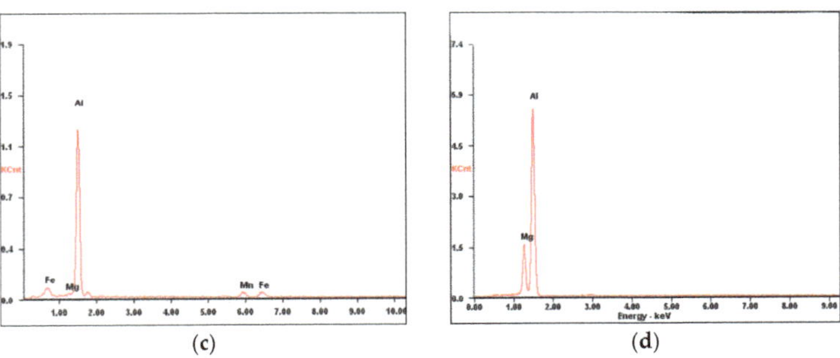

Figure 11. (**a**) Microstructure of the base metal; (**b**) energy-dispersive X-ray spectrometer (EDS) result of the phase particle A (Mg_2Si); (**c**) EDS result of the phase particle B (Al_6(Fe, Mn)); (**d**) EDS results of the phase particle C (Al_3Mg_2).

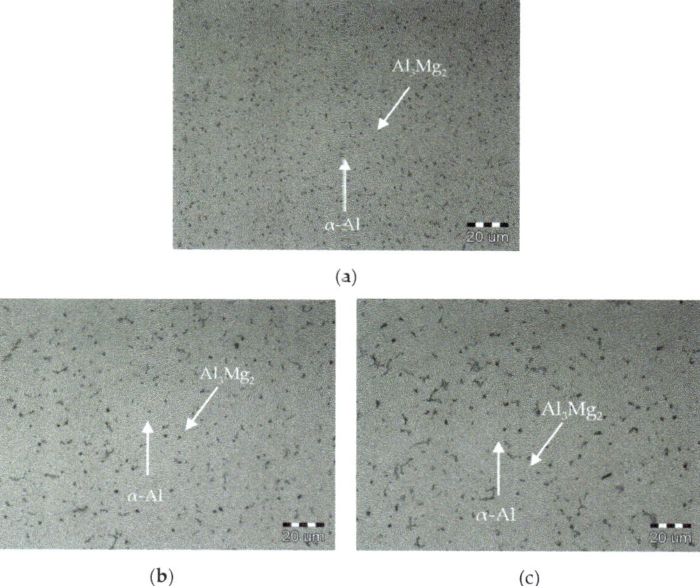

Figure 12. Microstructure of (**a**) the weld center from laser welding; (**b**) the weld center from 3.5 kA electric current aided laser welding; (**c**) the weld center from 5.0 kA electric current aided laser welding.

3.5. Microhardness

Microhardness tests were conducted on the cross-sections of the welded T-joints, and the measurement location was placed 1.5 mm below the skin surface of the welded T-joints, as illustrated in Figure 13. The microhardness distributions in the laser-welded and different electric current-aided laser-welded aluminum alloy T-joints (which failed across the faying surface and in the weld of the skin, respectively) are shown in Figure 14. For the lase-welded T-joint and for the electric current-aided laser-welded T-joints, the microhardness in the weld zone was the lowest because of the as-cast structure and growth of the strengthening phase β of decreased quantity. The microhardness of the base metal was the highest, which was attributed to the rolling deformation, initial annealing state and extremely fine phase β with a dispersed distribution. In the HAZ, the microhardness with an uneven distribution was higher than that in the weld zone. The hardness of the material reflects the ability to resist deformation; therefore, the hardness is directly relevant to the tensile shear loading capacity. Generally, with a decrease in the hardness, the tensile shear load decreases [30]. Because of the low hardness, the weld zone was the weakest area during the tensile shear tests and easily ruptured. Therefore, in the tensile shear tests, cracks all initiated from the edge of the weld at the faying surface. If the weld width at the faying surface (*W*) was small, the crack propagated across the faying surface between the skin and stringer until the tensile shear specimens failed, and if the weld width at the faying surface (*W*) was large, the fracture crack propagated in the weld zone of the skin until the tensile shear specimens failed, as shown in Figure 15.

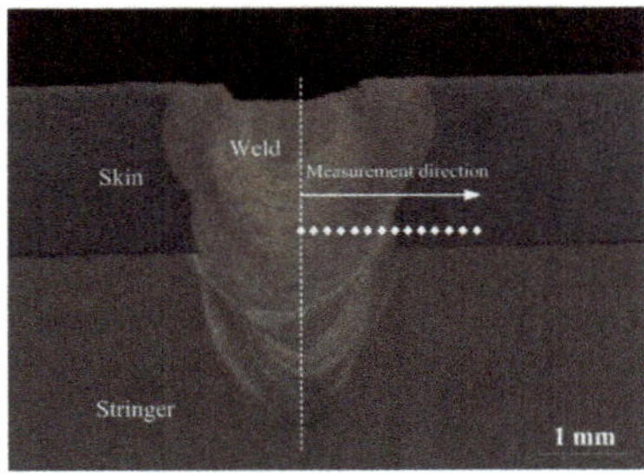

Figure 13. Schematic illustration of the microhardness measurement location and direction.

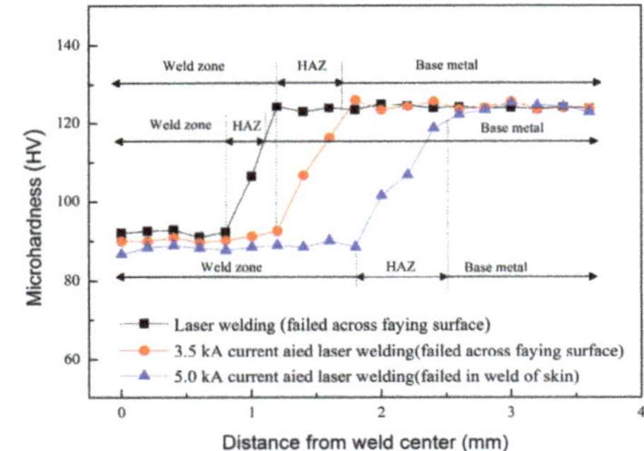

Figure 14. Microhardness distribution for different welded T-joints.

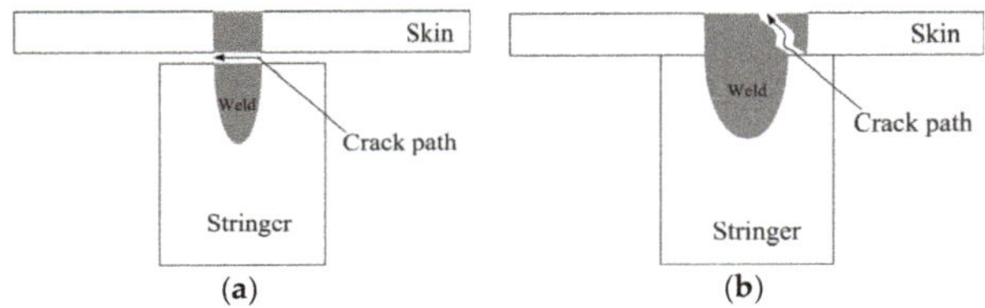

Figure 15. Schematic illustration of sample failure modes: (**a**) failed across the faying surface; (**b**) failed in the weld of skin.

As shown in Figure 14, the microhardness in the weld zone of the higher-electric-current-aided laser-welded T-joint was slightly lower than that of an even higher electric current. Thus, when the weld width at the faying surface (W) was sufficient and the tensile shear specimens failed in the weld zone of the skin, the welded T-joint tensile shear load increased to the maximum and then slightly decreased because the microstructure changed in the weld zone, as shown in Figure 12b,c, which is consistent with the tensile shear results shown in Figure 6. However, the reduction in the amount of tensile shear load was insignificant, which suggests that the tensile shear load of the aluminum alloy welded T-joint was mainly determined by the weld width at the faying surface (W).

4. Conclusions

Electric current-aided laser-welding technology was developed to stake weld aluminum alloy T-joints. The electric current used should be not more than 5.0 kA; otherwise, as in the present experimental investigation, the process stability reduces, resulting in a poor-quality weld.

With a higher electric current, the weld width at the faying surface (*W*) and penetration depth (*H*) both became larger; however, there was little change in the penetration depth (*H*). The weld width at the faying surface (*W*) visibly decreased, and the penetration depth had a small variation as the pressing force grew. The penetration depth (*H*) increased linearly with increasing laser power; however, the growth of the weld width at the faying surface (*W*) slowly decreased. The weld width and the penetration depth (*H*) at the faying surface (*W*) both clearly decreased when the welding speed increased. Consequently, the electric current and pressing force should be chosen to produce an expected weld width at the faying surface (*W*), whereas the laser power and welding speed should be primarily considered to obtain an optimal penetration depth (*H*).

The tensile shear load sharply increased at first and then reduced slowly as the weld width at the faying surface (*W*) increased. Some tensile shear specimens failed across the faying surface, and others failed in the weld zone of the skin. The specimens that failed in the weld of the skin could resist a higher tensile shear load compared with specimens that failed across the faying surface.

Because of the rolling deformation and extremely fine phase β with a dispersed distribution, the microhardness of the base metal was the highest. In the HAZ, the microhardness had an uneven distribution, which was higher than that in the weld zone. The weld zone was the weakest area during the tensile shear tests and was much more likely to rupture. The reduction in microhardness of the weld was only slight compared with that of the base metal and HAZ; therefore, the tensile shear load capacity of the aluminum alloy stake-welded T-joint was mainly determined by the weld width at the faying surface (*W*).

Acknowledgments: The work was financially supported by the Jilin Scientific and Technological Development Program (20160520055JH).

Author Contributions: Xinge Zhang designed the experiments, performed the experiments, analyzed the data and wrote the paper. Liqun Li and Yanbin Chen provided the experimental materials and laboratory equipment and directed the research. Zhaojun Yang and Xiaocui Zhu contributed to the discussion and interpretation of the results.

Conflicts of Interest: The authors declare no conflict of interest.

References

1. Romanoff, J.; Varsta, P.; Remes, H. Laser-welded web-core sandwich plates under patch-loading. *Mar. Struct.* **2007**, *20*, 25–48. [CrossRef]

2. Jelovica, J.; Romanoff, J.; Klein, R. Eigenfrequency analyses of laser-welded web-core sandwich panels. *Thin-Walled Struct.* **2016**, *101*, 120–128. [CrossRef]

3. Jelovica, J.; Romanoff, J.; Ehlers, S.; Varsta, P. Influence of weld stiffness on the buckling strength of laser-welded web-core sandwich plates. *J. Constr. Steel Res.* **2012**, *77*, 12–18. [CrossRef]

4. Janasekaran, S.; Tan, A.W.; Yusof, F.; Abdul Shukor, M.H. Influence of the overlapping factor and welding speed on T-Joint welding of Ti_6Al_4V and Inconel600 using low-power fiber laser. *Metals* **2016**, *6*, 134. [CrossRef]

5. Knox, E.M.; Cowling, M.J.; Winkle, I.E. Adhesively bonded steel corrugated core sandwich construction for marine applications. *Mar. Struct.* **1998**, *11*, 185–204. [CrossRef]

6. Silva, L.F.M.; Adams, R.D. The strength of adhesively bonded T-joints. *Int. J. Adhes. Adhes.* **2002**, *22*, 311–315. [CrossRef]

7. Grant, L.D.R.; Adams, R.D.; Silva, L.F.M. Effect of the temperature on the strength of adhesively bonded single lap and T joints for the automotive industry. *Int. J. Adhes. Adhes.* **2009**, *29*, 535–542. [CrossRef]

8. Frank, D. Fatigue strength assessment of laser stake-welded T-joints using local approaches. *Int. J. Fatigue* **2015**, *73*, 77–87. [CrossRef]

9. Meng, W.; Li, Z.G.; Huang, J.; Wu, Y.X.; Cao, R. Effect of gap on plasma and molten pool dynamics during laser lap welding for T-joints. *Int. J. Adv. Manuf. Technol.* **2013**, *69*, 1105–1112. [CrossRef]

10. Duley, W.W. *Laser Welding*, 1st ed.; John Wiley and Sons Ltd.: New York, NY, USA, 1998; pp. 1–9.

11. Köse, C.; Karaca, E. Robotic Nd:YAG fiber Laser welding of Ti-6Al-4V alloy. *Metals* **2017**, *7*, 221. [CrossRef]

12. Meng, W.; Li, Z.G.; Huang, J.; Wu, Y.X.; Chen, J.H.; Katayama, S. The influence of various factors on the geometric profile of laser lap welded T-joints. *Int. J. Adv. Manuf. Technol.* **2014**, *74*, 1625–1636. [CrossRef]
13. Romanoff, J.; Remes, H.; Socha, G.; Jutila, M.; Varsta, P. The stiffness of laser stake welded T-joints in web-core sandwich structures. *Thin-Walled Struct.* **2007**, *45*, 453–462. [CrossRef]
14. Li, K.; Lu, F.G.; Cui, H.C.; Li, X.B.; Tang, X.H.; Li, Z.G. Investigation on the effects of shielding gas on porosity in fiber laser welding of T-joint steels. *Int. J. Adv. Manuf. Technol.* **2015**, *77*, 1881–1888. [CrossRef]
15. Wang, M.; Jiang, M.L.; Wei, Q.; Gu, K.F. Technique of laser-TIG hybrid T-shape joint welding of titanium Alloy. *Adv. Mater. Res.* **2011**, *291–294*, 841–847. [CrossRef]
16. Hou, Z.L.; Li, C.B.; Liu, L.M. Laser-TIG Hybrid Welding of Magnesium Alloy T-Joint with Cold Filler Wire. *Mater. Trans.* **2015**, *56*, 1242–1247. [CrossRef]
17. Li, C.B.; Liu, L.M. Investigation on weldability of magnesium alloy thin sheet T-joints: Arc welding, laser welding, and laser-arc hybrid welding. *Int. J. Adv. Manuf. Technol.* **2013**, *65*, 27–34. [CrossRef]
18. Fratini, L.; Buffa, G.; Shivpuri, R. Influence of material characteristics on plastomechanics of the FSW process for T-joints. *Mater. Des.* **2009**, *30*, 2435–2445. [CrossRef]
19. Cui, L.; Yang, X.Q.; Zhou, G.; Xu, X.D.; Shen, Z.K. Characteristics of defects and tensile behaviors on friction stir welded AA6061-T4 T-joints. *Mater. Sci. Eng. A* **2012**, *543*, 58–68. [CrossRef]
20. Zhao, Y.; Zhou, L.L.; Wang, Q.Z.; Yan, K.; Zou, J.S. Defects and tensile properties of 6013 aluminum alloy T-joints by friction stir welding. *Mater. Des.* **2014**, *57*, 146–155. [CrossRef]
21. Urbikain, G.; Perez, J.M.; Lacalle, L.N.L.D.; Andueza, A. Combination of friction drilling and form tapping processes on dissimilar materials for making nutless joints. *Proc. Inst. Mech. Eng. Part B J. Eng. Manuf.* **2016**. [CrossRef]
22. Hong, K.M.; Shin, Y.C. Prospects of laser welding technology in the automotive industry: A review. *J. Mater. Process. Technol.* **2017**, *245*, 46–69. [CrossRef]
23. Caiazzo, F.; Alfieri, V.; Corrado, G.; Argenio, P.; Barbieri, G.; Acerra, F.; Innaro, V. Laser beam welding of a Ti–6Al–4V support flange for buy-to-fly reduction. *Metals* **2017**, *7*, 183. [CrossRef]
24. Hou, X.P.; Yang, X.Q.; Cui, L.; Zhou, G. Influences of joint geometry on defects and mechanical properties of friction stir welded AA6061-T4 T-joints. *Mater. Des.* **2014**, *53*, 106–117. [CrossRef]
25. Lee, H.S.; Yoon, J.H.; Yoo, J.T.; No, K. Friction stir welding process of aluminum-lithium alloy 2195. *Procedia Eng.* **2016**, *149*, 62–66. [CrossRef]
26. Krahmer, D.M.; Polvorosa, R.; Lacalle, L.N.L.D.; Alonso-Pinillos, U.; Abate, G.; Riu, F. Alternatives for specimen manufacturing in tensile testing of steel plates. *Exp. Tech.* **2016**, *40*, 1–11. [CrossRef]
27. Ertek Emre, H.; Kaçar, R. Resistance Spot Weldability of Galvanize Coated and Uncoated TRIP Steels. *Metals* **2016**, *6*, 299. [CrossRef]
28. Gao, S.; Wu, Z.S.; Jin, P.F.; Wang, J.J.; Shuai, P. Effect of deep cryogenic treatment on microstructure of 5A06 aluminum alloy MIG welded joint welding. *Mater. Sci. Forum* **2012**, *724*, 182–185. [CrossRef]
29. Wang, L.X.; Cong, B.Q.; Qi, B.J.; Li, W.; Yang, M.X.; Yang, Z.; Li, Y.L. Effect of ultra-high frequency pulse square-wave current on 5A06 aluminum alloy HPVP-GTAW joints. *Trans. China Weld. Inst.* **2013**, *34*, 61–64.
30. Chen, Y.B.; Miao, Y.G.; Li, L.Q.; Wu, L. Joint performance of laser-TIG double-side welded 5A06 aluminum alloy. *Trans. Nonferrous Met. Soc. China* **2009**, *19*, 26–31. [CrossRef]

 *metals*

Article

Control of Porosity and Spatter in Laser Welding of Thick AlMg5 Parts Using High-Speed Imaging and Optical Microscopy

Andrei C. Popescu [1,2,*], Christophe Delval [1] and Marc Leparoux [1]

[1]   Laboratory of Advanced Materials Processing, EMPA-Swiss Federal Laboratories for Materials Testing and Research, Feuerwerkerstrasse 39, 3602 Thun, Switzerland; christophe.delval@gmail.com (C.D.); marc.leparoux@empa.ch (M.L.)

[2]   National Institute for Lasers, Plasma and Radiation Physics, Atomistilor 409, 077125 Magurele, Romania

*   Correspondence: andrei.popescu@inflpr.ro; Tel.: +40-21-457-4550

Received: 13 September 2017; Accepted: 19 October 2017; Published: 26 October 2017

Abstract: We report on a feedback mechanism for rapid identification of optimal laser parameters during welding of AlMg5 coupons using real-time monitoring by high-speed imaging. The purpose was to constrain the liquid movement in the groove in order to obtain pore-free welds in this otherwise difficult-to-weld alloy. High-speed imaging of the welding process via an optical microscope allowed for recording at millimeter level, providing new information on liquid-metal dynamics during laser irradiation as well as plausible explanations for spatter occurrence and pores formation. The pore formation and especially the position of these pores had to be controlled in order to weld 3 mm thick samples. By tuning both laser power and pulse duration, pores were aligned on a single line, at the bottom of the weld. A laser pass of reduced power on that side was then sufficient for removing all pores and providing a suitable weld.

Keywords: laser welding; spatter; liquid metal; high-speed imaging; porosity control

---

1. Introduction

Weight and production cost reducing automotive body design are achieved by using laser butt welded semi products such as tailor blanks out of steel and combined materials [1,2]. Laser is preferred for welding, owing to to the high speed of the process, low distortion due to the small heat affected zone, manufacturing flexibility and ease of automation [3–5]. Further improvements in transportation can be achieved by replacing steel with lighter metals such as aluminum and its alloys [6–8]. However, obtaining reliable, reproducible flawless high-quality welds for these materials is challenging. Aluminum is also difficult to weld because it easily oxidizes. Moreover, due to its high thermal diffusivity, the weld fusion and penetration are more difficult to achieve in comparison to steel [9]. The fast heat exchange implies also a fast solidification that leads to high stresses, resulting in cracks, pores formation or even complete break of the welded area [10–14].

Another general drawback of laser welding, and especially in keyhole mode, consists of the ejection of liquid droplets, known as "spattering". It often causes defects such as craters, blowouts and underfills that seriously degrade the mechanical properties of the weld [15,16]. Categorization of spatter based on the physical phenomena that contribute to its formation has been proposed by Kaplan and Powell [17]. One cause of spatter formation is the vertical movement of liquid in the groove during irradiation caused by the vapor flow that expands in the keyhole. You et al. [18] attributed this type of spatter formation to the shockwave generated by the expanding vaporized matter from the laser molten pool. In order to understand and avoid this drawback, several research groups monitored in real time the laser welding of metals and alloys by fast imaging [16–20]. Thus, it was possible to extract relevant information about the keyhole (shape, size, liquid movement, pore formation), plume expansion (plume length, expansion speed, particles expulsion) and its effects over the molten pool (liquid spatter). By using a high-speed camera, Zhang et al. [19] visualized the influence of vapor plume on both the melt pool dynamics and transient keyhole cross-section, by applying a sophisticated sample preparation. They observed that for the full penetration of the melt, the recoil pressure of the plume induced a downward flow of the liquid in the keyhole. Because of this movement, the liquid on the keyhole bottom goes upwards with high speed, thus being the main cause for the occurrence of spatter.

High-speed imaging spatter detection contributed to distinguish the welding defects in case of laser welding of stainless steel, by identifying the spatter diameter together with the plume features and ejection direction [18,20]. The presence of spontaneous high amounts of spatter in the image sequences confirmed that a weld seam width reduction occurred. Based on this analogy, the image sequences could be used for evaluating in situ the weld seam quality. For the same material, Li et al. proposed to focus the laser beam in depth for achieving a more stable laser welding by producing a flow towards the bottom of the keyhole, as monitored by in situ X-ray imaging [21].

A method proposed in literature for reducing spatter and porosity during laser welding of dissimilar materials such as Al and Ti alloys is to offset the laser beam towards the Ti side close to the seam. Casalino et al. showed that this offset can produce a highly resistant weld, free of porosity and spatter, with no need for filler or groove preparation [22,23].

The aim of this research was to develop a method based on high-speed imaging and optical microscopy for rapid identification of the optimal parameters for laser welding of a thick sample. We aimed to obtain total heat penetration, to eliminate porosity and splatter and to reduce spatter. For this study, we selected the particular case of an AlMg5 alloy, where porosity is very difficult to be overcome by trial and error adjustments of laser parameters.

2. Materials and Methods

AlMg5 ingots were self-made by cast melting aluminum and magnesium (between 4.5 and 5.5 wt % Mg). Prismatic coupons of $20 \times 20 \times 3$ mm^3 were cut from the bulk material. Prior to laser irradiation, the pieces were sandpaper (P 180)-polished and rinsed with ethanol for contaminants and debris removal. The pieces were butt-welded using a pulsed Nd:YAG (λ = 1064 nm) laser source (SLS200-306 Lasag, Belp, Switzerland).

The characterization of laser beam using a MBS300 beam profiler (Ophir, North Logan, UT, USA) revealed a top hat pulse in the image plane of the final laser focusing lens. The spot size for a laser power of 1 kW and 2 ms pulse duration was found to be of 600 μm for a focusing of the beam on surface and of 850 μm for 2.5 mm in depth. The intensity profiles of the spots focused on surface and in depth are presented in Reference [16].

The AlMg5 specimens were pushed against each other on the work bench and held tight using a lock mechanism in order to minimize the gap.

The experimental set-up is schematically presented in Figure 1. The laser beam was perpendicular to the samples surface (Figure 1) and was delivered via optical fiber through an encased optical system mobile on the z axis, consisting of a collimator, guiding mirrors and a lens with 25 cm focal length. The laser beam remained in fixed position, while the samples were moved by an X–Y translation stage. The normal impinging laser beam has a spot diameter between 600 μm and 1.5 mm, for a 5.5 kW peak power and 1 or 5 ms pulse duration, respectively. For a full penetration butt weld on the 3 mm thick samples, the laser source had to be set to a maximum laser peak power of 5.5 kW with 850 μm spot diameter. The sample surface was kept above or below the beam focus at a distance of 2.5 mm. After establishing the full weld penetration, further process parameters could be tuned in order to improve the weld quality [24]. Pulse duration was varied between 1 and 6 ms. Working with a pulse repetition rate of 10 Hz, the pulse energy was stable for a chosen pulse peak power and all selected pulse durations. The samples were moved perpendicular to the impinging laser beam with stepper motor speeds between 7 and 40 cm/min. No filler material was used for welding the AlMg5 parts.

Figure 1. Schematic of experimental set-up for laser welding and monitoring. HS: High-speed camera.

In order to protect the weld pool against oxidation, high-purity Argon (Ar, 99.9999%) was used as shielding gas. The Ar gas was delivered through a 4 mm diameter polytetrafluoroethylene (PTFE) tubing placed in the xz plane, at 3 cm from the laser spot at an angle of approximately 50° in respect to the sample surface, while the sample was translated along the y direction (see Figure 1 for movement reference). An optimized Ar flow rate of 5 L/min was used for preventing oxidation. Thus, an Ar content of 99% was measured 1 cm ahead of the welding site by mass spectrometry.

Both single laser shot and full welding processes were recorded with a high-speed complementary metal-oxide-semiconductor (CMOS) camera (Motion Pro Y4-S3 mono; Videal AG, Niederoenz, Switzerland) with an image resolution of max. 1024 × 1024 pixels. To record the plume expansion through a 100 mm focal length Zeiss objective, the camera was mounted horizontally, parallel to the sample surface. Frame rates were chosen between 16,400 and 58,001 fps and the shutter speed was set between 100 ns and 20 μs.

For weld pool monitoring through a MZ10 microscope (Leica, Wetzlar, Germany) set to a 2× magnification and tilted to an angle of 45° with respect to the laser beam, the camera frame rate was chosen around 30,000 fps, with the shutter speed set to 100 ns. Molten pool illumination was achieved using cold white light delivered by a KL1500 Electronic halogen lamp source (Schott AG, Mainz, Germany) placed at 3 cm from the irradiation site, at an illumination angle of 45° in respect with the sample. Different color photographic filters were used in order to better distinguish either the features of the plume expansion or the spatter and weld pool. The recordings were processed with Pro Analyst motion analysis software (Version 1.5.9.8, Xcitex, Woburn, MA, USA). Knowing the number of acquired frames per second, the software can establish the occurring time of each frame and using the AlMg5 sample thickness as reference, it can calculate distances and travel speeds for plume expansion and flying particles.

The weld morphology and cross-sections of weld were studied by optical microscopy using an Axioplan microscope (Carl Zeiss, Oberkochen, Germany), equipped with a ProgResC14 plus camera (Jenoptik, Jena, Germany).

3. Results

3.1. Parameters Adjustment Using Optical Microscopy Data

First, it was identified that 5.5 kW laser peak power and over 2 ms pulse duration were mandatory in order to butt-weld 3 mm thick plates. Lower peak powers caused incomplete melting in depth.

Figure 2a–d corresponds to the welds conducted with pulses of 1, 2, 3 and 6 ms duration.

The droplets and particles deposited both around and inside the seam increased in number and dimensions with the pulse duration at constant power. For pulses of 1 ms, most particles had a diameter of 150 μm, increasing up to 1 mm for pulses of 6 ms. For pulse duration of 1 ms, most particles in the vicinity of the seam were polyhedral. By increasing the pulse duration, round particles started to be dominant (for clarity reasons, spherical particles will be thereon called "droplets"). Most of the droplets were surrounded by a halo, most probably part of the initial droplet that spread when it hit the surface. Figure 3 presents in detail a round particle that was still liquid when it collided with the sample surface, producing a splash of liquid metal around it (left side) and some examples of particles with irregular shape (right side).

As the laser peak power had to be maintained at 5.5 kW, the most obvious parameter to vary in order to improve the welds aspect was the laser pulse duration. An easy task in tuning the pulse duration was the study of spatter/splatter on samples surface. Evident signs of violent material expulsion were identified close to the seam in the form of splatter (massive molten metal spill) for 4 and 5 ms pulse durations. Considering these facts, for welding the 3 mm thick AlMg5 samples, a pulse duration of 2 ms was selected for further investigations.

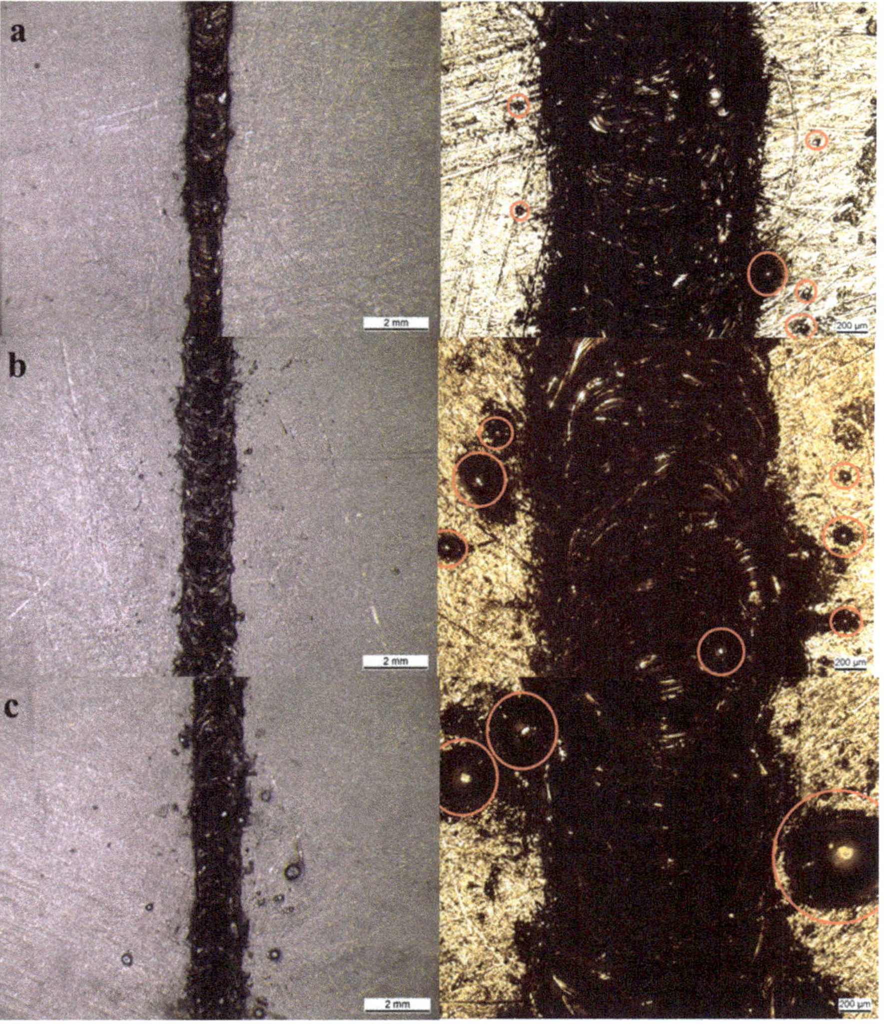

Figure 2. *Cont.*

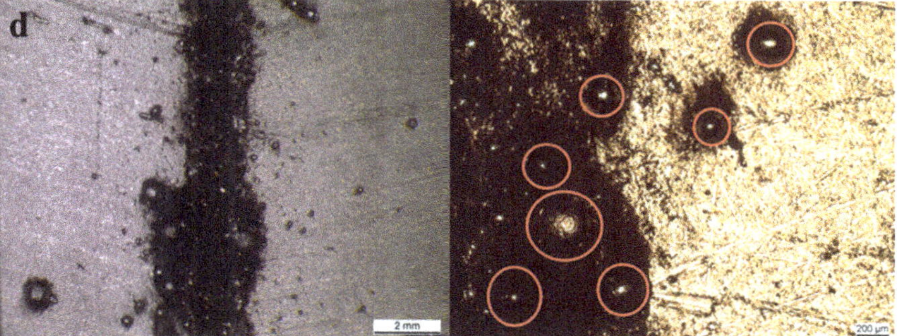

Figure 2. Optical microscopy images of seams in case of laser welding of AlMg5 parts under Ar. Laser peak power of 5.5 kW and pulse duration of (**a**) 1 ms; (**b**) 2 ms; (**c**) 3 ms; and (**d**) 6 ms, respectively. Red circles are marking droplets present around the seam.

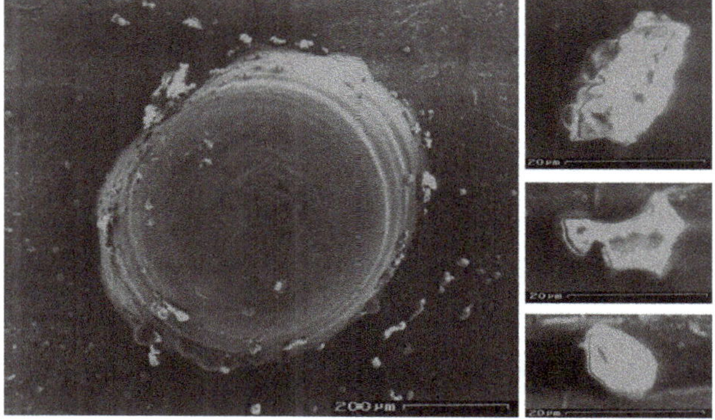

Figure 3. Optical microscopy images of particles found in the vicinity of the weld: left side-round droplet that was probably still liquid when it hit the surface. The impact produced a splash of molten metal around it; right side-particles of irregular shape.

By increasing the translation speed during welding from 7 to 14, 30 and finally 40 cm/min while keeping constant the laser peak power at 5.5 kW, for a pulse duration of 2 ms and a repetition rate of 10 Hz, a decrease of the amount of liquid spilled from the seam and drastically reduced density of droplets on the surface were observed. By doubling the speed from 7 to 14 cm/min, it was observed that the particles' diameters around the weld decreased from 60–200 μm to 20–100 μm. At 30 and 40 cm/min, no droplets could be found at the samples surface, close to the seam. Only rare droplets exclusively located inside the seam were observed. They exhibited diameters ranging from 200 to 500 μm. Moreover, the cross-sections of samples revealed full penetration welds. Therefore, 30 cm/min scan speed was kept as optimal for our welding conditions.

3.2. Parameters Adjustment Using High-Speed Imaging Data

For identifying the optimal experimental parameters for welding, high-speed imaging of the surface was used for monitoring the liquid phase behavior. This analysis can be of importance for welding from two perspectives: (i) It can provide information about spatter for a particular set of chosen experimental conditions. For welding, spatter can be chosen as one of the control parameters when searching for the optimal welding conditions, as its high amount could mean material loss which causes welding underfill, porosity, as well as a bad visual aspect of the welds; (ii) It allows for real-time visualization of liquid movement in the irradiated area, thus allowing the selection of irradiation parameters so that splatter (large liquid material downpour) can be ruled out.

First, to investigate the interaction between the laser beam and material, single irradiations onto the surface of a sandblasted AlMg5 sample were performed, with the beam focused 2.5 mm in depth. Figure 4 presents a subsequence of representative frames extracted from the real-time recording of an AlMg5 surface state evolution under the action of one laser pulse of 5.5 kW peak power and 2 ms duration.

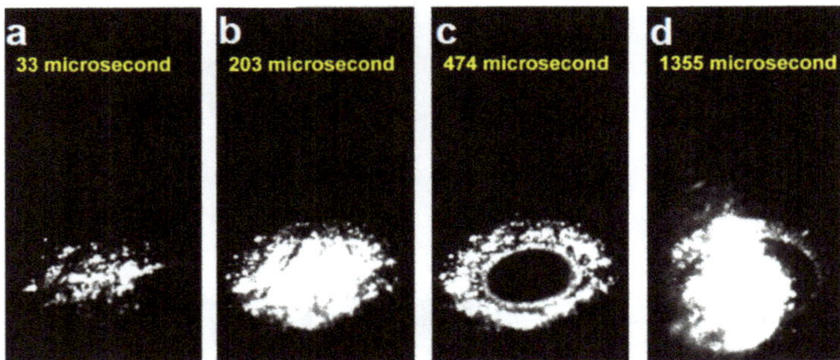

Figure 4. High-speed imaging of AlMg5 surface behavior under the action of a laser pulse of 5.5 kW power and 2 ms duration: (**a**) start of incandescence; (**b**) highest incandescence before melting; (**c**) expansion of the liquid phase starting from the center of the spot; (**d**) liquid expulsion from the molten pool.

In the following, for all set of frames, the time reference, "instant = 0 s", corresponds to the camera internal event trigger (i.e., "Motion Trigger"), defined as an increase from 2% to 5% of the light intensity in the area of irradiation site.

The purpose of this research was to identify the time for incandescence, melting and boiling on metal surface and further compare them with the times recorded for actual welds on the groove between two metal parts. The surface became incandescent 30 µs from the start of laser pulse irradiation (Figure 4a) and reached a maximum of brightness after approximately 200 µs of irradiation (Figure 4b). From this point on, a melting process that started from the center of the spot and expanded in a circular pattern towards the edges of the laser spot was observed. Figure 4c exhibits a representative frame for the liquid phase expansion. The entire irradiated area was completely melted after 800 µs. The molten pool continuously heated by the laser reached the boiling temperature. The first spatter was recorded after 1350 µs.

It has been also observed that the occurrence of events presented in Figure 4 remains the same for pulses with 5.5 kW peak power and longer pulse durations, but the recorded times decreased, while boiling and spatter (Figure 4d) intensified as pulse duration increased.

A basic analytical model that predicts temperature evolution on sample surface during the laser pulse action was used for comparison with the experimental observations [25]. The calculations were conducted on z direction only and surface absorption, conductivity, density, and heat capacity were considered as temperature-independent.

The simplified form of the heat equation becomes:

$$\frac{\partial T}{\partial t} = D\frac{\partial^2 T}{\partial z^2} \tag{1}$$

where T is temperature, t is time, D is the thermal diffusivity, $D = \frac{k}{\rho c_p}$, k = thermal conductivity, ρ = density, c_p = specific heat.

The solution of the one-dimensional heat equation is:

$$T(z, t < \tau) = \frac{2\alpha I_0(1 - R)}{k}(Dt)^{\frac{1}{2}} ierfc[\frac{z}{2(Dt)^{\frac{1}{2}}}]$$ (2)

where τ = pulse duration; α = optical absorption coefficient, I_0 = laser peak power, R = reflectivity, k = thermal conductivity; D = thermal diffusivity. *ierfc* = integral of complementary error function.

For the case of surface temperature ($z = 0$),

$$ierfc[\frac{z}{2(Dt)^{\frac{1}{2}}}] = \frac{1}{\sqrt{\pi}}$$ (3)

For AlMg5, the constants used in calculations were: k = 140 W/m·K, c_p =900 J/kg·K, $\alpha = 10^8$ m^{-1} while the reflectivity for 1064 nm wavelength was selected to be of 93% [26].

The simulated evolution of temperature during the laser pulse is given in Figure 5. At the end of the laser pulse, the surface temperature surpasses 3000 K. For the events recorded in Figure 4, the computed temperatures are T(33 µs) = 683 K, T(203 µs) = 1200 K, T(474 µs) = 1477 K, T(1.355 ms) = 2800 K. While the model accurately predicts what happens at 33 µs since the beginning of laser pulse, it overestimates the temperatures for the ms range. This is certainly due to model simplifications: Surface absorptivity would certainly increase with temperature, laser beam absorption in the plume and radiative losses are not computed. However, the model is useful in predicting the intervals where boiling can occur. AlMg5 boiling temperature is at ~2700 K [27], occurring after 1.355 ms since the start of the laser pulse, as caught on high-speed camera footage. The model obtained a value of 2800 K at 1.355 ms. With small adjustments, recordings can be optimized in order to capture the footage of interest only, thus reducing considerably the videos processing time and video storage space.

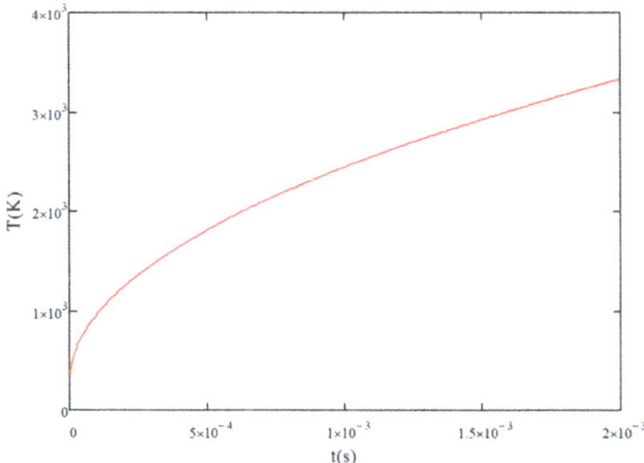

Figure 5. Simulation of AlMg5 surface temperature evolution during irradiation with a laser pulse of 5.5 kW peak power and 2 ms duration.

In a second step, a butt weld between two adjacent parts was performed and observed with the high-speed camera. Figure 6a shows that the metal heated by the laser spot first became incandescent. The joint borders progressively took a circular shape under the action of laser pulse (Figure 6b). Melting occurred after approximately 200 µs starting from the spot center that remained consistent with the previous observations obtained on flat surface irradiations. The liquid rose from the groove and produced liquid expulsion that began after approximately 830 µs (Figure 6c). At the end of the laser pulse, the liquid fell back into the groove forming a whirlpool (Figure 6d). We believe that the liquid follows the crater walls shaped conically when returning into the groove, thus producing this whirlpool. During this upwards/downwards movement, the liquid loses heat and becomes denser than the liquid remaining on the bottom of the groove. The pressure put by the returning liquid upon the bottom pool can cause a spill between at the joint between the two pieces to be welded (see area marked with a red circle in Figure 7a).

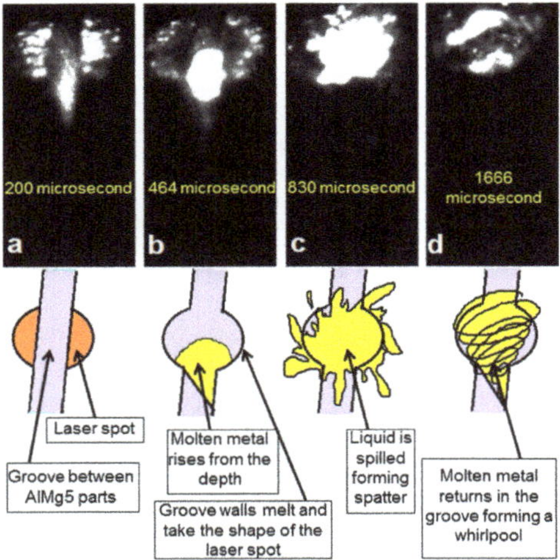

Figure 6. High-speed imaging of phenomena occurring during laser irradiation with a pulse of 5.5 kW power and 2 ms duration of the joint between two AlMg5 pieces: (**a**) metal reaches incandescence; (**b**) molten metal rises from the groove; (**c**) expulsion of liquid phase; (**d**) metal returns in the groove forming a whirlpool.

In case of butt welding, spatter occurs even before surface boiling. We believe this to be the cause of gas bubbles that emerge from the bottom of the molten pool causing liquid spatter. To confirm this assumption, the joint between two AlMg5 coupons was irradiated with single laser pulses (power = 5.5 kW, frequency = 1 Hz, in-depth focusing at 2.5 mm), and subsequently, a cross-section of the weld was studied by optical microscopy (Figure 7). Indeed, we found pores that seemed to be trapped during solidification, close to the bottom of the weld. The increase of pulse duration generated larger pores in the solidified keyhole. In the case of a 2 ms pulse, ~50 μm pore was formed (Figure 7a), while increasing the pulse duration to 3 ms resulted in a pore with a diameter of ~300 μm (Figure 7b). This pore diameter ascending trend continued up to a pulse duration of 6 ms, when most of the keyhole was devoid of metal (Figure 7c).

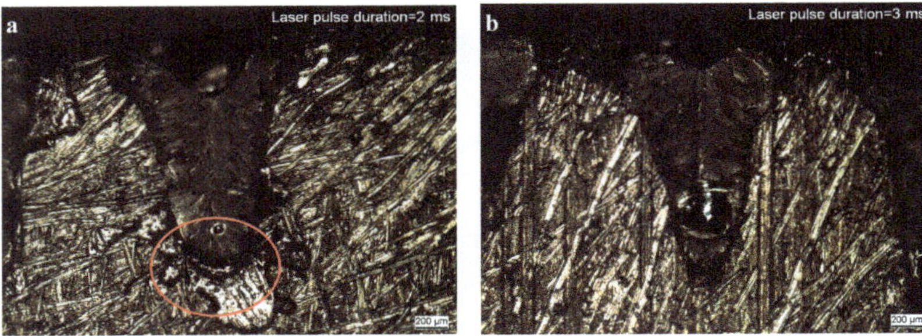

Figure 7. *Cont.*

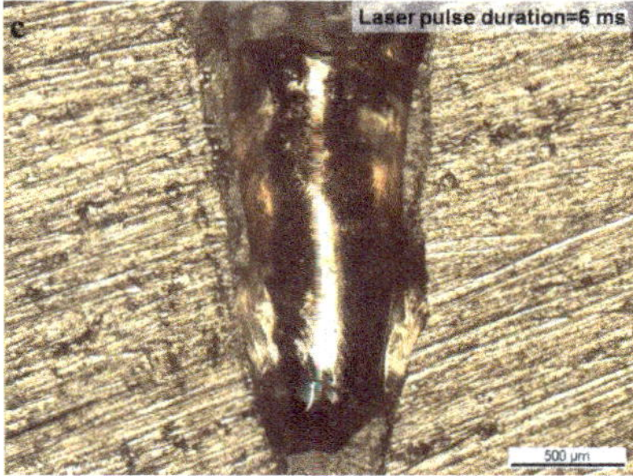

Figure 7. Optical microscopy images of cross-sections of laser welds conducted with discrete pulses, power = 5.5 kW, frequency = 1 Hz, in-depth focusing of 2.5 mm: (**a**) weld with reduced porosity and spilled material; (**b**) weld with large pore; and (**c**) keyhole devoid of metal.

3.3. Obtaining Pore-Free Welds in AlMg5 Alloy Using an Infrared Millisecond Laser Source

The molten pool manifested very little boiling and spatter for a 3 kW regime. However, cross-sections revealed that the weld was not complete, 3 kW peak power proving to be insufficient to melt 3 mm thick AlMg5 samples. The weld also displayed root porosity (Figure 8a). A peak power of 5.5 kW was necessary to weld 3 mm thick samples, but in this case, numerous large pores randomly distributed in the weld were observed in cross-section.

Weld experiments were also conducted with laser beams focused in depth from 0 to 5 mm, with a step of 0.5 mm between samples. Beam focusing on surface was ruled out, as it induced too much spatter and material loss, visible both in the high-speed images and also visually, in the weld aspect. Spatter reduced progressively with beam defocusing in depth and pores were more confined towards the bottom of the weld as the defocusing increased. The optimal defocus distance was found to be 2.5 mm in depth, corresponding to a circular spot of 850 μm diameter. In this irradiation regime total heat penetration was achieved and pores were isolated at the bottom of the weld, close to the bottom edge of coupon (Figure 8b). By increasing the laser power even more, the spatter and pores will increase, while continuing defocusing to 4 or 5 mm in depth would reduce the molten pool depth. It was clear that a pore-free single-step welding with our laser source on this alloy was not possible under the investigated parameters. A second welding step on the other side, where the pores were confined, was necessary. A 3 kW pulse peak power focused 2.5 mm in depth, at the same scan speed and pulse repetition rate as in the previous case, it was sufficient enough to remove the pores and obtain a pore-free weld (Figure 8c).

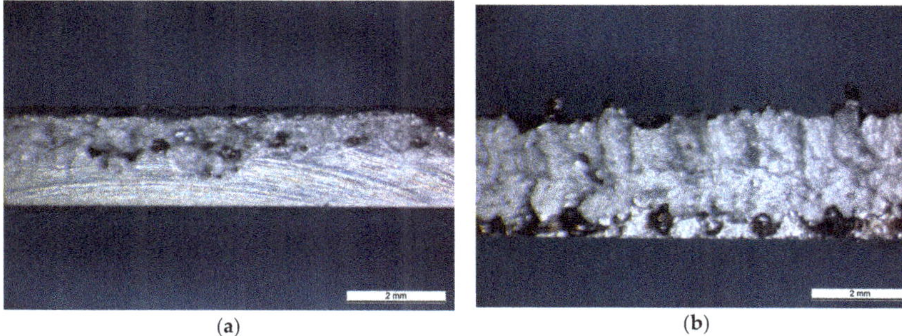

(**a**) (**b**)

Figure 8. *Cont.*

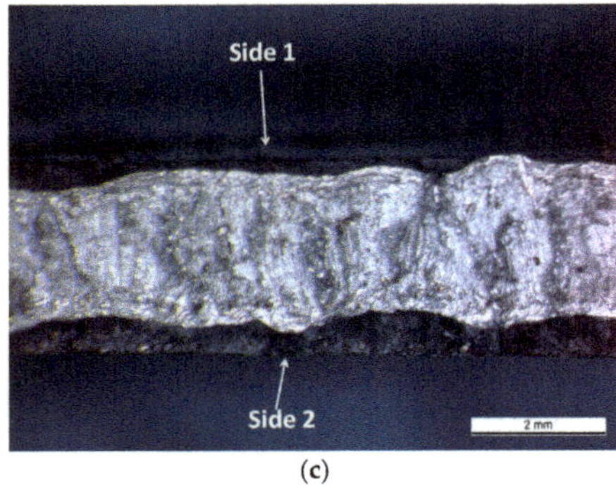

(c)

Figure 8. Optical microscopy image of welded joints in case of casted AlMg5 parts: (**a**) Ar, peak power = 3 kW, frequency = 10 Hz, pulse duration = 2 ms, 1 pass; (**b**) Ar, peak power = 5.5 kW, frequency = 10 Hz, pulse duration = 2 ms, 1 pass; (**c**) Side 1: Ar, peak power = 5.5 kW, frequency = 10 Hz, pulse duration = 2 ms, 1 pass; Side 2: Ar, peak power = 3 kW, frequency = 10 Hz, pulse duration = 2 ms, 1 pass.

4. Discussion and Conclusions

The laser welding process of AlMg5 parts with a millisecond Nd:YAG laser source emitting in infrared was monitored by high-speed imaging and correlated with optical microscopy analyses for a faster selection of optimal parameters in view of obtaining a pore-free and splatter-free weld.

High-speed imaging via a microscope objective of an actual laser welding process was conducted and we found that in the case of beams focused in depth, liquid metal is pushed upwards by the vaporized metal expanding from the bottom of the molten pool. It caused spill and spatter and was partially impeded to refill the groove by the gas metal cloud emerging from the bottom of the groove.

While providing useful information, it must be stated that this high-speed imaging investigation in static regime can differ from the actual welding process in dynamic regime. Indeed, in a static regime, a laser beam at normal incidence on a relatively smooth surface is partly reflected in a specular manner, while a small part of it is absorbed by the medium. In a dynamic high-power regime, a keyhole containing mostly metal vapour is formed. The laser beam can penetrate deeper into the material by means of this cavity and it is reflected multiple times on the keyhole walls, a part of it being absorbed at each contact with the walls. Thus, the multiple Fresnel absorption/reflection can significantly increase absorption efficiency for the laser radiation in a dynamic regime, and therefore the involved temperatures could be more elevated while event times could be shorter as compared to the static regime [28,29].

Compared to the laser irradiation of coupons surface, spatter is formed earlier in the case of irradiation of the joint between two coupons. One potential explanation for the spatter occurring at different times (830 µs vs. 1350 µs, respectively) might be that the mechanism of spatter formation differs: in the case of surface irradiation, metal has to reach first the melting temperature and then the boiling temperature for spatter to occur, while for the joint irradiation case, molten metal pouring from the edges in contact, is lifted upwards by the gases rising from the groove.

Optical microscopy images of cross-sections after single-pulse irradiations (Figure 7) showed that the pores could originate from the bottom of the molten pool and their size is increasing as the laser pulse power increases. The metallic mirror-like shine of the pores together with their spherical shape point towards the hypothesis that they are a result of hot metallic vapors that rose from the bottom of the molten pool in a liquid environment that later solidified and trapped them. The vapors condensed on the solidified spherical walls during cooling, providing the mirror-like aspect.

The sample thickness which required maximum laser power together with the information on liquid dynamics provided by high-speed imaging and optical microscopy reduced the options of parameters to be varied to a minimum, making the optimization process quite fast. As high power was needed for welding the 3 mm thick samples, the pores presence could not be ruled out. A solution was to contain them at the bottom of the weld by tuning the beam focusing. As laser power decreased by defocusing, the temperature of the molten metal decreased, meaning that it solidified faster after the pulse ending. An experiment involving high-speed camera for monitoring the molten pool together with optical microscopy of the cross-sections revealed that 2.5 mm defocusing in depth of the beam was the optimal regime for containing the pores close to the bottom of the weld, immediately after their genesis. In order to ensure a pore-free weld, the sample side with the aligned pores had to be melted in order for the gases to be released and the liquid metal to rearrange. After a few tests, the optimal pulse power that ensured a pore-free weld was found to be of 3 kW.

High-speed imaging monitoring supported by optical microscopy cross-section investigations should allow users to adjust laser parameters for any type of high-power processing laser source in order to be able to weld even the most problematic metals and alloys.

Acknowledgments: The authors acknowledge the support of this research by SCIEX-NMS.CH (Scientific Exchange Programme—New Member States of the European Union–Switzerland) under the contract 12.346 PLASDIAMET (Diagnostic method for liquid droplets in ablation plasma generated during high resolution laser processing of materials). Andrei C. Popescu acknowledges the CNCS/CCCDI–UEFISCDI for financial support during writing of this manuscript under the Project PN-III-P2-2.1-PED-2016-1309.

Author Contributions: Andrei C. Popescu designed and performed the experiments, conducted optical and electron microscopy analysis and wrote the manuscript. Christophe Delval performed the high speed imaging monitoring and extracted necessary data from the recordings using the ProAnalyst software. Marc Leparoux provided the analysis tools for this research, synthesized the AlMg5 samples, supervised the experiments, corrected and structured the manuscript in the final form.

Conflicts of Interest: The authors declare no conflict of interest.

References

1. Arcelor Mittal. Available online: http://automotive.arcelormittal.com/tailoredblanks (accessed on 31 August 2017).
2. Wisco Tailored Blanks. Available online: http://www.tailored-blanks.com/en/about.php (accessed on 31 August 2017).
3. Lu, J.; Kujanpää, V. Review study on remote laser welding with fiber lasers. *J. Laser Appl.* **2013**, *25*, 052008. [CrossRef]
4. Kose, C.; Karaca, E. Robotic Nd:YAG fiber laser welding of Ti-6Al-4V Alloy. *Metals* **2017**, *7*, 221. [CrossRef]
5. Wu, Q.; Gong, J.K.; Chen, G.Y.; Xu, L. Research on laser welding of vehicle body. *Opt. Laser Technol.* **2008**, *40*, 420–426. [CrossRef]
6. Lanza, M.; Lauro, A.; Scanavino, S. Fabrication and weldability in structures. *AL Alumin. Alloys* **2001**, *13*, 80–86.
7. Gao, X.; Zhang, L.; Liu, J.; Zhang, J. A comparative study of pulsed Nd:YAG laser welding and TIG welding of thin Ti6Al4V titanium alloy plate. *Mater. Sci. Eng. A* **2013**, *599*, 14–21. [CrossRef]
8. Leo, P.; D'Ostuni, S.; Casalino, G. Hybrid welding of AA5754 annealed alloy: Role of post weld heat treatment on microstructure and mechanical properties. *Mater. Des.* **2016**, *90*, 777–786. [CrossRef]
9. Good Fellow. Available online: http://www.goodfellow.com/E/Aluminum.html;http://www.goodfellow.com/E/Stainless-teel-AISI-304.html (accessed on 31 August 2017).
10. Jeffus, L.F. *Welding: Principles and Applications*, 6th ed.; Cengage Learning: New York, NY, USA, 2007.
11. Harvey, J.P.; Chartrand, P. Modeling the hydrogen solubility in liquid aluminum alloys. *Metall. Mater. Trans. B* **2010**, *41*, 908–924. [CrossRef]
12. Ola, O.; Doern, F. Keyhole-induced porosity in laser-arc hybrid welded aluminum. *Int. J. Adv. Manuf. Technol.* **2015**, *80*, 3–10. [CrossRef]
13. Scintilla, L.D. Continuous-wave fiber laser cutting of aluminum thin sheets: Effect of process parameters and optimization. *Opt. Eng.* **2014**, *53*, 66–113. [CrossRef]

14. Leo, P.; Renna, G.; Casalino, G.; Olabi, A.G. Effect of power distribution on the weld quality during hybrid laser welding of an Al-Mg alloy. *Opt. Laser Technol.* **2015**, *73*, 118–126. [CrossRef]
15. Karlsson, J.; Kaplan, A.F.H. Analysis of a fibre laser welding case study, utilising a matrix flow chart. *Appl. Surf. Sci.* **2011**, *257*, 4113–4122. [CrossRef]
16. Popescu, A.C.; Delval, C.; Shadman, S.; Leparoux, M. Investigation and in situ removal of spatter generated during laser ablation of aluminium composites. *Appl. Surf. Sci.* **2016**, *378*, 102–113. [CrossRef]
17. Kaplan, A.F.H.; Powell, J. Spatter in laser welding. *J. Laser Appl.* **2011**, *23*, 032005. [CrossRef]
18. You, D.; Gao, X.; Katayama, S. Monitoring of high-power laser welding using high-speed photographing and image processing. *Mech. Syst. Signal Process.* **2013**, *20*, 39–52. [CrossRef]
19. Zhang, Y.; Chen, G.; Wei, H.; Zhang, J. A novel "sandwich" method for observation of the keyhole in deep penetration laser welding. *Opt. Lasers Eng.* **2008**, *46*, 133–139. [CrossRef]
20. You, D.; Gao, X.; Katayama, S. Visual-based spatter detection during high-power disk laser welding. *Opt. Laser Eng.* **2014**, *54*, 1–7. [CrossRef]
21. Li, S.; Chen, G.; Katayama, S.; Zhang, Y. Relationship between spatter formation and dynamic molten pool during high-power deep-penetration laser welding. *Appl. Surf. Sci.* **2014**, *303*, 481–488. [CrossRef]
22. Casalino, G.; Mortello, M.; Peyre, P. Yb-YAG laser offset welding of AA5754 and T40 butt joint. *J. Mater. Process. Technol.* **2015**, *223*, 139–149. [CrossRef]
23. Casalino, G.; Mortello, M. Modeling and experimental analysis of fiber laser offset welding of Al-Ti butt joints. *Int. J. Adv. Manuf. Technol.* **2016**, *83*, 89–98. [CrossRef]
24. Xue, X.; Amorim, B.P.J.; Liao, J. Effects of pulsed Nd:YAG laser welding parameters on penetration and microstructure characterization of a DP1000 steel butt joint. *Metals* **2017**, *7*, 292. [CrossRef]
25. Sands, D. Pulsed laser heating and melting. In *Heat Transfer—Engineering Applications*; Vikhrenko, S.V., Ed.; InTech: Rijeka, Croatia, 2011; pp. 47–70.
26. Gill, D.H.; Newnam, B.E. Picosecond-pulse damage studies of diffraction gratings. In *Damage in Laser Materials*; Bennet, H.E., Guenther, A.H., Milam, D., Newnam, B.E., Eds.; National Bureau of Standards Special Publication: Gaithersburg, MD, USA, 1986; Volume 727, pp. 154–161.
27. Haynes, W.M. *CRC Handbook of Chemistry and Physics*, 97th ed.; CRC Press: New York, NY, USA, 2017.
28. Svenungsson, J.; Choqueta, I.; Kaplan, A.F.H. Laser welding process—A review of keyhole welding modeling. *Phys. Procedia* **2015**, *78*, 182–191. [CrossRef]
29. Jin, X.; Cheng, Y.; Zeng, L.; Zou, Y.; Zhang, H. Multiple reflections and Fresnel absorption of Gaussian laser beam in an actual 3D keyhole during deep-penetration laser welding. *Int. J. Opt.* **2012**, *361818*, 1–8. [CrossRef]

Article

FEM Simulation of Dissimilar Aluminum Titanium Fiber Laser Welding Using 2D and 3D Gaussian Heat Sources

Sonia D'Ostuni [1], Paola Leo [1] and Giuseppe Casalino [2,*]

[1] Innovation Engineering Department, University of Salento, Via per Arnesano s.n., 73100 Lecce, Italy;
    sonia.dostuni@unisalento.it (S.D.); paola.leo@unisalento.it (P.L.)
[2] DMMM Politecnico di Bari, Viale Japigia, 182, 70126 Bari, Italy
*   Correspondence: giuseppe.casalino@poliba.it; Tel.: +39-080-5962753

Received: 1 June 2017; Accepted: 8 August 2017; Published: 10 August 2017

Abstract: For a dissimilar laser weld, the model of the heat source is a paramount boundary condition for the prediction of the thermal phenomena, which occur during the welding cycle. In this paper, both two-dimensional (2D) and three-dimensional (3D) Gaussian heat sources were studied for the thermal analysis of the fiber laser welding of titanium and aluminum dissimilar butt joint. The models were calibrated comparing the fusion zone of the experiment with that of the numerical model. The actual temperature during the welding cycle was registered by a thermocouple and used for validation of the numerical model. When it came to calculate the fusion zone dimensions in the transversal section, the 2D heat source showed more accurate results. The 3D heat source provided better results for the simulated weld pool and cooling rate.

Keywords: dissimilar welding; fiber laser; finite element analysis

---

1. Introduction

Laser welding is recognized as an effective process to weld metals with a laser beam of high-power, high-energy density. In fact, the power density of a laser beam is much higher than that of arc or plasma. Consequently, a deep narrow penetration weld can be effectively produced. These properties have made laser welding a suitable technology for weldments that are made from metals of different compositions and properties [1,2]. A dissimilar joint is as strong as the weaker of the two metals being joined, i.e., possesses sufficient tensile strength and ductility so that the joint will not fail in the weld, has good fatigue behavior [3].

Among them, Al/Ti dissimilar joints are of major interest in aeronautics and automotive applications, where weight reduction, coupled with high mechanical strength and corrosion resistance, are paramount. Between the different Al/Ti welding processes laser welding offers numerous advantages. Especially in aluminum alloys when used in keyhole mode improves the absorption of the beam due to the multiple reflections in the cavity [4]. Moreover, high energy density, high cooling and heating rate allow for reducing the importance of mixing and diffusion phenomena, and thus reduce the formation of intermetallic compounds in the case of dissimilar joints. An Al/Ti joint has a remarkably lower elongation due to the high residual stresses, which facilitate the crack ignition and propagation. Therefore, the quality depends heavily on the process parameters, which determine the magnitude of thermal stresses [5].

The selection of the welding parameters is crucial for obtaining a satisfactory quality weld. Residual stresses and temperature field in laser welding joints can be predicted by numerical analysis such as a finite element one. The Finite Element Method has been one of the performing techniques to predict the joint properties in the welding process, which involves thermal, metallurgical and mechanical phenomena.

The computation of thermal field relies strongly on the heat source model. Rosenthal was the first researcher who proposed a model for the heat source in welding [6]. He proposed an analytical solution considering a punctual or a line heat source. Since then, other more realistic models have been proposed. For arc welding, several heat source configurations have also been proposed. Two and three-dimensional approaches can be used. Zeng et al. described the thermal elastic-plastic analysis using finite element techniques to analyze the thermos-mechanical behavior and evaluate the residual stresses and welding distortion on the AZ31B magnesium alloy and 304L steel butt joint in laser-TIG hybrid welding [7]. A modified three-dimensional conical heat source was used for performing the simulation in arc welding [8,9]. In certain cases, the finite element model is integrated with other computational techniques like artificial intelligence to establish an automated and iterative optimization algorithm [10]. Zeng et al. calculated the thermal cycles and temperature distribution of MIG welding of 5A06 aluminum alloy structure during discontinuous welding. The finite element method transient heat transfer analysis was used to save computing time and improve calculation accuracy [11].

For the fiber laser, Casalino et al. [12] developed and applied a stationary process with a surface heat source model based on thermal load through several specific elements next to the welding line. Other researchers [13] proposed the combination of Gaussian distribution on the surface and distribution along the thickness to consider 3D distribution by applying the conical Gaussian heat source model. They found that 3D conical Gaussian heat distribution can obtain better results with high depth to width ratio (defined as the ratio between the weld penetration evaluated on the axis of fused zone, and the width of the welded seam in the horizontal direction on the sample surface). Nagel et al. [14] proposed some strategies for the optimization of the laser welding of high alloy steel sheets using two different heat sources.

In this paper, two-dimensional and three-dimensional heat distribution was used for welding laser simulation of dissimilar Al/Ti button joint. The objective of this study is to compare the two approaches to establish the best model. The simulated fusion zone was compared with the macrograph obtained from the experiment to calibrate the model. Then, the validation is based on the comparison of the temperature profile measured with thermocouples during the welding cycle.

2. Experimental Setup

A laser butt joint has been produced from two plates of aluminum and titanium (3 and 2 mm, respectively), according to the scheme of Figure 1. Tables 1 and 2 report the chemical composition of the two alloys and the mechanical and thermo-physical properties at room temperature. Particularly, the thermal conductivity has been considered as a function of temperature (Tables 3 and 4).

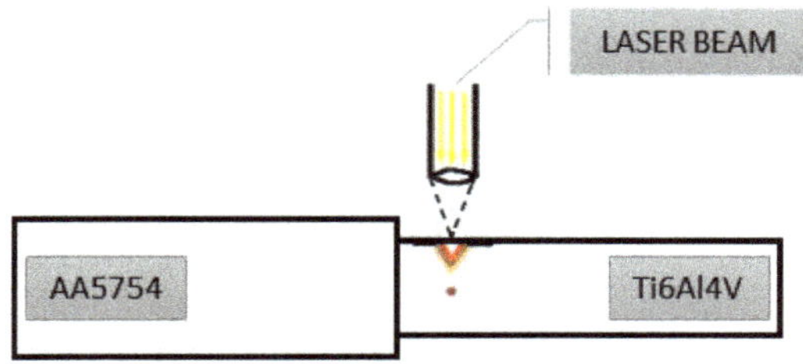

Figure 1. Scheme of laser welding on AA5754/Ti6Al4V.

Table 1. Chemical composition of AA5754 aluminum and Ti6Al4V titanium alloys (wt %).

AA5754								
Si	Fe	Cu	Mn	Mg	Cr	Zn	Ti	Al
0.40	0.40	0.10	0.50	2.6–3.6	0.30	0.20	<0.15	balance

Ti6Al4V								
C	Fe	N_2	O_2	Al	V	H_2	Ti	C
<0.08	<0.25	<0.05	<0.2	5.5	3.5	<0.0375	balance	<0.08

Table 2. Mechanical and thermo-physical properties of the two alloys.

Property	AA5754	Ti6Al4V
Young modulus [GPa]	70	114
Poisson ratio	0.3	0.3
Density [g/cm^3]	2.7	4.4
Liquidus Temperature [K]	870	1923
Solidus Temperature [K]	856	1880

Table 3. Temperature dependent thermal conductivity of AA5754.

AA5754	
Temperature	Thermal Conductivity [W/mK]
293	138
373	147.2
473	152.7
573	162.7
673	152.7
773	158.75
873	138
1773	138

Table 4. Temperature dependent thermal conductivity of Ti6Al4V.

Ti6Al4V	
Temperature	Thermal Conductivity [W/mK]
293	6.01
773.15	14.78
793.15	15
823.15	15.15
953.15	17.20
993.15	17.80
1013.15	18.30
1053.15	18.80
1093.15	19.50
1113.15	20
1133.15	20.50
1153.15	21
1173.15	21.60
1273.15	23.91
1933.15	34.3

The experimental trials were carried out using an Ytterbium Fiber Laser System (IPG YLS-4000), with a maximum output power equal to 4 kW (IPG Laser, Barbuch, Germany). The laser beam was delivered through a 200 μm optical fiber with a Beam Parameter Product (BPP) equal to 6.3 mm·mrad, the product of a laser beam's divergence angle (half-angle) and the radius of the beam at its narrowest point. The laser beam, whose wavelength was 1070.6 nm, has been focused continuously through a lens with focal distance of 250 mm producing a spot diameter of 0.4 mm on the workpiece surface.

Argon and helium were employed as shielding gas with 10 L/min volumetric flow rate, particularly Argon has been employed on the upper surface and helium on the bottom surface. The laser beam axis was placed on the titanium side, some 1 mm far from the interface (laser offset). The welding parameters have been 1200 W power at 1000 m/min welding rate.

The microstructure of the weld was analyzed by optical microscopy (OM; Nikon Epiphot 200, Nikon, Tokyo, Japan) and Zeiss EVO scanning electron microscope (SEM, Zeiss, Oberkochen, Germany). For optical microscopy observation, the transversal sections of the samples were cut and prepared using the standard metallographic grinding and polishing techniques and attached using Keller reagent (95 mL H_2O, 2.5 mL HNO_3, 1.5 mL HCl, HF 1 mL). The dimension of the fusion zone (FZ) was evaluated using NIS-Element software (version 4.5, Nikon, Tokio, Japan, 2016)) for the image analysis. NIS-Elements is a Nikon software supplied with Epiphot 200 OM. The software is tailored to facilitate image capture, object measurement and counting. Vickers microhardness profile (0.3/15) was collected using a Vickers Affri Wiky 200JS2 microhardness tester (Affri, Wood Dale, IL, USA) at half of weld cross section thickness. The distance between indentations was equal to 300 μm. The hardness at the Al/Ti fusion zones interface, where an intermetallic compound layer was observed, had been identified with nanoindentation (0.01/15) using Leica VMHT (Leica Microsystems Wetzlar GmbH, Wetzlar Germany) due to the reduced size of the layer.

Two thermocouple recording systems were placed in the middle of the plates, 2 mm distant from the weld centre line in both titanium and aluminum side. The calibration of the two models was carried out by comparing the size and shape of the fusion zone of the numerical model and the experimental one. The validation of the model was made by comparing the experimental and numerical thermal cycle.

3. Numerical Model

3.1. Model for the Plates

The plates of titanium and aluminum (200 mm × 50 mm) were joined along the long side. The plates have different thickness, i.e., 2 mm for titanium one and 3 mm for Aluminum one. The adopted mesh is the same for the two models. To obtain accurate results, a fine mesh was adopted close to the welding line. Mesh size was determined rapidly by trial-and-error; likewise, it is generally done in the literature. Mesh size is equal to $1 \times 0.5 \times 0.5$ mm^3 inside of 10 mm from the interface and the number of the nodes is equal to 20 along the x-axis. At a distance from the interface higher than 10 mm, the mesh size increases along the x-axis. Particularly from 10 to 50 mm far from the interface (the limit of the Al sheet), the number of the nodes is equal to 20, but the ratio in size between the last element and first element in the distribution is 0.1 (arithmetic sequence). The mapped mesh has 40,000 elements and 49,692 nodes. Figure 2 shows the results of the meshing procedure.

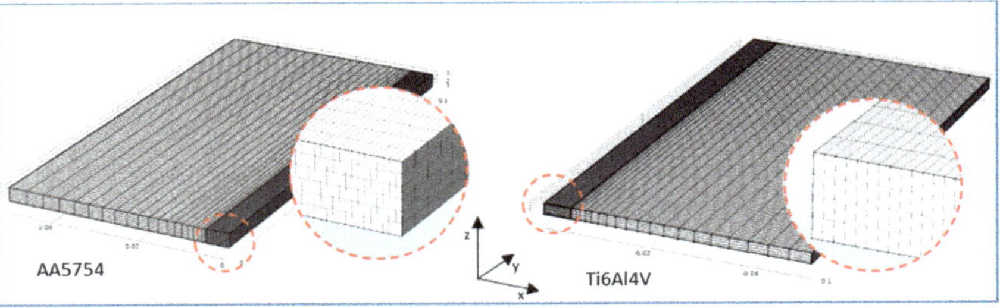

Figure 2. Mesh outlook.

The numerical simulation was performed using the finite element code COMSOL Multiphysics. Comsol Multiphysics is a Multiphysics modeling tool that solves all types of problems based on Finite Element [15].

During the welding process, the transfer of heat is governed by the general equation heat flow (1a):

$$\rho C_p \frac{\delta T}{\delta t} + \rho C_p v \nabla T + \nabla q = hf \tag{1a}$$

where

ρ [g/mm^2] is the density of the metal as a function of temperature,

C_p [J/(g·K)] is specific heat of the metal as a function of temperature,

v [m/s] is the velocity field,

q [W/mm^2] is the heat lost to the surroundings by combination of radiation and convection and conduction,

hf [W/mm^2] is the heat source.

The heat lost is given by Equation (1b):

$$-q = \varepsilon \sigma \left(T_r^4 - T^4 \right) + h(T_r - T) \tag{1b}$$

where ε is the emissivity of the surface and is taken as 0.5 for titanium and 0.3 for aluminum. σ is the Stefan–Boltzmann constant and is taken as 5.67×10^{-8} W/(m^2·K^4), T_r is the room temperature and was taken at 293 K. h is the heat transfer coefficient assumed equal to 20 W/(m^2·K) for air and 200 W/(m^2·K) for the bottom surface in contact with the workspace.

When the contact is formed by pressing two similar or dissimilar metallic materials together, only a small fraction of the nominal surface area is in contact because of the roughness and irregularities of the contacting surfaces. When a heat flux is imposed across the junction, there are only a limited number and size of the contact spots that results in an actual contact area. The actual contact area is significantly smaller than the apparent contact area and causes a thermal contact resistance. There are several analytic expressions for predicting the contact conductance, and several values in literature [16]. Based on these values, by acting on a contact resistance value, the calibration has been carried out by comparing the fusion zones of the numerical model with the experimental results. Finally, the value of contact conductance is assumed to be equal to 30,000 W/(m^2·K), for both of the two models [17]. Mechanical constraints were imposed for the simulation of the clamping system. Constraints were imposed to the four exterior nodes of the plates so degrees of freedom were zero (no displacement is permitted).

3.2. 2D Heat Source

Since the pioneering work of Rosenthal [6] that proposed punctual and linear heat sources, several more realistic sources have been proposed. When the distribution along the thickness is not important like in thin plates, the surface Gaussian heat source model is a good proposal for bed-on-plate cases when both TIG and conductive laser welding must be simulated [18]. For the surface laser heat source, the radius (R) should be calculated first by the formula written as [19]:

$$R = \frac{2M_0^2 \lambda f}{\pi D_0}, \tag{2}$$

where M_0^2 is the beam quality equal to 1.1 for fiber laser, f is the focal length of the focusing lens and D_0 is the diameter of the lens.

For laser welding, the heat source of the laser beam was simulated by a traveling two-dimensional distribution of heat source (Figure 3a). Particularly, the heat surface distribution was built by combining two Gaussian distributions one in the plane XZ and one moving with welding rate v in plane YZ. Those two Gaussian distributions were obtained, and, with an energy of about 99.7% of the total laser energy, a fusion of the titanium alloys was obtained. The heat source radius (R) was assumed equal to three standard deviations of the Gaussian pulse (Figure 3b):

$$gp(s) = \frac{1}{\sigma\sqrt{2\pi}}e^{-\frac{t^2}{2\sigma^2}} = \frac{3}{R\sqrt{2\pi}}e^{-\frac{9s^2}{2R^2}} \tag{3}$$

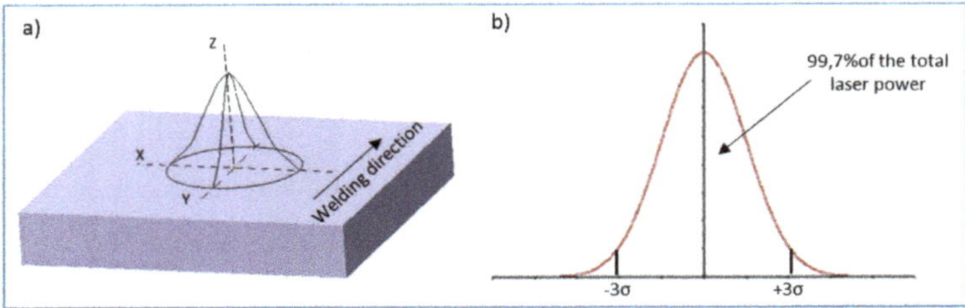

Figure 3. (a) Combining of the two Gaussian distribution in plane *XZ* and *XY*, and (b) Gaussian distribution.

The 2D total heat source (W) (Figure 4) is expressed in Equation (4):

$$hs_{2D}(x,y)_{2D} = P_{\text{laser}}[gp(x) \times gp(y - vt)] \tag{4}$$

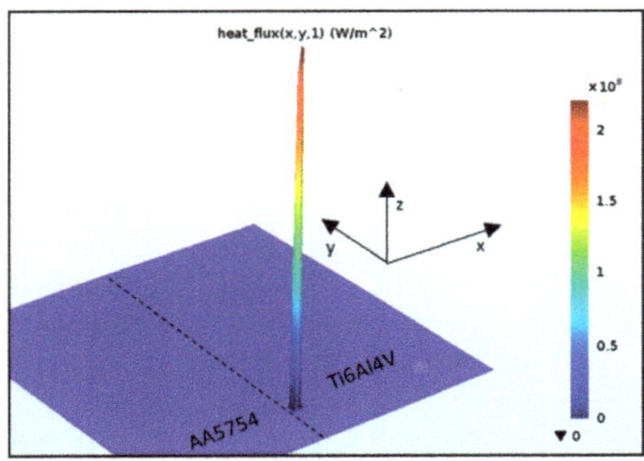

Figure 4. Heat flux in 2D Gaussian heat distribution.

3.3. 3D Heat Source

For the three-dimensions laser source, the heat flux was described in the numerical model as a volume and surface distributed heat flux:

$$hf_{3D} = (1 - \varphi)[hf_{\text{sur}}(x,y)]_{\text{load}} + \varphi[hf_{\text{vol}}(z)]_{\text{load}} \tag{5}$$

where the coefficient φ is the energy fraction that will be introduced through the cylinder, and the remaining energy will be introduced via the surface heat source (the value being used is 0.9). For the surface, a Gaussian distribution was used, the heat source radius was equal to two standard deviations of the Gaussian pulse under the assumption that 95.44% of the total fusion energy of the titanium alloys was applied. In this case, considering the surface heat distribution proposed by Goldak and Akhlagi [20], the surface heat flux hf_{sur} [W/mm^2] can be expressed as shown in Equation (6):

$$hf_{\mathrm{sur}}(x,y) = \frac{2\eta P_{\mathrm{laser}}}{\pi R^2}\mathrm{e}^{-\frac{2r^2}{R^2}} \tag{6}$$

where η is the process efficiency (the value being used is 1), P_{laser} is the laser power in W and R is the heat source radius.

For volume heat source, a constant heat distribution cylinder was considering, with a radius R_{FZ} equal to those of the molten cylinder (Figure 5). The numeric value for volumetric heat source is given by the ratio between the laser power and the volume of cylinder with a radius R_{FZ}.

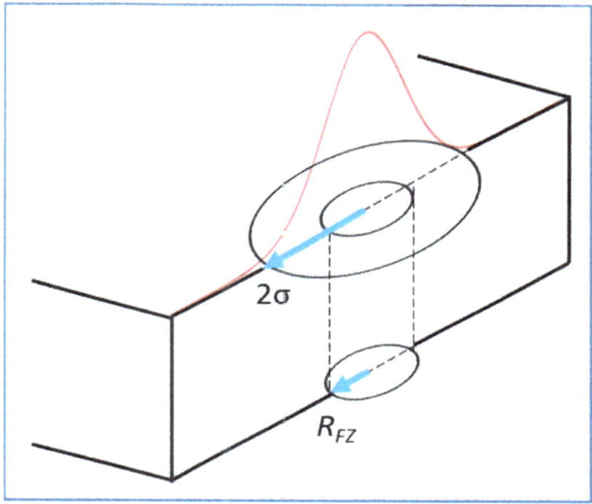

Figure 5. Total heat flux in 3D heat distribution.

Thus, finally, the keyhole is modelled in the software using the heat source radius, the radius of the molten cylinder, measured by experimental results, and the energy fraction.

4. Results and Discussion

4.1. Metallurgical Characterization of Weld

The appearance of Ti6Al4V/AA5754 laser welded joint after chemical etching is shown in Figure 6. Both titanium and aluminum alloy melted at the joint interface and separate fusion zones were observed (Al side and Ti side) as well as two heat affected zone (HAZ) between FZ and the base material. Good weld appearance of the cross section was obtained with full penetration and low level of porosity. It was demonstrated that in aluminum-titanium dissimilar weld, porosity tended to be produced in the fusion line. It happens that gases in the seam are hard to escape and concentrated in the middle of fusion line during the solidification process [21]. In Figure 6, some gas trapping occurred on both sides of the intermetallic layer but not in the intermetallic layer.

Figure 7 shows the Ti microstructure of the different zones showed in Figure 6, i.e., base material, heat affected zone and fusion zone. The basic mental (BM) is composed of dark β phase in the dominating bright α matrix. Particularly, the β phase is distributed at the boundary of the α grains. This is a typical microstructure for α-β titanium alloys in mill-annealed conditions [22]. The microstructure within the joint depends on the heat received from the laser beam, and varies according to the distance from it. In FZ, a predominantly martensitic microstructure of acicular type (α') is present. In fact, as reported in literature [22–24], the microstructure of the laser fusion zone of Ti-6Al-4V alloy is completely martensitic due to the high cooling speed from β field. The heat affected zone is a mixture of martensitic and primary α grain. During the heat thermal cycle due to the laser heating, an increase of β phase results. However, due to the rapid cooling, that transformation is never complete, so the microstructure of the HAZ is mixed, formed from martensitic grains and grain α.

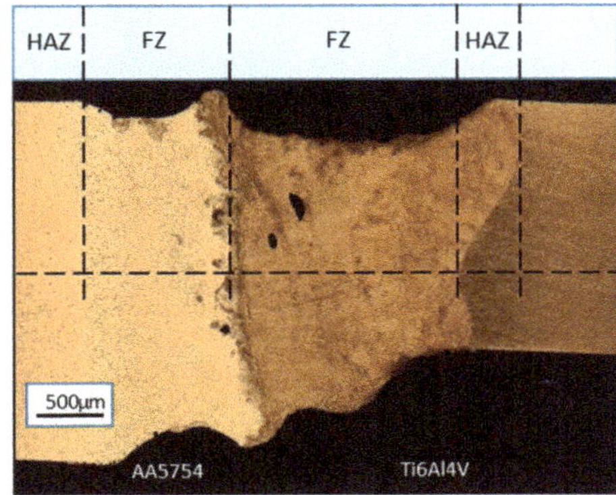

Figure 6. Appearance of Ti6Al4V/AA5754 laser welded joint after chemical etching.

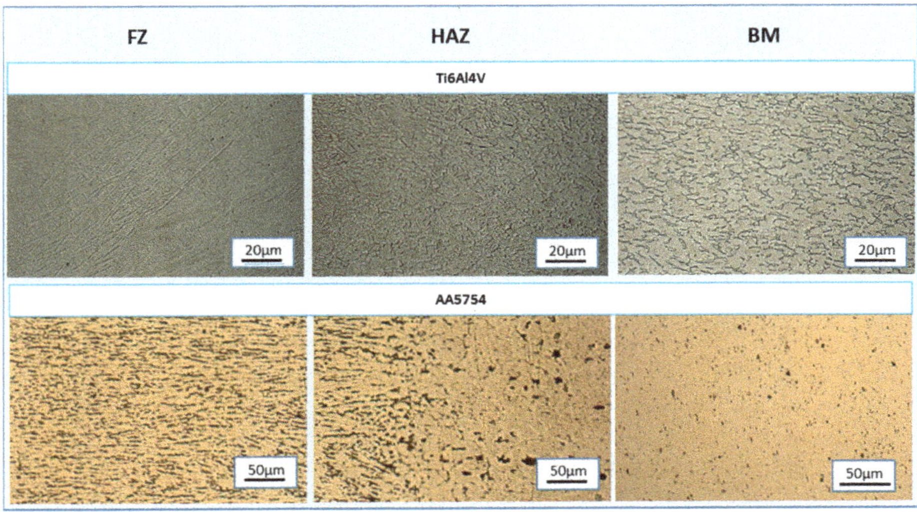

Figure 7. Microstructure of the various zones in Ti6Al4V and AA5754.

The base material aluminum alloy AA5754 was supplied in the annealed condition (Figure 7). In the aluminum matrix, there are some second phases. These phases have been identified in literature as (Fe,Mn)Al$_6$, (Fe,Mn)$_3$SiAl$_{12}$, Mg$_2$Si and Mg$_2$Al$_3$ [25–27].

The microstructure of the fusion zone exhibits a very fine dendritic microstructure due to the high cooling rate. The dendrites grow in direction parallel to the direction of heat flow giving to columnar grains [28,29]. In the HAZ microstructure, no detectable changes are evident to the OM; however, in this zone, solubilization of magnesium based has been reported in the literature [30]. Figure 8 shows the micro hardness profile (0.3/15) in the transverse section of the weld. The microhardness was very high in the titanium FZ where the microstructure was martensitic. In the HAZ, the value diminished with the lower amount of martensitic microstructure. The rise in the microhardness of the FZ of aluminum in the weld was caused by the rapid cooling that produced a very fine solidification structure and solid solution strength. The increment in the HAZ of aluminum was due to the dissolution of magnesium compounds during the welding cycle [30].

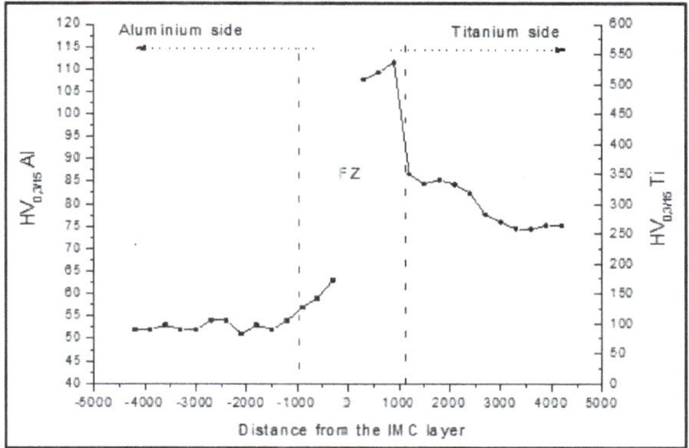

Figure 8. Microhardness profile at half thickness of the weld cross section.

A layer of intermetallic compounds (IMC) formed between the two fusion zones (Figure 9), whose stoichiometry was clarified in some paper [26,31,32]. This layer is due to the reaction in the temperature of the two alloys, and the size is variable as a function of the process parameters [32]. Particularly for the welding process parameters concerning this study, the average thickness of the IMC layer was equal 50 ± 5 μm to and the average nanohardness value (0.01/15) was equal to 2485 ± 232.

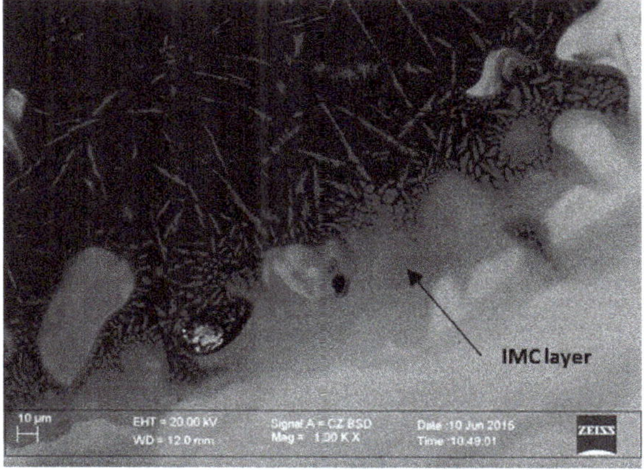

Figure 9. SEM Micrograph of intermetallic compounds (IMC) layer at the Al/Ti joint interface.

4.2. Calibration of the Model

After the simulation, the fusion zone was compared with the experimental one for calibration purposes. Figure 10 shows the comparison between the simulated and experimental fusion zone profile for aluminum and titanium. For both 2D and 3D heat source modeling, a complete penetration of keyhole was obtained, and, particularly for the 2D heat source, the boundary of the fusion zone obtained numerically had similar shapes in comparison with the experimental ones.

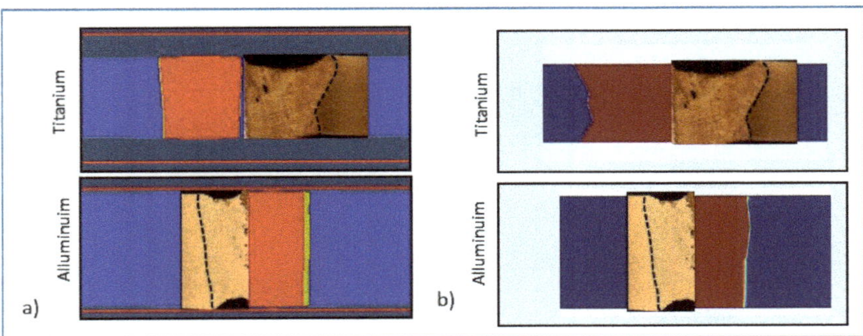

Figure 10. Calibration of (**a**) 3D and (**b**) 2D heat source models by the cross section of the weld.

In Tables 5 and 6, the dimensions of the fusion zone profile, respectively, in the top, middle and bottom, are reported both for 2D and 3D numerical simulation and the experimental measure. A better matching between the numerical and experimental results for the size of fusion zone has been obtained using a 2D source.

Table 5. Dimension (mm) of the fusion zone profile in the aluminum side.

Aluminum Fusion Zone	2D Heat Source	3D Heat Source	Experimental Data
Top	118	136	116
Middle	120	135	112
Bottom	114	135	108

Table 6. Dimension (mm) of the fusion zone profile in the titanium side.

Titanium Fusion Zone	2D Heat Source	3D Heat Source	Experimental Data
Top	232	204	225
Middle	207	196	198
Bottom	226	187	196

In fact, by correctly adjusting keyhole parameter in 3D heat source, only the top width of the weld was close to the experimental result, while the 2D heat source could almost precisely calculate the top, middle and bottom width. The observed performance can be explained as follows. It is well known that the behavior of heat flow depends on welding conditions [14,33]. For higher welding powers and thinner plates, the heat flow is predominantly 2D, whereas for lower welding powers and thicker plates, the heat flow was 3D. Therefore, there must be a transition thickness in which the heat flow has a behavior, which is a mix of the 2D and 3D ones. Probably for the thin plates examined in this study, the heat flow was predominantly 2D, and that is why 2D heat flow modelling is more appropriate.

Figures 11 and 12 show the temperature distribution on the top when the weld bead profile reached the quasi-steady state. The thermal conductivity of the two materials influenced the position of the maximum temperature in the melt pool and away from the melt pool. For both numeric simulation, the maximum temperature was recorded within the titanium side, where the beam laser was directed. The width of the heat affected zone in the aluminum side was greater due to the high thermal conductivity of the aluminum alloy. On the contrary, the low thermal conductivity of the titanium alloy leads to accumulating the heat over the metal by reducing the area interested from metallurgical transformation.

Figure 11. Isometric view of temperature distributions (in Kelvin) using 2D heat source modelling.

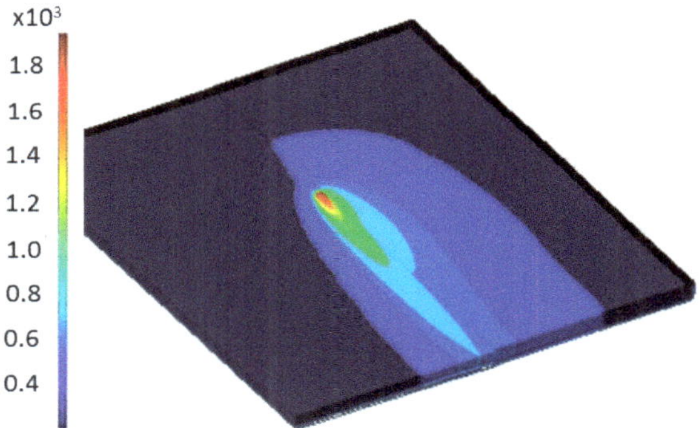

Figure 12. Isometric view of temperature distributions (in Kelvin) using 3D heat source modelling.

The shape of the weld pool longitudinal plane (Figure 13) showed a teardrop shape for the 2D heat source. Thus, this means that probably the 2D heat source cannot perfectly simulate the laser heat source. However, better results have been obtained using the 3D heat source; in this case, the resulting molten puddle had an elliptical shape, which is more similar to the experimental pool shape as shown by microstructure evolution of the grains (Figure 14) in the fusion zone. This shape of the grains derives from the elliptical molten pool [24,34].

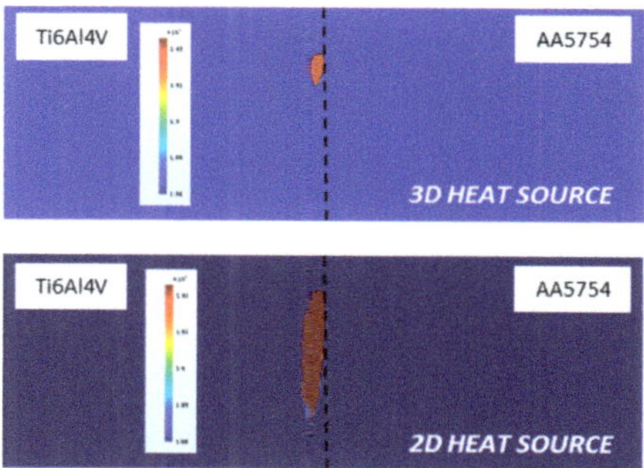

Figure 13. Weld pool numerical results for the numerical simulations.

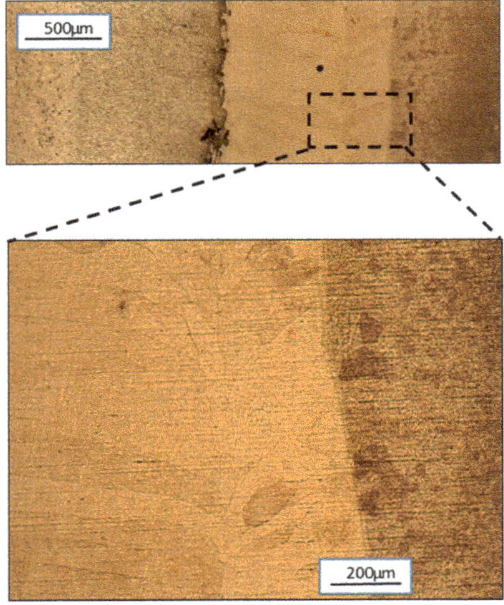

Figure 14. Interface zoom-up in the weld cross section ns.

4.3. Validation of the Model

The temperature cycle was measured in two different points that were placed 2 mm away from the weld centerline on both sides of the weld (Figures 15 and 16). The temperature profiles were compared with those obtained from the thermocouples. High thermal gradients, and fast cooling rate were present, due to the laser process. When it concerns the temperature peak, the numerical results matched the experimental one for both of the heat sources. Nonetheless, regarding the cooling rate, the 3D heat source was more precise than the 2D one.

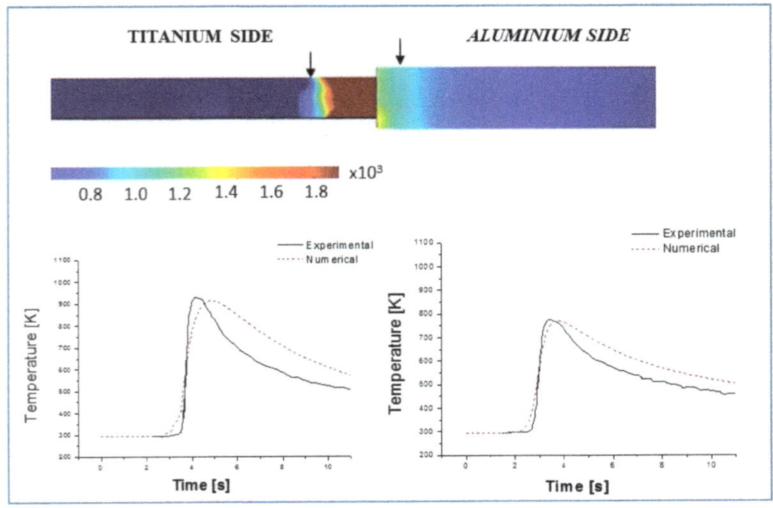

Figure 15. Experimental and numerical thermal cycle using 2D heat source modelling.

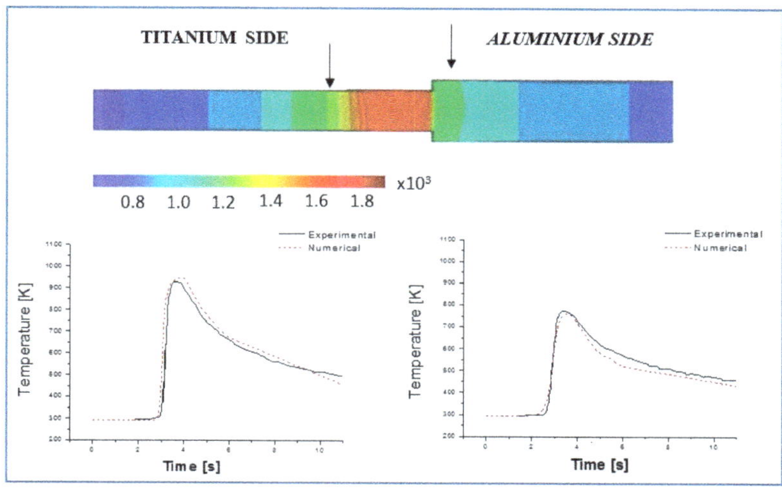

Figure 16. Experimental and numerical thermal cycle using 3D heat source modelling.

5. Conclusions

In this paper, the numerical models to simulate the laser welding process of butt dissimilar Al/TI joints were developed. Two different 2D and 3D heat source modelling processes have been utilized to simulate the proper heat flux during the welding. The numeric results were compared with the experimental ones to calibrate and to validate the two models. The metallurgical analyses showed that the titanium fusion zone was principally martensitic, and the heat affected zone was a mixture of martensitic and primary α grain. In the aluminum fusion zone, a dendritic structure was present and the heat affected zone was characterized by the solubilization of the magnesium compounds.

The FEM simulation of the thermal cycle of fiber offset welding was satisfactory. The following points were demonstrated:

(1) The calculations for the fusion zone dimensions were accurate both for the 2D and the 3D heat source. By using that 2D heat source, a better matching of numeric and experimental results was obtained at the three levels at which the molten zone sizes were taken.

(2) In the longitudinal section, the numerical results were not as accurate for both of the heat sources. For the 2D one, a teardrop shape of the molten weld pool formed while the 3D heat source produced an elliptical one. It is possible to conclude that the 3D heat source can better approximate the heat flux during laser welding and the maximum temperature gradients, which determined the change in the grain growth direction in the titanium side.

(3) The overall thermal cycle accuracy was good for 2D and 3D heat sources. However, the 3D heat source provided better results for the cooling rate simulation.

Author Contributions: The welds were fabricated in the TISMA ((Innovative Welding for Advanced Materials) laboratory of Bari, which is led by Giuseppe Casalino. Paola Leo looked after the metallographic preparation and analysis of the microstructure and Sonia D'Ostuni built the numerical model. Discussion and conclusions were written with the contribution of all authors.

Conflicts of Interest: The authors declare no conflict of interest.

References

1. Oliveira, J.P.; Zeng, Z.; Andrei, C.; Braz Fernandes, F.M.; Miranda, R.M.; Ramirez, A.J.; Omori, T.; Zhou, N. Dissimilar laser welding of superelastic NiTi and CuAlMn shape memory alloys. *Mater. Des.* **2017**, *128*, 166–175. [CrossRef]

2. Casalino, G.; Guglielmi, P.; Lorusso, V.D.; Mortello, M.; Peyre, P.; Sorgente, D. Laser offset welding of AZ31B magnesium alloy to 316 stainless steel. *J. Mater. Process. Technol.* **2017**, *242*, 49–59. [CrossRef]

3. Zeng, X.; Oliveira, J.P.; Yang, M.; Song, D.; Peng, B. Functional fatigue behavior of NiTi-Cu dissimilar laser welds. *Mater. Des.* **2017**, *114*, 282–287. [CrossRef]

4. Katayama, S. *Handbook of Laser Welding Technologies*; Woodhead Publishing Limited: Sawston, UK, 2013.

5. Casalino, G.; Mortello, M.; Peyre, P. Yb–YAG laser offset welding of AA5754 and T40 butt joint. *J. Mater. Process. Technol.* **2015**, *223*, 139–149. [CrossRef]

6. Rosenthal, D. Mathematical theory of heat distribution during welding and cutting. *Weld. J.* **1941**, *20*, 220–234.

7. Zeng, Z.; Li, X.; Miao, Y.; Wu, G.; Zhao, Z. Numerical and experiment analysis of residual stress on magnesium alloy and steel butt joint by hybrid laser-TIG welding. *Comput. Mater. Sci.* **2011**, *50*, 1763–1769. [CrossRef]

8. Pham, S.M.; Tran, V.P. Study on the Structure Deformation in the Process of Gas Metal Arc Welding (GMAW). *Am. J. Mech. Eng.* **2014**, *2*, 120–124.

9. Dhinakaran, V.; Suraj Khope, N.S.S.; Sankaranarayanasamy, K. Numerical Prediction of Weld Bead Geometry in Plasma Arc Welding of Titanium Sheets Using COMSOL. In Proceedings of the 2014 COMSOL Conference in Bangalore, Bangalore, India, 13–14 October 2014.

10. Casalino, G.; Hu, S.J.; Hou, W. Deformation prediction and quality evaluation of the gas metal arc welding butt weld. *Proc. Inst. Mech. Eng. Part B* **2003**, *217*, 1615–1622. [CrossRef]

11. Zeng, Z.; Wang, L.; Wang, Y.; Zhang, H. Numerical and experimental investigation on temperature distribution of the discontinuous welding. *Comput. Mater. Sci.* **2009**, *44*, 1153–1162. [CrossRef]

12. Casalino, G.; Michelangelo, M. A FEM model to study the fiber laser welding of Ti6Al4V thin sheets. *Int. J. Adv. Manuf. Technol.* **2016**, *86*, 1339–1346. [CrossRef]

13. Azizpour, M.; Ghoreishi, M.; Khorram, A. Numerical simulation of laser beam welding of Ti6Al4V sheet. *J. Comput. Appl. Res. Mech. Eng.* **2015**, *4*, 145–154.

14. Falk, N.; Flaviu, S.; Benjamin, K.; Jean, P.B.; Jorg, H. Optimization strategies for laser welding high alloy steel sheets. *Phys. Proced.* **2014**, *56*, 1242–1251.

15. *Comsol Multiphysics*, version 5.2a; Software for Multiphysics Modeling; COMSOL: Stockholm, Sweden, 2016.

16. Contuzzi, N.; Campanelli, S.L.; Casalino, G.; Ludovico, A.D. On the role of the Thermal Contact Conductance during the Friction Stir Welding of an AA5754-H111 butt joint. *Appl. Therm. Eng.* **2016**, *104*, 263–273. [CrossRef]

17. Casalino, G.; Mortello, M.; Peyre, P. FEM Analysis of Fiber Laser Welding of Titanium and Aluminum. *Procedia CIRP* **2016**, *41*, 992–997. [CrossRef]

18. Teixeira, P.R.D.F.; Araújo, D.B.D.; Cunda, L.A.B.D. Study of the gaussian distribution heat source model applied to numerical thermal simulations of tig welding processes. *Ciênc. Eng.* **2014**, *23*, 115–122. [CrossRef]

19. Steen, W.M.; Mazumder, J. *Laser Material Processing*; Springer: Berlin, Germany, 2010.

20. Goldak, J.; Akhlagi, M. *Computational Welding Mechanics*; Springer: Ottawa, ON, Canada, 2005; pp. 26–27.

21. Lv, S.X.; Jing, X.J.; Huang, Y.X.; Xu, Y.Q.; Zheng, C.Q.; Yang, S.Q. Investigation on TIG arc welding–brazing of Ti/Al dissimilar alloys with Al based fillers. *Sci. Technol. Weld. Join.* **2012**, *17*, 519–524. [CrossRef]

22. Lutjering, G.; Williams, J.C. *Titanium (Engineering Materials and Processes)*, 2nd ed.; Springer: Berlin, Germany; New York, NY, USA, 2007.

23. Akman, E.; Demir, A.; Canel, T.; Sınmazcelik, T. Laser welding of Ti6Al4V titanium alloys. *J. Mater. Process. Technol.* **2009**, *209*, 3705–3713. [CrossRef]

24. Xu, P.; Li, L.; Zhang, C. Microstructure characterization of laser welded Ti-6Al-4V fusion zones. *Mater. Charact.* **2014**, *87*, 179–185. [CrossRef]

25. Kumar, S.; Nadendla, H.B.; Scamans, G.M.; Eskin, D.G.; Fan, Z. Solidification behavior of an AA5754 alloy ingot cast with high impurity content. *Int. J. Mater. Res.* **2012**, *103*, E1–E7. [CrossRef]

26. Raghavan, V. Phase Diagram Evaluations: Section II. *J. Ph. Equilib. Diffus.* **2005**, *26*, 171–172. [CrossRef]

27. *ASM Specialty Handbook: Aluminum and Aluminum Alloys*; Davis, J.R. (Ed.) ASM International: Novelty, OH, USA, 1993.
28. Porter, D.A.; Easterling, K.E. *Phase Transformations in Metals and Alloys*; Chapman & Hall: London, UK, 1992.
29. Verhoeven, J.D. *Fundamentals of Physical Metallurgy*; Wiley: Hoboken, NJ, USA, 1975.
30. Leo, P.; D'Ostuni, S.; Casalino, G. Hybrid welding of AA5754 annealed alloy: Role of post weld heat treatment on microstructure and mechanical properties. *Mater. Des.* **2016**, *90*, 777–786. [CrossRef]
31. Bailey, N. *Welding Dissimilar Metals*; The Welding Institute: Cambridge, UK, 1986.
32. Gao, M.; Chen, C.; Gu, Y.; Zeng, X. Microstructure and tensile behavior of laser arc hybrid welded dissimilar Al and Ti alloys. *Materials* **2014**, *7*, 1590–1602. [CrossRef]
33. Sorensen, M.B. Simulation of Welding Distortion in Ship Section. Ph.D. Thesis, University of Denmark, Lyngby, Denmark, 1999.
34. Messler, R.W. *Principles of Welding*; Wiley-VCH, John Wiley & Sons, Inc.: Singapore, 2004.

Article

Effects of Pulsed Nd:YAG Laser Welding Parameters on Penetration and Microstructure Characterization of a DP1000 Steel Butt Joint

Xin Xue [1,2], António B. Pereira [2], José Amorim [2] and Juan Liao [1,*]

[1] School of Mechanical Engineering and Automation, Fuzhou University, Fuzhou 350116, China; xin@fzu.edu.cn or xin@ua.pt

[2] Centre for Mechanical Technology and Automation, Department of Mechanical Engineering, University of Aveiro, 3810-193 Aveiro, Portugal; abastos@ua.pt (A.B.P.); josepamoria@ua.pt (J.A.)

* Correspondence: jliao@fzu.edu.cn or jliao@ua.pt; Tel.: +86-0591-2286-6793

Received: 13 June 2017; Accepted: 27 July 2017; Published: 1 August 2017

Abstract: Of particular importance and interest are the effects of pulsed Nd:YAG laser beam welding parameters on penetration and microstructure characterization of DP1000 butt joint, which is widely used in the automotive industry nowadays. Some key experimental technologies including pre-welding sample preparation and optimization design of sample fixture for a sufficient shielding gas flow are performed to ensure consistent and stable testing. The weld quality can be influenced by several process factors, such as laser beam power, pulse duration, overlap, spot diameter, pulse type, and welding velocity. The results indicate that these key process parameters have a significant effect on the weld penetration. Meanwhile, the fusion zone of butt joints exhibits obviously greater hardness than the base metal and heat affected zone of butt joints. Additionally, the volume fraction of martensite of dual-phase steel plays a considerable effect on the hardness and the change of microstructure characterization of the weld joint.

Keywords: pulsed Nd:YAG laser beam welding; DP1000 steel; penetration; hardness; phase transformation

---

1. Introduction

The need to reduce fuel consumption and greenhouse gas emissions motivates vehicle manufacturers to utilize lightweight metals with better ductility and strength [1,2]. Nowadays, dual-phase (DP) have already built a good reputation for performance and safety [3,4]. Meanwhile, the application of DP steels unavoidably involves welding in the manufacturing process and the properties of welded joints due to the integrity and safety requirements. Although such traditional welding methods as resistance spot welding (RSW) [5], tungsten inert gas welding (TIG) [6], metal inert-gas welding (MIG) [7], and gas arc [8,9] have been widely used, the advantages of laser welding are also well known: energy efficiency, slight heat-affected zone, little heat input per unit volume, and deep penetration [10]. The chief limitation, i.e., cost per watt, is swiftly falling with new advances in laser technology [11]. Although in some cases a proper filler material may be demanded when an interfacial gap attends to be filled [12], autogenous welding is still one of the most common forms of laser welding.

Recently, many efforts have been performed to understand the sensitivity of the laser welding process in advanced high-strength steels. Pulsed Nd:YAG laser welding has especially attracted attention recently in many academic research and industry applications [13–19]. Tzeng [20] used zinc-coated steel and attempted to make a laser welding joint without gas-formed porosity. Xia et al. [21] used laser welding of DP steels and investigated the sensitivity of heat input and martensite on adjacent heat-affected zone (HAZ) softening. Several Nd:YAG and diode laser welds were made on DP450, DP600, and DP980 steels over a wide range of heat input. Dong et al. [22] studied the rate-dependent properties by means of a wide strain rate range, deformation, and fracture behavior of DP600 steel and its welded joint (WJ) subjected to the Nd:YAG laser welding. Their results showed that the DP600 WJ exhibits continuous yielding at quasi-static strain rates and develops a yield point at higher strain rates. Baghjari et al. [23] paid attention to the Nd:YAG laser welding of AISI420 martensitic stainless steel as well as the sensitivities of voltage, laser beam diameter, energy frequency, pulse duration, and welding velocity to the size and deformation of the welded joint. However, previous studies [24–28] indicated that the mechanical properties of DP steel butt joint have effects evident in the process operation parameters. This may be due to the soft zone formed in the adjacent heat-affected zone (HAZ). Then the question arises how the welding parameter manipulates the mechanical properties of DP steel butt joints by using laser welding. Although many efforts have been made to improve the tensile properties of DP steel butt joints, investigations into the penetration and microstructure properties of this kind of weld are limited. Therefore, it is essential to characterize the weld penetration and microstructure in terms of multiple welding parameters.

The purpose of this article is to experimentally investigate the influences of key process parameter on weld penetration of the fusion zone during pulsed Nd:YAG laser beam welding, and to analyze the hardness and microstructure evolution of DP steel butt joint. In Section 2, the base material used, DP1000, was characterized by some general mechanical tests. Section 3 introduces the main experimental procedures including cutting surface preparation for pre-welding sample, pulsed Nd:YAG laser beam welding, optimization design of welding sample fixture and the microstructure, and hardness testing method. Section 4 provides some corresponding process parameter dependent theory for a better understanding of the interaction between weld quality and process parameters. In Section 5, the influences of laser welding parameters on depth penetration and micro-hardness are analyzed and discussed, as well as the microstructure evolution before and after welding process.

2. Base Material Characterization: DP1000 Steel

The studied dual-phase steel (DP1000) sheet, a type of advanced high-strength steel (AHSS), has been widely utilized in automotive or shipping structures such as frames, wheels, and bumpers. DP steel contains a soft ferrite matrix with hard martensitic "islands". It exhibits good performance during the forming processes [29]. Figure 1 shows the microstructure of the experimental base material by means of a Scanning Electron Microscope (SEM) (Hitachi, Tokyo, Japan). The selected DP1000 steel with sheet thickness of 1.0 mm was supplied by the SSAB (Stockholm, Sweden) in as-received state. Two identical sheets with dimensions of 40 mm × 14 mm (length × width) were in one butt joint configuration after welding. The chemical composition is presented in Table 1.

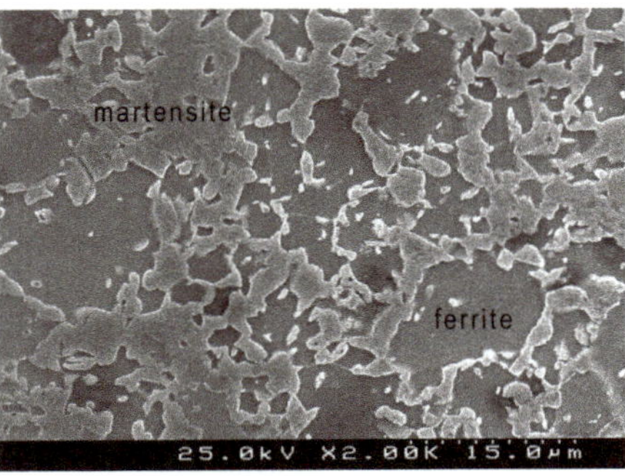

Figure 1. Microstructure observation of dual-phase DP1000 steel used.

Table 1. Chemical composition of the studied DP 1000 steel in wt %.

C	Si	Mn	P	S	N	Cr	Ni	Cu	Al	V	B	Nb	Cekv
0.14	0.20	1.46	0.013	0.003	0.002	0.03	0.03	0.01	0.051	0.01	0.0004	0.014	0.39
Cekv = C + Mn/6 + (Cr + Mo + V)/5 + (Ni + Cu)/15													

In order to obtain the basic mechanical properties, as listed in Table 2, uniaxial tension tests along the rolling direction (RD) were carried out with a nominal strain rate of 10^{-4} s^{-1}. A Shimadzu universal tensile testing machine (Schimadzur, Kyoto, Japan) was used with a top capacity 100 kN. All tension samples were cut by a water-jet and the dimensions complied with the ASTM E8 standard [30]. The test was repeated four times to verify the reproducibility of the tested results. In order to accurately measure the strain, a digital image correlation (DIC) system was adopted to capture the displacement fields. The ARAMIS-5M commercial code [31] was utilized to obtain the strain from numerous optical images. More detailed descriptions on this measurement technology can be found in previous works [9,32–34]. The macroscopic hardness of the base material was obtained using a Vickers hardness tester, namely Shimadzu HMV-2000 (Schimadzu, Kyoto, Japan).

Table 2. Basic mechanical properties of the studied DP1000 (RD: Rolling direction).

Direction	Elastic Modulus, E (GPa)	Poisson' Ratio υ	Yield Stress σ_y (MPa)	Ultimate Stress σ_u (MPa)	Elongation e (%)	Hardness HV0.5
RD	210	0.3	779	1125	9.35	382

3. Experimental Procedures

3.1. Cutting Surface Preparation for Pre-Welding Sample

The sheet cutting processes usually involve heating and melting of a material. This can lead to inhomogeneous microstructure changes and influence the performance of the materials used in the subsequent welding process. For instance, Figure 2 illustrates a typical cross-sectional surface cut by a guillotine, mainly including a cutting area, deformation area, rough area, and butt. This leads to a significant anomalistic gap at the pre-welding sample inspection. In fact, laser beam welding is seriously dependent on the position and surface quality of the prepared samples, especially when a narrow laser beam is adopted. Thus, it is vital to cut precise edges for almost no gap between the samples to be welded.

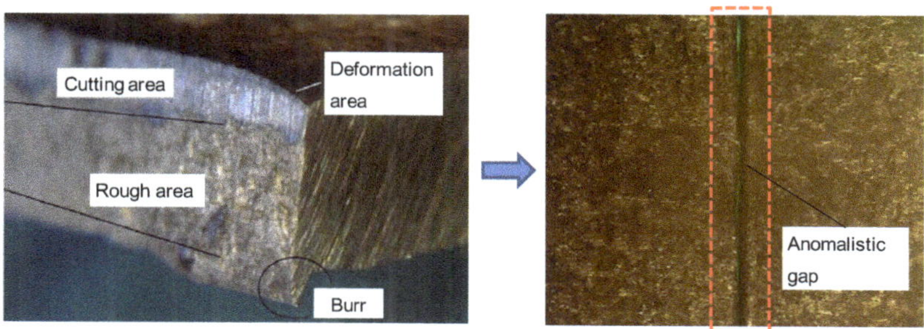

Figure 2. Illustration of a typical cross-sectional surface cut by guillotine and the anomalistic gap at the pre-welding inspection.

In this work, to ensure all the samples are welded under the same configuration, the inhomogeneous cross section of the cut sample was firstly machined flat using computer numerical control (CNC) and then the burr was carefully removed with sandpaper. To remove oil or other impurities, the machined samples were cleaned with acetone. Figure 3 illustrates this additional and important experimental preparation work, namely cutting surface preparation. It can be observed that there is almost no gap, which leads to consistency in pre-welding inspection.

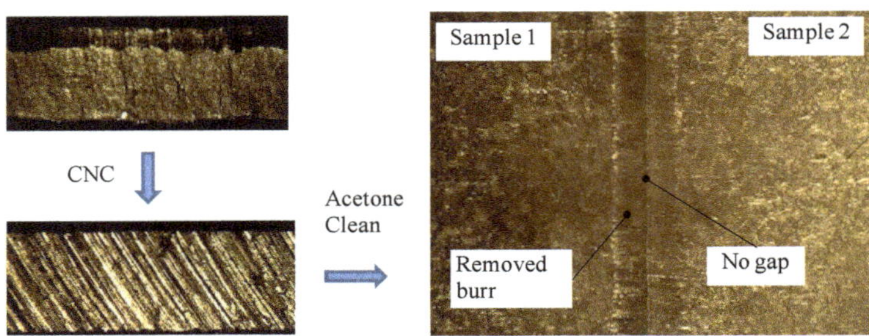

Figure 3. Cutting surface preparation for pre-welding inspection by using computer numerical control (CNC).

3.2. Pulsed Nd:YAG Laser Welding Set-Up

In this work, pulsed Nd:YAG laser tested system, i.e., SISMA SWA300, was utilized to perform experiments. Figure 4 shows the experimental apparatus used. The main parameters of the control system are as follows: maximum mean power of 300 W, wave length of 1064 nm, maximum peak laser power of 12 kW, maximum pulse energy of 100 J, and the range of pulse range from 0.2 ms to 25 ms. To ensure the pool of the butt joint against the oxidation, argon shielding gas with high purity (99.99%) was supplied to both bottom and top surfaces of the specimens at a gas flow rate of 10 L/min. A solid state pulsed Nd:YAG laser with the above parameters is appropriate for fiber passing from the resonator to the laser beam head. This ensures that little welding occurs far from the laser source. It is particularly useful for large or complex structures. In this work, the distance between the laser exit and the sample was set at a value of 105 mm with optimal resolution.

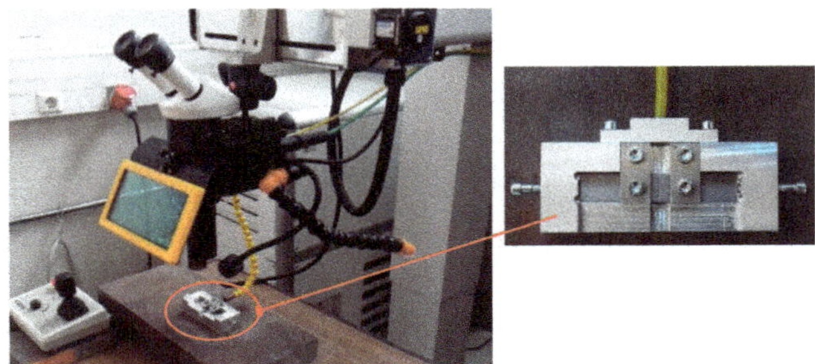

Figure 4. Experimental apparatus of pulsed Nd:YAG laser welding.

3.3. Design Optimization of Welding Sample Fixture

In order to avoid accidental misalignment or movement during laser welding, a characteristic welding sample fixture has been designed to ensure stable and consistent experimental tests. It is a relatively simple but good enough sheet metal sample fixture. It includes three steps to assemble the welding samples, as shown in Figure 5. The samples snap into place and clamp with screws along the longitudinal and normal directions.

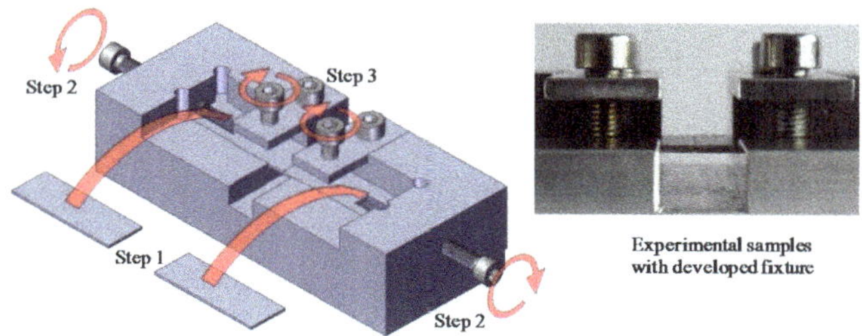

Figure 5. Design of welding sample fixture.

For laser welding, the flow ways of assisted gas inlet are usually ignored or at casual placement during the pre-welding preparation. In this work, a two-hole inlet structure design (see Figure 6) of assisted gas flow was developed to ensure that both the top and bottom surfaces of the welding sample have entire shielding gas during the welding process. Figure 7 compares three types of assisted gas solutions, i.e., without assisted gas, one-hole inlet structure, and two-hole inlet structure. It indicates that the design with a two-hole inlet structure allows the possibility of reducing the oxidation on both the top and bottom weld surfaces.

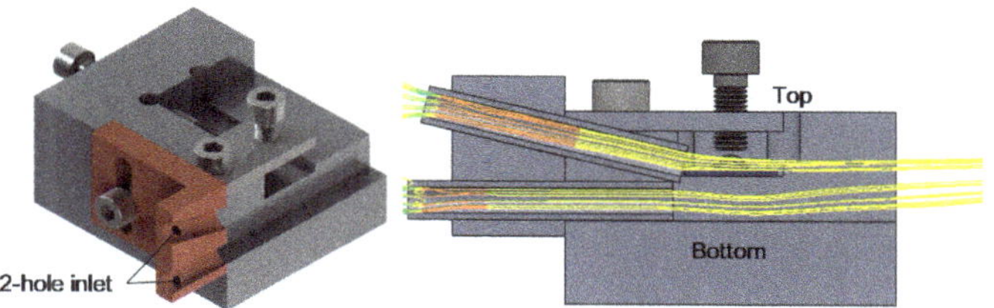

Figure 6. Optimization design of assisted gas flow solution with two-hole inlet structure.

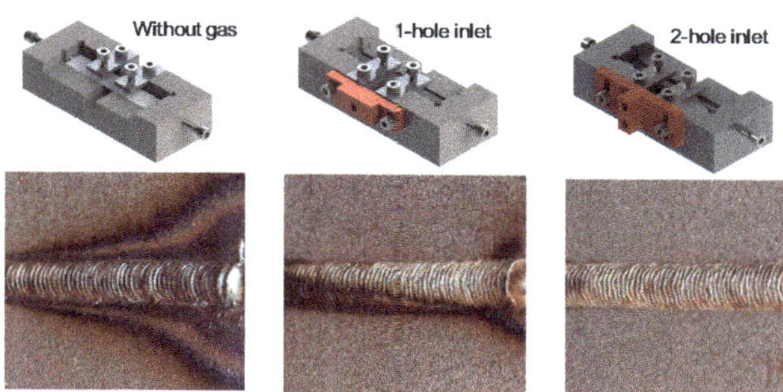

Figure 7. Comparison of welding top surface quality with different assisted gas inlet structures.

3.4. Microstructure and Micro-Hardness Testing

The microstructure examination was conducted by means of scanning electron microscope (SEM). The etching condition was carried out through the immersion of the samples with 5% nital solution. The first macro-hardness tests were performed with a holding time 10 s using a Vickers hardness tester, namely Shimadzu HMV-2000 (Shimadzu, Kyoto, Japan). The test force of 20 mN can be used as well as the application of 115° triangular pyramid indenter. The distance between the test positions is 0.25 mm. The second is a CSM Nano-indenter (TTX-NHT, CSM Instruments, Neuchatel, Switzerland) used for the small-scale micro-hardness measurement. The tested samples were polished by emery particle paper TTP2000 (Zebra Technologies, Lincolnshire, IL, USA). For the measurement of weld penetration, a microscope, namely Mitutoyo^{TM} (Kanagawa-ken, Tokyo, Japan), was used. All the corresponding experimental tests were sufficiently spaced to prevent any underlying obstruction of strain fields giving rise to contiguous indentations.

4. Process Parameter Dependent Theory

4.1. Process Parameters Relationship of General Laser Welding

To obtain exhaustive penetrated high-quality joint after laser beam welding, it is essential to study the optimal process conditions. Generally, they consist of three main groups: laser beam parameters (diameter, wavelength, power, and divergence), sample capabilities (material density, thermal conductivity, thickness, potential heat of melting and cooling), and key welding parameters (velocity, shielding gas direction, surface absorptivity, and focusing element length). Compared to absorptivity about 10–20% at room temperature for most steels, it is able to increase to 80–90% during heating and leaps when the material's melting point is reached. The basic relationship of necessary heat Q and mass m can be explained by [35]:

$$Q = m[c(T_m - T_0) + L_m],\tag{1}$$

where c, T_m, T_0, and L_m are the particular heat, melting and initial temperatures, and potential heat of melting, respectively. For continuous laser beam welding, the above relationship can be transformed as follows:

$$Q/t = \rho Dhv[c(T_m - T_0) + L_m],\tag{2}$$

where Q/t denotes requested power for melting. ρ, D, h, and v are the density, laser diameter, penetration, and welding velocity, respectively. The depth of penetration is obtained by:

$$h = \frac{Q/t}{\rho Dv[c(T_m - T_0) + L_m]}.\tag{3}$$

Including the absorption of weld surface and the losses of heat conduction, the calculation of penetration is estimated as:

$$h = \kappa \frac{P}{Dv},\tag{4}$$

where P and κ are the power and a constant associated with surface absorption and latent energy losses. P/v defines the ratio of heat input to a unit length. The penetration depth is related to the adopted power and is proportionally opposite to laser spot diameter and welding velocity. More detailed descriptions of pulsed laser parameters can be seen in the next section.

4.2. Pulse Welding Parameters

For a pulsed laser, the main conditions, i.e., flash lamp charging voltage U (V), frequency f (Hz), and pulse duration t (ms) are three key process parameters in terms of laser source. The practical pulse energy E (J) can be defined by the above parameters. Peak power P_{peak} (kw) is given by a portion of laser beam energy and pulse duration as follows:

$$P_{peak} = E/t.\tag{5}$$

For a given spot size, the laser beam intensity is dependent on peak power. Average laser power P (W) is defined as a product of practical energy and frequency:

$$P = Ef,\tag{6}$$

and gives the laser welding velocity. Nowadays, the first general pulsed laser welding method is spot welding. It is also generally applied for rude fixing of structures to be subsequently seam process. One typical advantage of this procedure is to mitigate dimensional distortions caused by high thermal gradients with respect to power density. To obtain successive strict welds, pulse overlap (P_O) must be adopted as well as the appropriate combination of welding parameters. Thus, pulse overlap can be given by:

$$P_O = 1 - \frac{v}{Df}. \tag{7}$$

Compared to conventional and continuous laser beam welding, higher pulse densities result in higher thermal transformation rates. This could induce weld defects and inhomogeneous microstructure. Thus, another important parameter, namely effective peak power density (D_{epp}), should be considered for the important contribution of weld quality. It can be defined by a product of peak power density (D_{pp}) and overlapping factor O_f:

$$D_{epp} = O_f \times D_{pp}, \text{ with } O_f = \frac{1}{1 - P_O}. \tag{8}$$

4.3. Design of Experiments for Laser Welding

In order to explore the effects of each single process factor on penetration and microstructure properties of DP steel, experimental design is performed. First, determine the parameters that ensure fairly good laser quality. In other words, the full weld penetration should be obtained. Then, based on the full weld penetration, further process parameters optimization can be performed to analyze the effect of microstructure properties, e.g., micro-hardness. Additionally, the optimal weld should ensure no cracks, high resistance, a suitable size and shape of melting fusion, and so on.

Table 3 presents the selected welding parameters of the experimental plan. More detailed lists of each analyzed welding parameter are given in the Appendix A, Tables A1–A6. In each set of one-factor experiments, only one parameter was varied, keeping all the other parameters fixed to identify the sensitivity of the one tested parameter. Seven different types of pulse, i.e., a simple rectangular pulse, a modular pulse with three sections, a rectangular pulse with a ramp start, a rectangular pulse with a ramp end, a rectangular pulse with a ramp start and end, the emission of laser beam pulses with three consecutive equal intervals of 1 ms, and a pulse with simple power divided to be better resolution at low power, were analyzed and discussed. It should be noted that the first simple rectangular pulse is the best and most widely used in real or experimental tests.

Table 3. Design of experiments for welding parameters.

Series	Parameters	Variation of Single Factor
A	Power percent (%)	20, 40, 60, 80, 86
B	Duration (ms)	0.3, 3, 6, 9, 12, 15, 18, 21, 23
C	Overlap (%)	0, 20, 40, 60, 80, 95
D	Laser beam diameter (mm)	0.6, 0.8, 1.0, 1.2, 1.4, 1.6, 1.8, 2.0
E	Pulse type	[pulse type icons]
F	Velocity (mm/s)	0.1, 1.0, 1.5, 2.0, 2.7

5. Results and Discussions

5.1. Effects of Laser Beam Parameters on Weld Penetration

5.1.1. Effects of Laser Beam Power on Weld Penetration

Laser beam power was changed in the interval from 20% to 80%. Figure 8 shows the weld surfaces on the top and bottom and at the beginning (left side) and end (right side) of the perpendicular longitudinal section. It can be observed that the weld penetrations reach almost full when the power percentages are higher than 60%. However, the 40% laser power is obviously insufficient. In addition, there are some spatters on the top weld surface when the laser power percent is larger than 60%. This might be due to the clamping force on the samples. The potential question of thermodynamic instability in the welding pool arises.

As shown in Figure 9, it is possible to estimate that the ideal power for the welding of the studied case should be between 40% and 60%. Weld penetration has a sharp rise in this range of laser power. However, the sensitivity of the weld penetration to laser beam power greater than 60% becomes less pronounced. In a word, the effects of power on penetration are significant and must be taken into account during laser welding applications.

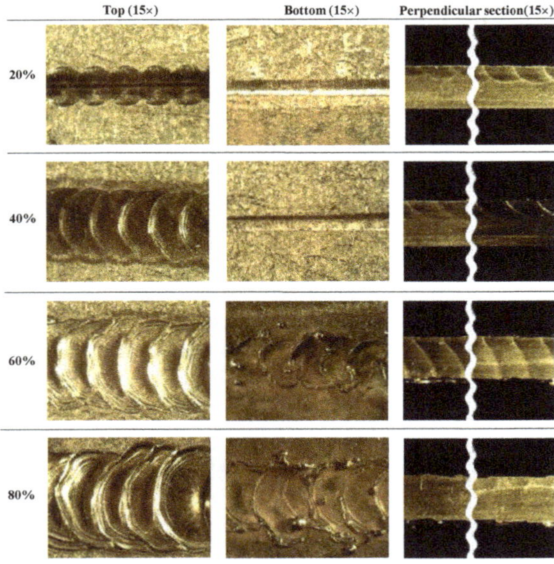

Figure 8. Comparison of weld surfaces under different laser beam powers.

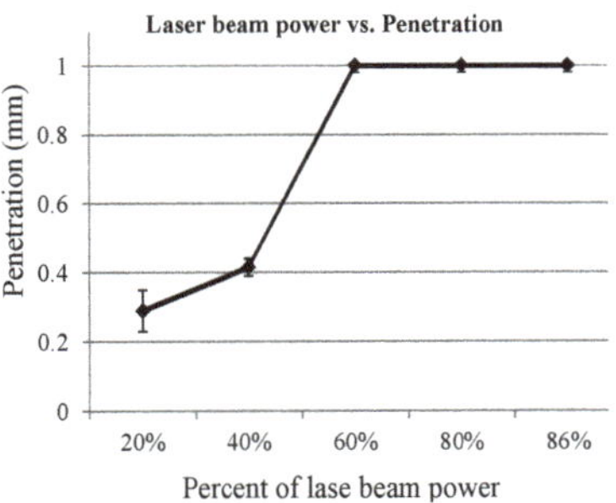

Figure 9. The effects of laser beam power on penetration.

5.1.2. Effects of Pulse Duration on Weld Penetration

Figure 10 reveals that the greater the pulse duration the deeper the weld penetration. The longest pulse duration of 21 ms and/or 23 ms induces full penetration. By varying the pulse duration and keeping the energy constant, the longer the pulse duration, the lower the energy received by the part, leading to a decrease in penetration. However, in reality, the high pulse duration is not recommended. This is because too high a pulse duration can cause the pulse overlap to not be constant, especially for continuous laser beam welding. Therefore, the experimental results of these samples over a long time are not reliable, but only provide a trend or sensitivity reference. For the analyzed DP1000 steel, the optimal duration of the pulse should be in the middle range, from 9 ms to 15 ms.

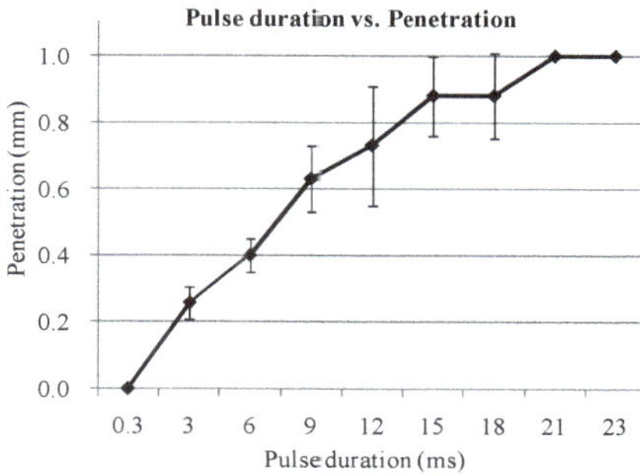

Figure 10. The effects of pulse duration on weld penetration.

5.1.3. Effects of Overlap on Weld Penetration

It can be understood from Equations (7) and (8) that overlapping is related to the selection of pulse duration, spot size (diameter), and traverse velocity for a specific mean power. In this work, a series of overlap values were tested experimentally to assess the effects on weld penetration. Figure 11 shows that the overlap seems not to affect the weld penetration, but only up to the value of 80%. For the last sample applied to an overlap of 95%, the laser pulse is similar to the continuous laser as the pulses are almost constant. For this value of overlap, the samples are completely melted and crossed by the beam. The suitable overlap should be less than 80%, although it hardly affects weld penetration for pulsed Nd:YAG laser beam welding.

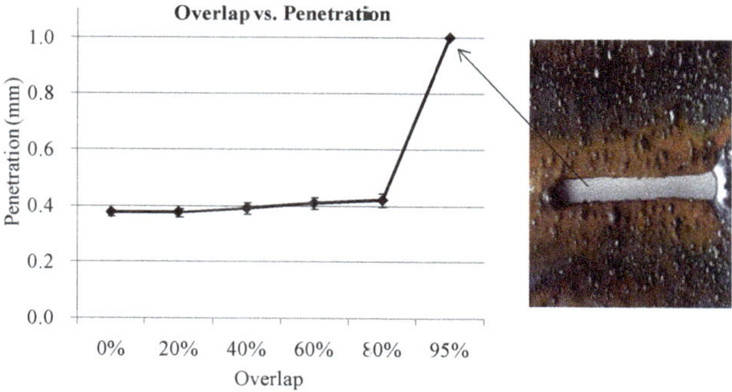

Figure 11. Effects of overlap on weld penetration.

5.1.4. Effects of Spot Laser Diameter on Weld Penetration

Figure 12 shows that an increase of diameter decreases the weld penetration. This experimental result is consistent with the theoretical relationship in Equation (4). It is known that a greater diameter leads to smaller power density due to the larger area in the same laser beam power. The small power density induces a decrease in the laser impact in the piece. In other words, when the diameter decreases or the power is channeled to a smaller point, the laser beam should be more concentrated and potent, causing an increase in weld penetration.

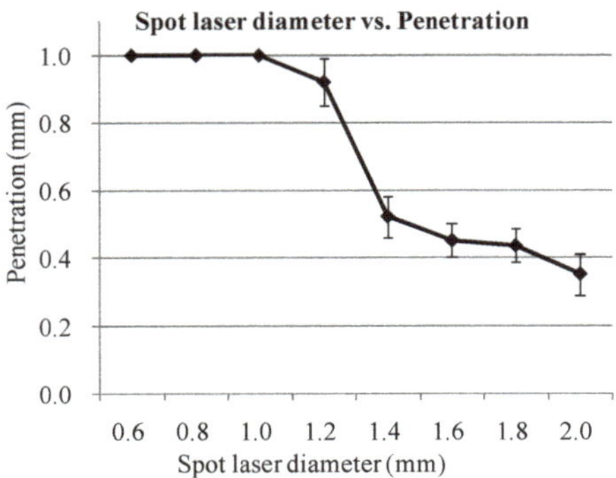

Figure 12. Effects of spot laser diameter on weld penetration.

Figure 12 indicates that the penetration decreases sharply when the applied spot diameter increases from 1.2 mm to 1.4 mm. To further observe the distinction, the weld surfaces under different spot diameters are compared in Figure 13. Meanwhile, there are some significant fusion spatters on the top and bottom weld surfaces when the spot diameter is smaller than 1.0 mm. This might be because the thermodynamic behavior in the welding pool manifests shrinkage when a small laser beam spot size is applied. These additional spatters imply that the weld penetrations are sufficient.

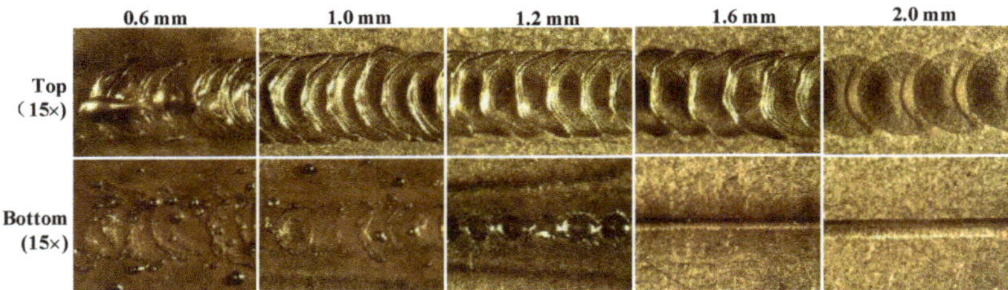

Figure 13. Comparison of weld surfaces under different spot diameters.

5.1.5. Effects of Pulse Type on Penetration

Figure 14 compares the effects of seven different pulse types on weld penetration. It indicates that the pulse type has no significant effect on penetration, with the exception of the last type, namely scale expanded pulse. This special pulse type results in a low penetration (Height h = 0.3 mm) compared to the others (between 0.5 mm and 1.0 mm). This is because the pulse with simple power is divided to be better resolution at low power. However, this kind of pulse maintains a penetration that is virtually constant throughout the cord. It should be noted that the "scale expanded" pulse is used mainly as heat treatment.

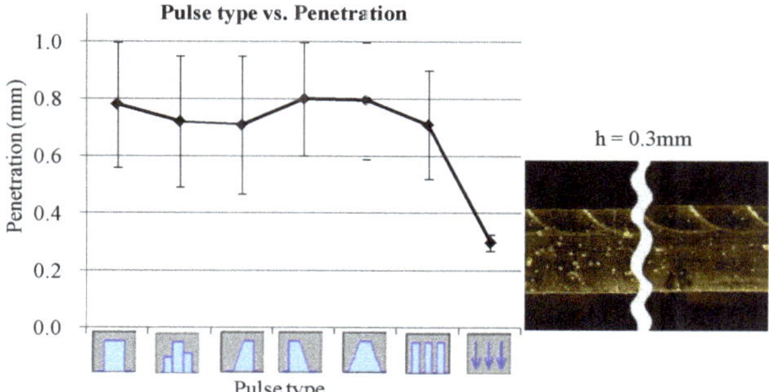

Figure 14. Effects of pulse type on weld penetration.

5.1.6. Effects of Welding Velocity on Weld Penetration

As shown in Figure 15, the penetration seems to be directly influenced by the welding velocity: The higher the welding velocity, the deeper the weld penetration. In the case of DP1000 laser beam welding, weld penetration has an obvious increase for the low bound of the acceptable welding velocity that the welding velocities are less than 1.5 mm/s. The relationship between welding velocity and penetration may be explained as follows: The greater the welding speed, the less time the material has to cool, so the temperature of the material increases until the end of the weld. The fusion material becomes "softer" and facilitates the penetration of the laser beam, especially for a thin sheet. In addition, the cycle of heating and cooling makes a different change in the microstructure of the material, e.g., causing an increase in the hardness of the material, hindering further penetration.

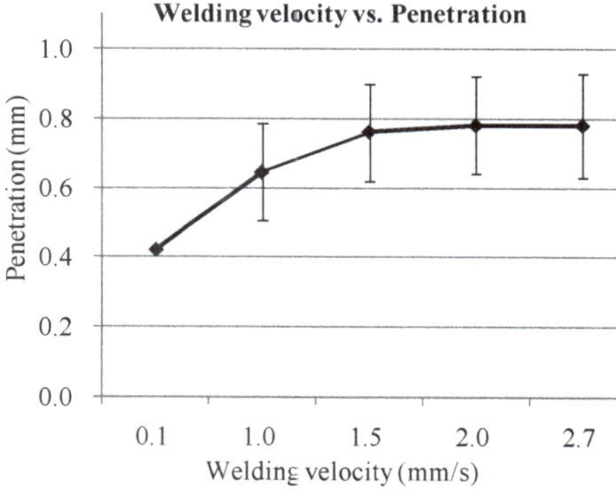

Figure 15. Effects of welding velocity on penetration.

5.1.7. ANOVA for Quadratic Model of Weld Penetration

According to the design of experiments for welding parameters listed in Table 3, the second-order polynomial (regression) equation used to represent the response surface $\hat{R}(x)$ is given by:

$$\hat{R}(x) = \beta_0 + \sum_{i=1}^{n} \beta_i x_i + \sum_{i=1}^{n} \beta_{ii} x_i^2 + \sum_{i}\sum_{j>1} \beta_{ij} x_i x_j + \eta, \tag{9}$$

where β is the polynomial coefficient, η is the minor error, and n is the number of design variables (six factors in this work). Statistical testing of the empirical models has been done by Fisher's statistical test for Analysis of Variance—ANOVA. Table A7 shows that the ANOVA tests applied to the individual coefficients in the model are significant. The *F*-value is the ratio of the mean square due to regression to the mean square due to residual. The Model *F*-value of weld penetration response is 12.94, which implies that the models are significant. There is only 0.36% chance in the weld penetration response model that the "Model *F*-Value" could occur due to noise. The determination coefficient of the regression analysis results is $R^2 = 0.966$, indicating a high degree of correlation between the experimental values and empirical calculated values obtained from the response models.

5.2. Effects of Laser Parameters on Hardness and Its Microstructure Observation

In order to evaluate the microstructure evolution of the material before and after laser beam welding, one of the optimal welding parameters, as listed in Table 4, can be obtained through the above experimental sensitivity analyses. The tested sample had full penetration with almost no oxidation and spatter on both top and bottom surfaces. Figure 16 shows the weld surfaces of fusion zone using these optimal parameters.

Table 4. Optimal welding parameters of DP 1000 steel used.

Material	Power (%)	Pulse Duration (ms)	Overlap (%)	Diameter (mm)	Velocity (mm/s)	Energy (J)	Pulse Type
DP1000	57	9.0	60	0.6	0.3	45	Rec.

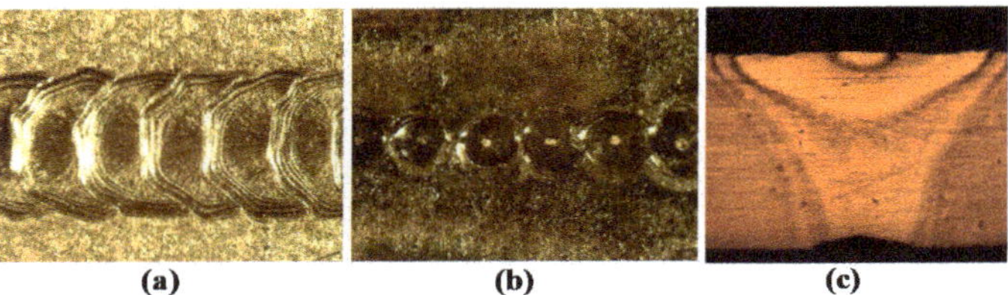

Figure 16. Weld surfaces of fusion zone using the optimal parameters: (**a**) Top surface (15×); (**b**) bottom surface (15×); (**c**) perpendicular longitudinal section.

5.2.1. Microscopic and Macroscopic Hardness

The hardness is considered an important indicator of welded joint properties and performance. Too much hardness inside the fusion and the heat-affected zones require higher pre-heat temperatures, slower cooling, and more complicated hydrogen controlled welding procedures [16]. The laser Nd:YAG process gives higher hardness values compared to other processes [20]. Thus, the mechanism of formation of softening zones HAZ and fusion zone (FZ) of the welded joint should account for the high capability of the welds. In this work, the metallurgical properties of the welded joint of DP1000, including the variations of microscopic and macroscopic hardness of FZ and HAZ under optimal welding parameters, are observed experimentally. According to Taylor et al. [31], the microscopic and macroscopic hardness can be transformed into the same unit for comparison, as listed in Table 5.

Table 5. Comparison of average measured hardness.

Material Tested	Micro-Indentation	Vickers
Base DP 1000 steel	425 HV (10.7 GPa)	382 HV
Heat-affected zone	372 HV (9.36 GPa)	316 HV
Fusion zone of weld	504 HV (14.5 GPa)	449 HV

Figure 17 shows the micro-indentation hardness distribution and measured macroscopic hardness of laser welded DP1000 steel butt joints. The fusion zone (FZ) of welded joints shows higher hardness than the base metal and HAZ of butt joints. Table 5 shows that the hardness in the HAZ is lower compared to the base metal. The previous study indicates that the soft zone in the HAZ of the dual-phase butt joints is caused by the tempering of pre-existing martensite [20].

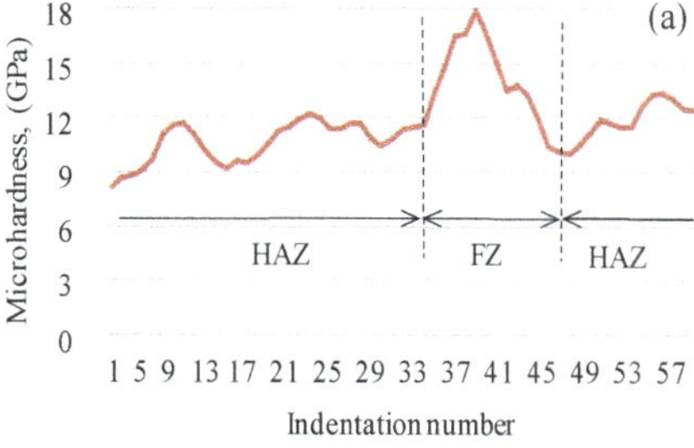

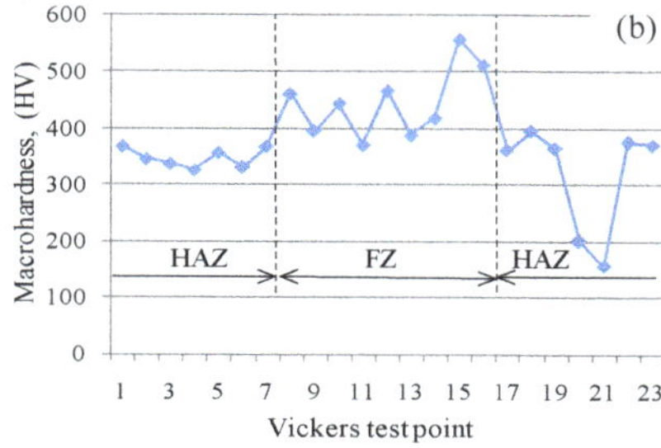

Figure 17. (**a**) Micro-indentation hardness; (**b**) measured macroscopic hardness (HAZ: Heat-affected zone; FZ: Fusion zone).

5.2.2. Microstructure Characterization of Nd:YAG Laser Welded Joint

The SEM examinations of HAZ and FZ regions, as shown in Figure 18, indicate that the microstructure of DP1000 steel butt joints in FZ region is martensitic together with a few encompassing ferrite and bainite molecules. The changeable hardness in the FZ of DP1000 butt joints reveals the existence of a so-called multiple-ingredient microstructure. The quick-cooling process of the weld butt pool during the laser welding should lead to the production of martensite in the FZ region. Thus, the martensite volume fraction of dual-phase steels should play a considerable effect on the change of microstructure characterization during the laser beam welding.

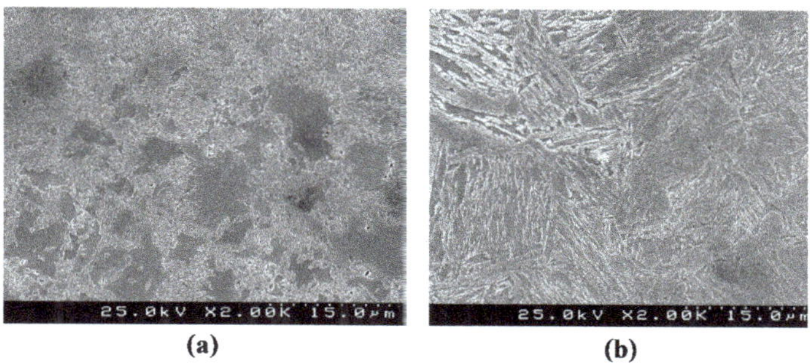

Figure 18. SEM micrographs showing the microstructure change of laser welded DP1000 steel butt joints: (**a**) HAZ; (**b**) FZ.

This may be explained as follows: Firstly, the peak temperature experienced in the soft zone during laser beam welding seems to result in a partial disappearance of pre-existent martensite. It is likely that the peak temperature in the soft zone can be high enough to stimulate partial martensite-to-austenite solid state transformation. Meanwhile, the subsequent cooling phase might not provide a trigger of the backward austenite-to-martensite solid state transformation, or full martensite-to-austenite transformation in the heating phase and partial austenite-to martensite transformation in the cooling phase, or both partial transformations in the heating and cooling phases during the welding thermal cycle. Actually, the occurrence of the microstructure change depends on the location within the welded joint, in terms of the complexity of the temperature field changes. Secondly, if there was no occurrence of the abovementioned solid-state phase transformations, the presence of a higher amount of martensite in the DP steels might cause greater tempering of pre-existing martensite in the HAZ, giving rise to more severe softening in this material.

6. Conclusions

The pulsed Nd:YAG laser beam welding of dual-phase DP1000 steel is addressed using theoretical and experimental methods. Some key experimental technologies, including cutting surface preparation for pre-welding sample and optimization design of welding sample fixture for a sufficiently shielding gas flow, are performed to ensure consistent and stable testing. For a better understanding of the interaction between weld quality and process parameters, a parameter-dependent theory is introduced. The main conclusions are as follows.

1. The ideal laser beam power for welding in the studied case should be between 40% and 60% with a sharp rise. However, the sensitivity of the weld penetration to the greater than 60% laser beam power becomes less pronounced.
2. The greater the pulse duration, the deeper the weld penetration.
3. The overlap seems not to affect the weld penetration, but only up to the value of 80%.
4. The weld penetration has a sharp decrease when the applied spot diameter increases from 1.2 mm to 1.4 mm.
5. The pulse type has no significant effect on penetration, with the exception of the scale expanded pulse.
6. The penetration seems to be directly influenced by the welding velocity, i.e., the higher the welding velocity, the deeper the weld penetration.
7. The ANOVA tests applied to the individual coefficients in the regression model show that the Model F-value of weld penetration response is 12.94, which implies that the models are significant. The determination coefficient of the regression analysis results indicates that a high degree of correlation between the experimental values and empirical calculated values is obtained from the models.

Acknowledgments: The authors would like to thank Engineer António Festos for support concerning the experimental tool machining. This work was funded by the Portuguese Foundation of Science and Technology (UID/EMS/00481/2013 and SFRH/BPD/114823/2016 and CENTRO-01-0145-FEDER-022083) and the Scientific Program of Fuzhou University (0020-650421 and 0020-650425).

Author Contributions: António B. Pereira and José Amorim conceived and designed the experiments; José Amorim performed the experiments; Xin Xue and José Amorim analyzed the data; Juan Liao and António B. Pereira contributed reagents/materials/analysis tools; Xin Xue, José Amorim, and Juan Liao wrote the paper.

Conflicts of Interest: The authors declare no conflict of interest.

Appendix A

The detailed design of experiments on the effects of each analyzed welding parameter on weld penetration are presented in this section.

Table A1. Experimental plan for the effects of laser beam power on weld penetration.

Series	Power (%)	Duration (ms)	Overlap (%)	Diameter (mm)	Pulse Type	Energy (J)	Velocity (mm/s)	Max. Velocity (mm/s)
A1	20					24.0		6.3
A2	40					43.5		3.5
A3	60	-	-	-	-	66.1	-	2.2
A4	80					91.6		1.6
A5	86					100.0		1.4

Table A2. Experimental plan for the effects of pulse duration on weld penetration.

Series	Power (%)	Duration (ms)	Overlap (%)	Diameter (mm)	Pulse Type	Energy (J)	Velocity (mm/s)	Max. Velocity (mm/s)
B1		0.3				1.3		10
B2		3				13.0		10
B3		6				26.1		5.8
B4		9				39.1		3.9
B5	50	12	50	1.3	Simple rectangular	52.2	1.4	2.9
B6		15				65.2		2.3
B7		18				78.3		1.9
B8		21				91.3		1.6
B9		23				100.0		1.4

Table A3. Experimental plan for the effects of overlap on weld penetration.

Series	Power (%)	Duration (ms)	Overlap (%)	Diameter (mm)	Pulse Type	Energy (J)	Velocity (mm/s)	Max. Velocity (mm/s)
C1			0					5.5
C2			20					4.4
C3	50	12.5	40	1.3	Simple rectangular	54.3	0.2	3.3
C4			60					2.2
C5			80					1.1
C6			95					0.2

Table A4. Experimental plan for the effects of spot laser diameter on weld penetration.

Series	Power (%)	Duration (ms)	Overlap (%)	Diameter (mm)	Pulse Type	Energy (J)	Velocity (mm/s)	Max. Velocity (mm/s)
D1				0.6				1.2
D2				0.8				1.7
D3				1.0				2.1
D4	50	12.5	50	1.2	Simple rectangular	54.3	1.2	2.5
D5				1.4				3.0
D6				1.6				3.4
D7				1.8				3.8
D8				2.0				4.3

Table A5. Experimental plan for the effects of pulse type on weld penetration.

Series	Power (%)	Duration (ms)	Overlap (%)	Diameter (mm)	Pulse Type	Energy (J)	Velocity (mm/s)	Max. Velocity (mm/s)
E1						54.3		2.7
E2						54.5		2.7
E3	50	12.5	50	1.3		59.0	2.6	2.6
E4						59.0		2.6
E5						56.9		2.6
E6						53.4		2.8
E7						28.5		5.3

Table A6. Experimental plan for the effects of velocity on weld penetration.

Series	Power (%)	Duration (ms)	Overlap (%)	Diameter (mm)	Pulse Type	Energy (J)	Velocity (mm/s)	Max. Velocity (mm/s)
F1							0.1	
F2							1.0	
F3	50	12.5	50	1.3	Simple rectangular	54.3	1.5	2.7
F4							2.0	
F5							2.7	

Table A7. ANOVA for quadratic model of weld penetration.

Source	SS	DF	MS	F-Value	p-Value	Evaluation
Model	1.446	14	0.13	12.94	0.0036	Significant
A	0.115	1	0.12	10.37	0.0181	
B	0.010	1	0.010	0.88	0.3839	
C	0.379	1	0.379	34.13	0.0011	
D	0.014	1	0.014	1.30	0.2976	
E	0.003	1	0.003	1.02	0.3505	
F	0.008	1	0.008	2.59	0.1587	
AB	0.074	1	0.074	6.65	0.0419	
AC	0.138	1	0.138	12.40	0.0125	
AD	0.003	1	0.003	0.28	0.6188	
AE	0.060	1	0.060	5.36	0.0599	
AF	0.045	1	0.045	4.02	0.0919	
BC	0.099	1	0.099	8.91	0.0245	
BD	0.073	1	0.073	23.42	0.0029	
BE	0.020	1	0.020	6.48	0.0437	
BF	0.065	1	0.065	20.66	0.0039	-
CD	0.079	1	0.079	25.35	0.0024	
CE	0.040	1	0.040	12.73	0.0118	
CF	0.009	1	0.009	2.81	0.1447	
DE	0.007	1	0.007	2.40	0.1724	
DF	0.019	1	0.019	5.93	0.0508	
EF	0.154	1	0.154	49.40	0.0004	
A2	0.001	1	0.001	0.09	0.7744	
B2	0.001	1	0.001	0.05	0.8341	
C2	0.013	1	0.013	1.13	0.3293	
D2	0.002	1	0.002	0.17	0.6938	
E2	0.080	1	0.080	25.61	0.0023	
F2	0.007	1	0.007	2.30	0.1798	
Residual	0.067	6	0.011	-	-	
Total	1.512	20	-	-	-	

$R^2 = 0.966$; SS—sum of square; DF—degree of freedom; MS—mean square.

References

1. Rossinia, M.; Spena, P.R.; Cortese, L.; Matteis, P.; Firrao, D. Investigation on dissimilar laser welding of advanced high strength steel sheets for the automotive industry. *Mater. Sci. Eng. A* **2014**, *628*, 288–296. [CrossRef]
2. Blacha, S.; Weglowski, M.S.T.; Dymek, S.; Kopuscianski, M. Microstructural characterization and mechanical properties of electron beam welded joint of high strength grade S690QL. *Arch. Metall. Mater.* **2016**, *61*, 1193–1200. [CrossRef]
3. Grouve, W.J.B.; Warnet, L.; Akkerman, R.; Wijskamp, S.; Kok, J.S.M. Weld strength assessment for tape placement. *Int. J. Mater. Form.* **2010**, *3*, 707–710. [CrossRef]
4. Hernandez, V.H.B.; Panda, S.K.; Kuntz, M.L.; Zhou, Y. Nanoindentation and microstructure analysis of resistance spot welded dual phase steel. *Mater. Lett.* **2010**, *64*, 207–210. [CrossRef]

5. Ramazani, A.; Mukherjee, K.; Abdurakhmanov, A.; Abbasi, M.; Prahl, U. Characterization of microstructure and mechanical properties of resistance spot welded DP600 Steel. *Metals* **2015**, *5*, 1704–1716. [CrossRef]

6. Huang, Y.M.; Wu, D.; Zhang, Z.F.; Chen, H.B.; Chen, S.B. EMD-based pulsed TIG welding process porosity defect detection and defect diagnosis using GA-SVM. *J. Mater. Process. Technol.* **2017**, *239*, 92–102. [CrossRef]

7. Narwadkar, A.; Bhosle, S. Optimization of MIG welding parameters to control the angular distortion in Fe410WA Steel. *Mater. Manuf. Process.* **2015**, *31*, 2158–2164. [CrossRef]

8. Ahmadzadeh, M.; Farshi, B.; Salimi, H.R.; Fard, A.H. Residual stresses due to gas arc welding of aluminum alloy joints by numerical simulations. *Int. J. Mater. Form.* **2013**, *6*, 233–247. [CrossRef]

9. Ramazani, A.; Mukherjee, K.; Abdurakhmanov, A.; Prahl, U.; Schleser, M.; Reisgen, U.; Bleck, M. Micro-macro-characterisation and modelling of mechanical properties of gas metal arc welded (GMAW) DP600 steel. *Mater. Sci. Eng. A* **2014**, *589*, 1–14. [CrossRef]

10. Mei, L.F.; Chen, G.Y.; Jin, X.Z.; Zhang, Y.; Wu, Q. Research on laser welding of high-strength galvanized automobile steel sheets. *Opt. Lasers Eng.* **2009**, *47*, 1117–1124. [CrossRef]

11. Sharma, R.S.; Molian, P. Yb:YAG laser welding of TRIP780 steel with dual phase and mild steels for use in tailor welded blanks. *Mater. Des.* **2009**, *30*, 4146–4155. [CrossRef]

12. Xu, W.; Westerbaan, D.; Nayak, S.S.; Chen, D.L.; Goodwin, F.; Zhou, Y. Tensile and fatigue properties of fiber laser welded high strength low alloy and DP980 dual-phase steel joints. *Mater. Des.* **2013**, *43*, 373–383. [CrossRef]

13. Ghaini, F.M.; Hamedi, M.J.; Torkamany, M.J.; Sabbaghzadeh, J. Weld metal microstructural characteristics in pulsed Nd:YAG laser welding. *Scr. Mater.* **2007**, *56*, 955–958.

14. Mahamood, R.M.; Akinlabi, E.T. Scanning speed and powder flow rate influence on the properties of laser metal deposition of titanium alloy. *Int. J. Adv. Manuf. Technol.* **2017**, in press. [CrossRef]

15. Hazratinezhad, M.; Arab, N.B.M.; Sufizadeh, A.R.; Torkamany, M.J. Mechanical and metallurgical properties of pulsed neodymium-doped yttrium aluminum garnet laser welding of dual phase steels. *Mater. Des.* **2012**, *33*, 83–87. [CrossRef]

16. Mirakhorli, F.; Malek Ghaini, F.; Torkamany, M.J. Development of weld metal microstructures in pulsed laser welding of duplex stainless steel. *J. Mater. Eng. Perform.* **2012**, *21*, 2173–2176. [CrossRef]

17. Seang, C.; David, A.K.; Ragneau, E. Effect of Nd:YAG laser welding parameters on the hardness of lap joint: Experimental and numerical approach. *Phys. Proc.* **2013**, *41*, 38–40. [CrossRef]

18. Sun, Q.; Di, H.S.; Li, J.C.; Wang, X.N. Effect of pulse frequency on microstructure and properties of welded joints for dual phase steel by pulsed laser welding. *Mater. Des.* **2016**, *105*, 201–211. [CrossRef]

19. Wang, J.F.; Yang, L.J.; Sun, M.S.; Liu, T.; Li, H. A study of the softening mechanisms of laser-welded DP1000 steel butt joints. *Mater. Des.* **2016**, *97*, 118–125. [CrossRef]

20. Tzeng, Y.F. Effects of operating parameters on surface quality for the pulsed laser welding of zinc-coated steel. *J. Mater. Process. Technol.* **2000**, *100*, 163–170. [CrossRef]

21. Xia, M.S.; Biro, E. Effects of heat input and martensite on HAZ softening in laser welding of dual phase steels. *ISIJ Int.* **2008**, *48*, 809–814. [CrossRef]

22. Dong, D.Y.; Liu, Y.; Yang, Y.L.; Li, J.F.; Ma, M.; Jiang, T. Microstructure and dynamic tensile behavior of DP600 dual phase steel joint by laser welding. *Mater. Sci. Eng. A* **2014**, *594*, 17–25. [CrossRef]

23. Baghjari, S.H.; Akbari Mousavi, S.A.A. Effects of pulsed Nd:YAG laser welding parameters and subsequent post-weld heat treatment on microstructure and hardness of AISI 420 stainless steel. *Mater. Des.* **2013**, *43*, 1–9. [CrossRef]

24. Farabia, N.; Chen, D.L.; Zhou, Y. Microstructure and mechanical properties of laser welded dissimilar DP600/DP980 dual-phase steel joints. *J. Alloys Compd.* **2011**, *509*, 982–989. [CrossRef]

25. Lee, J.H.; Park, S.H.; Kwon, H.S.; Kim, G.S.; Lee, C.S. Laser, tungsten inert gas, and metal active gas welding of DP780 steel: Comparison of hardness, tensile properties and fatigue resistance. *Mater. Des.* **2014**, *64*, 559–565. [CrossRef]

26. Reisgen, U.; Schleser, M.; Mokrov, O.; Ahmed, E. Shielding gas influences on laser weldability of tailored blanks of advanced automotive steels. *Appl. Surf. Sci.* **2010**, *257*, 1401–1406. [CrossRef]

27. Schemmann, L.; Zaefferer, S.; Raabe, D.; Friedel, F.; Mattissen, D. Alloying effects on microstructure formation of dual phase steels. *Acta Mater.* **2015**, *95*, 386–398. [CrossRef]

28. Wang, X.N.; Chen, C.J.; Wang, H.S.; Zhang, S.H.; Zhang, M.; Luo, X. Microstructure formation and precipitation in laser welding of microalloyed C-Mn steel. *J. Mater. Process. Technol.* **2015**, *226*, 106–114. [CrossRef]
29. Liao, J.; Sousa, J.A.; Lopes, A.B.; Xue, X.; Barlat, F.; Pereira, A.B. Mechanical, microstructural behaviour and modelling of dual phase steels under complex deformation paths. *Int. J. Plast.* **2017**, *93*, 269–290. [CrossRef]
30. Xue, X.; Liao, J.; Vincze, G.; Pereira, A.B.; Barlat, F. Experimental assessment of nonlinear elastic behaviour of dual-phase steels and application to springback prediction. *Int. J. Mech. Sci.* **2016**, *117*, 1–15. [CrossRef]
31. ARAMIS—3D Motion and Deformation Sensor. Available online: http://www.gom.com/metrology-systems/aramis.html (accessed on 30 July 2017).
32. Tasan, C.C.; Diehl, M.; Yan, D.; Zambaldi, C.; Shanthraj, P.; Roters, F.; Raabe, D. Integrated experimental-simulation analysis of stress and strain partitioning in multiphase alloys. *Acta Mater.* **2014**, *81*, 386–400. [CrossRef]
33. Ghadbeigi, H.; Pinna, C.; Celotto, S. Quantitative strain analysis of the large deformation at the scale of microstructure: Comparison between digital image correlation and micro grid techniques. *Exp. Mech.* **2012**, *52*, 1483–1492. [CrossRef]
34. Alharbi, K.; Ghadbeigi, H.; Efthymiadis, P.; Zanganeh, M.; Celotto, S.; Dashwood, R.; Pinna, C. Damage in dual phase steel DP1000 investigated using digital image correlation and microstructure simulation. *Model. Simul. Mater. Sci.* **2015**, *23*, 085005. [CrossRef]
35. Duley, W.W. *Laser Welding*; John Wiley & Sons: Hoboken, NJ, USA, 1998; ISBN 978-0-471-24679-4.

Article

Effects of Laser Offset and Hybrid Welding on Microstructure and IMC in Fe–Al Dissimilar Welding

Giuseppe Casalino [1,*], Paola Leo [2], Michelangelo Mortello [3], Patrizia Perulli [1] and Alessandra Varone [4]

[1]   DMMM Politecnico di Bari, 70124 Bari, Italy; patrizia.perulli@poliba.it

[2]   Dipartimento di Ingegneria dell'Innovazione, Universita del Salento, 73100 Lecce, Italy; paola.leo@unisalento.it

[3]   Welding Engineering Research Centre, Building 46, Cranfield University, Bedfordshire MK43 0AL, UK; M.Mortello@cranfield.ac.uk

[4]   Department of Industrial Engineering, University of Rome "Tor Vergata", Via del Politecnico 1, 00133 Rome, Italy; Alessandra.varone@uniroma2.it

*   Correspondence: Giuseppe.Casalino@poliba.it; Tel.: +39-080-5962753

Received: 27 June 2017; Accepted: 20 July 2017; Published: 25 July 2017

Abstract: Welding between Fe and Al alloys is difficult because of a significant difference in thermal properties and poor mutual solid-state solubility. This affects the weld microstructure and causes the formation of brittle intermetallic compounds (IMCs). The present study aims to explore the weld microstructure and those compounds over two different technologies: the laser offset welding and the hybrid laser-MIG (Metal inert gas) welding. The former consists of focusing the laser beam on the top surface of one of the two plates at a certain distance (offset) from the interfaces. Such a method minimizes the interaction between elevated temperature liquid phases. The latter combines the laser with a MIG/MAG arc, which helps in bridging the gap and stabilizing the weld pool. AISI 316 stainless steel and AA5754 aluminum alloy were welded together in butt configuration. The microstructure was characterized and the microhardness was measured. The energy dispersive spectroscopy/X-ray Diffraction (EDS/XRD) analysis revealed the composition of the intermetallic compounds. Laser offset welding significantly reduced the content of cracks and promoted a narrower intermetallic layer.

Keywords: laser offset welding; hybrid welding; microstructure; intermetallic layer

---

1. Introduction

Lightweight metals and their alloys are increasingly used as efficient structural materials. The reduction of the overall weight of a vehicle decreases the fuel consumption and carbon emissions. This accomplishment is highly requested for automotive, aerospace, and shipbuilding industries [1]. Aluminum is one of the most popularly used lightweight metal, thanks to its low density, good corrosion resistance, and excellent workability [2]. Replacing conventional steel components with hybrid dissimilar Al–Fe assemblies is beneficial to improve flexibility, vehicle energy efficiency, and cut down the manufacturing costs.

Achieving a reliable fusion welded joint between Al and Fe alloys is challenging, due to the low mutual solid solubility and the large difference in thermal properties. This includes the melting points (660 versus 1535 $^\circ$C), the thermal conductivities (238 versus 77.5 W·m^{-1}·K^{-1}), and thermal expansion coefficients (23.5 $\times$ 10^{-6} and 11.76 $\times$ 10^{-6}/K). Additionally, the nearly-zero solid state solubility of Al in Fe, and the zero solubility of Fe in Al result in the formation of brittle intermetallic compounds (IMCs), which deteriorate the mechanical properties and form cracks [3,4]. According to Fe–Al phase diagram [5], IMCs include Fe-rich compounds (FeAl and Fe$_3$Al) and Al-rich compounds (FeAl$_2$, Fe$_2$Al$_5$, and FeAl$_3$). The crystalline arrangement suffers from low mobility of dislocations and slip systems. Thus, the formation of Al–Fe phases plays a key role in achieving an effectual connection between the two metals, but the excessive formation of IMCs results in brittleness and degrades the strength of the joint [3]. Furthermore, a "strength-mismatch effect" was observed in between two different steels [6,7]. The mismatch characteristic of the weld is to ensure that the welded joint withstands in-service constraints and provides good weld quality [8].

Some authors demonstrated the feasibility of using friction stir welding (FSW) for producing tough Fe–Al welds [9–14]. However, laser welding is potentially a more attractive technology for dissimilar joining, mostly in automotive and aerospace applications, thanks to its high versatility, precision, and productivity. It exhibits locally focused energy delivery in a small spot size and high process speed, which provides a smaller weld pool, higher cooling rate, and a shorter reaction time for IMC growth.

Since the thermal gradients and the reaction time determines the width of the IMC zone, one of the main concerns of researchers is the investigation of the welding conditions which minimize the growth of brittle phases. The control of temperature at the interfaces is fundamental for the right growth of the IMC layer. Meco et al., analyzed the transient thermal cycle of the laser–material interactions' fundamental parameters. The growth and distribution of IMCs correlated with the thermal cycle [15]. Qin et al., developed a finite element model to investigate the temperature fields in the weld zone during laser-arc welding. They found that the distribution of the temperature and the reaction time is not uniform along the thickness, which leads to non-uniform IMC layers [16]. Gao et al., proved that the non-uniformity in the thickness and the irregular shape of the interface increase with the heat input [17]. Recent studies showed that the intermetallic bond is altered by post-welding heat treating the joint [18].

Apart from the extension of the reaction area, the properties of IMCs are hugely responsible for the weld resistance. Chen et al., proved that the interposition of a Ni-foil interlayer improves the toughness, even if additional IMCs form [19]. The composition of the filler material is highly influential on the seam properties. Mathieu et al., adopted a hot wire to improve the adhesion of the filler material at the interfaces and demonstrated how the welding conditions are responsible for the zone in which the fracture occurs [20]. Dharmendra et al., used a low-melting point Zn-15Al filler alloy, which exhibited good wetting properties [21]. Song et al., explored how the dynamics of Al-Si filler material at liquid state affect the morphology of the weld [22]. Finally, the interposition of Al-Si filler material during resistance spot welding was used in overlap configuration [23].

The combination between laser and arc into a hybrid process improves the process robustness and stability, enables deeper penetration, provides compensation for the geometrical defects and misalignments, and ensures tolerance in joint fit-up and wider control of weld metallurgy [24,25]. Among process parameters, the power distribution has proved to have a deep impact on the weld quality for similar and dissimilar welding [26–28].

The laser offset welding (LOW) was used to weld dissimilar assemblies. The beam focused on the top surface of one of the two materials to promote a liquid–solid interaction, which produces the bond. Such a method limited the growth of an IMC layer and promoted the diffusion through the liquid-solid boundary, without mixing liquid phases at elevated temperature. This approach was used for several dissimilar welds [29–33]. Pardal et al., reduced the mutual inter-diffusion and the thickness of IMCs by carrying out the dissimilar Al–Fe joining process in conduction mode. Aluminum melted but the steel did not, so a liquid–solid interface formed in the volume of the weld [34].

In this work, a comparative study between laser offset and hybrid laser welding of AA5754 and 316 stainless steel is presented. The morphology of the weld was revealed by visual inspection. The microstructure and interlayer were characterized by optical and electron microscopy. The phases and their distribution were identified by microhardness, which was compared with data available in the literature. The EDS/XRD analysis revealed the composition of the intermetallic compounds.

The overall result showed an effective bond between aluminum and steel, if complex interactions are controlled and limited to a small amount.

2. Experimental Procedures

2.1. Material Properties and Weld Configuration

In this study, the butt weld geometry was used for testing the weldability of a dissimilar Al–Fe weld. The dimensions of the sheets (length × width × thickness) were $100 \times 50 \times 3$ mm^3 for the aluminum sheet and $100 \times 50 \times 2$ mm^3 for the steel one. The difference in plate thickness was chosen to improve the wettability of aluminum on the steel. 0.8 mm diameter AISI 316 steel filler wire was used for the hybrid laser-arc process. Tables 1 and 2 show the characteristics of the as-received materials.

Table 1. Chemical composition of the as-received materials (wt %).

Metal	C	Al	Cr	Mn	Mo	Mg	Ni	Ti	P	S	Si	Fe
AISI 316	0.08	-	18	2	3	-	14	-	0.045	0.03	1	balance
AA5754	-	balance	0.30	0.50	-	3.6	-	0.15	-	-	0.40	0.40

Table 2. Properties of the as-received materials: ultimate tensile strength (UTS), yield stress (YS), Young module (E), elongation to fracture %(A%), Vickers microhardness (HV), thermal conductivity (K), Liquidus Temperature (T_L), density (ρ), specific heat capacity (c).

Metal	UTS (MPa)	YS (MPa)	E (GPa)	A%	HV	K (W/(m·K))	T_L (°C)	ρ (g/cm^3)	c (J/(g·°C))
AISI 316	580	290	193	50	178	16.3	1400	8	0.5
AA5754	230	80	68	17	62	147	600	2.66	0.9

2.2. The Welding Procedures

The welds were produced using two different technologies: the laser offset welding (LOW) and the hybrid laser-MIG welding. When using LOW, the laser source was focused on the steel side at a certain distance (offset) from the bimetal interface (Figure 1). In this investigation, the off-set value was about 1 mm from the laser beam axis. The keyhole moved along a linear path, parallel to the interface. The heat transferred to the aluminum side through the steel heat affected zone, as shown in Figure 2. The thermal energy spreading from the keyhole caused the fusion of the aluminum. In this way, the steel fusion zone (FZ) separated the steel molten pool from the aluminum-fused zone, which avoided the excessive growth of the IMC layer.

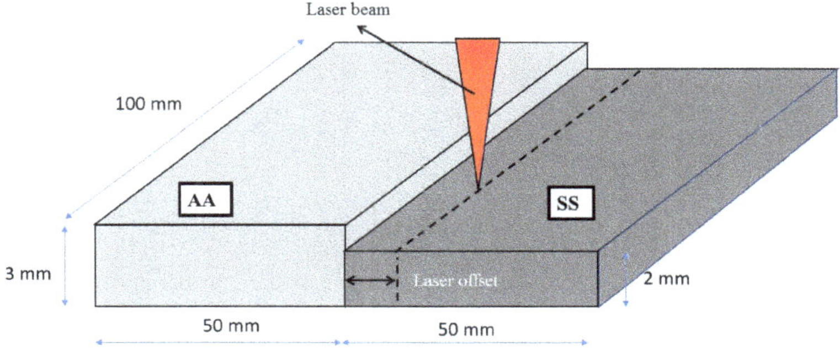

Figure 1. Schematic illustration of the laser offset welding (LOW) technique procedure.

Focusing the beam on the steel plate instead of the aluminum one is advantageous because it enables better process control. The stability of the keyhole depends on the balance of different force contributions, which include recoil pressure of the vaporized material, pressure inside the cavity, surface tension, pressure, and weight of molten metal. If the process is not largely robust and stable, small deviations of the process conditions from the design point, might compromise the stability of the keyhole, leading to collapse and uncontrolled liquid and thermal flows. Aluminum plate is highly reflective to beam radiation and exhibits high diffusivity, making it difficult to keep the process stable during the weld. Thus, even if the melting point of aluminum is lower than steel one, the process has been conducted by focusing the beam on the steel side, in opposition to brazing principles.

On the other hand, during laser-MIG hybrid welding (Figure 3) the laser-arc coupled source was directed straight to the weld centerline and it moved along that line. After preliminary trials, the wire was positioned at a 1 mm distance from the laser focus. The laser and arc combination promotes a stable wire deposition, without any spatter generation.

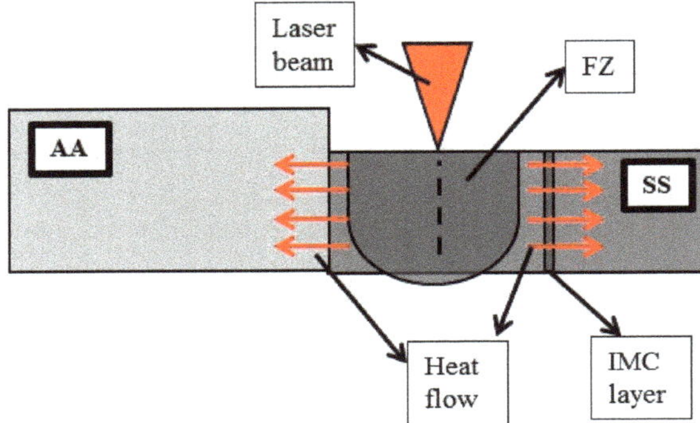

Figure 2. Schematic drawing for describing the joining mechanism by the LOW technique.

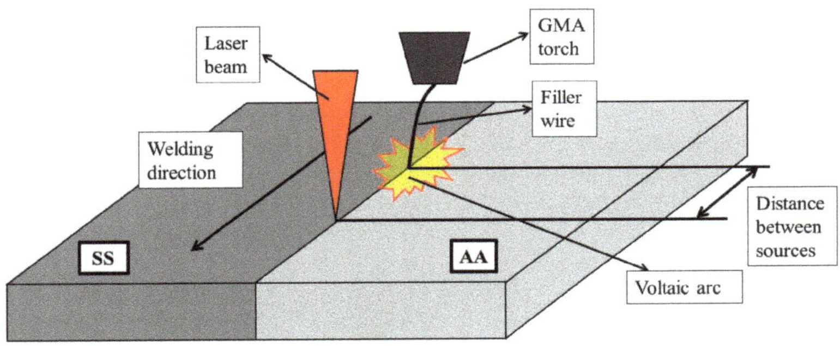

Figure 3. Schematic illustration of laser-MIG (Metal inert gas) technique procedure.

Before welding, the sheets were machined at low milling speed, which reduced the thermal contact resistance. Dust and contaminants were removed by cleaning with acetone.

2.3. Set-Up of the Welding Systems

Two different laser systems were used to perform the trials:

For laser offset welding, it consisted of a 6-axis robot, a stationary shielding box system and a workbench equipped with clamps and supporting table (Figure 4). A YLS-4000 Yb-doped fiber laser with a wavelength of 1070 nm and a maximum power of 4 kW (IPG Laser GmbH, Barbuch, Germany) was used in continuous wave regime. The fiber had 200 μm diameter, while the optics (focal lens and collimator) provided a magnification factor of 2, resulting in roughly a 0.4 mm beam diameter, which was calculated by the $1/e^2$ width method, near-Gaussian distribution.

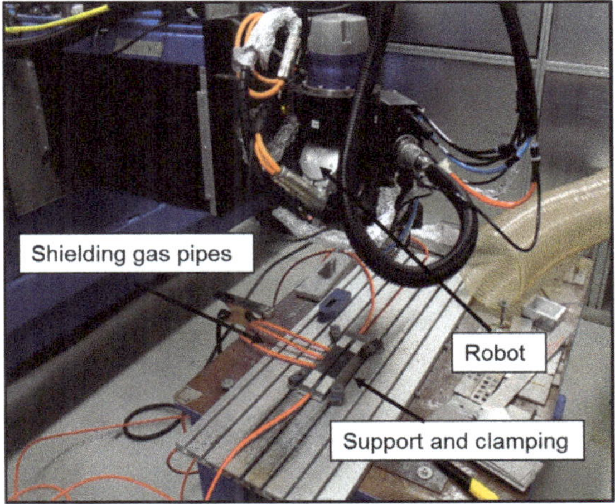

Figure 4. Overall experimental set-up.

For hybrid laser welding, a CO_2 laser (Rofin, Hamburg, Germany) whose maximal power was 3 kW was used operating in continuous wave mode. The focusing mirror was parabolic and it had a 160 mm radius. A laser beam coaxial argon shielding gas is used, which is advantageous for arc stability. Figure 5 shows the laser-MIG coupling.

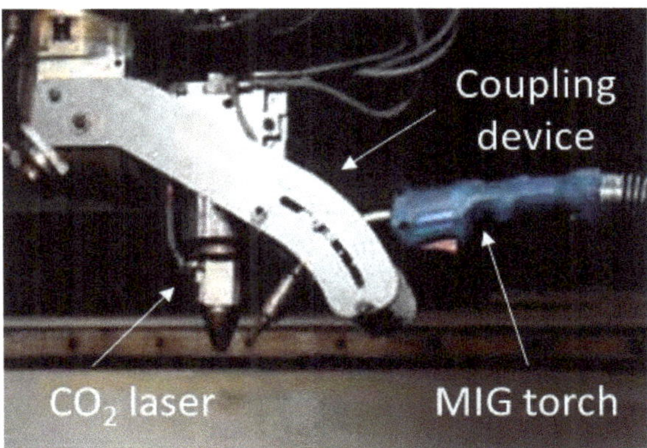

Figure 5. Laser-MIG coupling.

2.4. Process Parameters

Preliminary bead-on-plate tests were conducted to find out the process parameters that enabled the weld formation. The process parameters and their levels are summarized in Table 3. The overall power used for the hybrid welding (laser power + MIG power) was 3420 W (versus 2500 W used for laser welding).

Table 3. Experimental plan.

	Laser Power (kW)	Welding Speed (m/min)	Wire Feed Speed (m/min)	Current (A)	Voltage (V)
Laser Offset Welding	2.5	2	-	-	-
Hybrid Laser-MIG Welding	3.42	2	1.2	80	24

2.5. Metallographic Analysis and Mechanical Testing

Weld cross sections were cold mounted and then they were grinded and mechanically polished using a variable speed. The microstructure was analyzed by Epiphot 200 Optical Microscope (OM, Nikon, Tokyo, Japan) and EVO scanning electron microscope (SEM, Zeiss, Oberkochen, Germany), equipped with an energy-dispersive X-ray spectrometer (EDS, Bruker AXS Inc, Madison, WI, USA). The samples were prepared by a standard metallographic procedure, which involved etching with the following reagents:

- Keller's solution (1 mL HF, 1.5 mL HCl, 2.5 mL HNO_3, and 95 mL H_2O) for aluminum microstructure.
- Vilella's solution (1 g picric acid, 5 mL HCl, 100 mL ethanol) for steel microstructure.

Vickers micro-hardness tests with a load of 0.1 Kg (AffriWiky 200JS2) were carried out to estimate local mechanical properties of welds and intermetallic phases created at the steel/aluminum interface.

X-ray diffraction measurements were carried out on a Rigaku diffractometer with $CuK\alpha$ radiation ($\lambda = 0.154$ nm). The X-ray diffraction data were collected at a scanning rate of $0.02°/s$ in 2θ ranging from $20°$ to $100°$ with count time 1.0 s in the fusion zone of both the aluminum and steel sheet. In the thin intermetallic layer, the X-ray diffraction data were collected at a scanning rate of $0.02°/s$ in 2θ ranging from $20°$ to $55°$ with count time 6.0 s.

3. Base Material Characterization

AA5754 Al–Mg alloy was supplied in annealed and recrystallized state. The optical micrograph (Figure 6) shows the aluminum matrix (solid solution phase) together with a series of intermetallic precipitates. Based on previous works [35–37], it can be concluded that the acicular shape, light gray particles are $(Fe,Mn)Al_6$ (Figure 6), while the rounded shape dark gray particles consist of fragile $(Fe,Mn)_3SiAl_{12}$ (Figure 6). The larger black particles are Mg_2Si (Figure 6), while the smaller ones are Mg_2Al_3 [35,36].

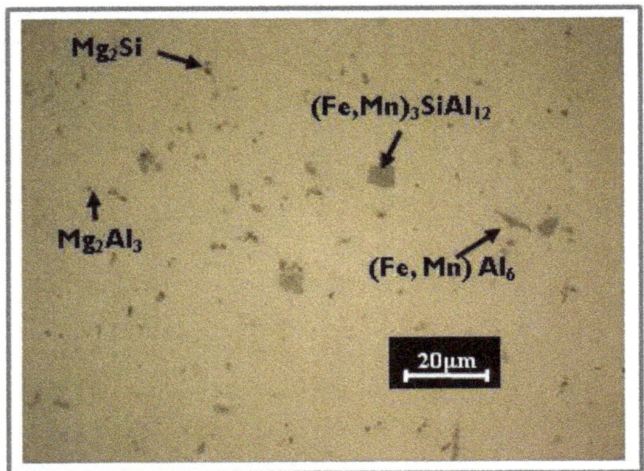

Figure 6. Optical micrograph of AA5754 base material. Light gray acicular shape particles $(Fe,Mn)Al_6$, rounded shape dark gray particles $(Fe,Mn)_3SiAl_{12}$, black particles Mg_2Si, the smaller ones are Mg_2Al_3.

The microstructure of the 316L base material is presented in Figure 7, showing an equiaxed, twinned microstructure. Annealing twins (induced by heat treatment) and deformation twins are typical of austenitic stainless steels, which are characterized by low stacking fault energy (SFE). The low SFE austenitic steel induces a planar array of dislocations during the deformation, promoting deformation twinning. The twin boundaries are barriers to the dislocation slipping, which increases the strain-hardening rate [36].

Figure 7. Polarized light micrograph base material of 316L stainless steel showing austenitic structure with twins.

4. Laser Off-Set Welding Results

The morphology of the cross section of the joint produced by laser offset welding is shown in Figure 8. At this stage, the sample was weakly etched, which permitted focusing the analysis on the bead shape. Detailed microstructural features were resolved later.

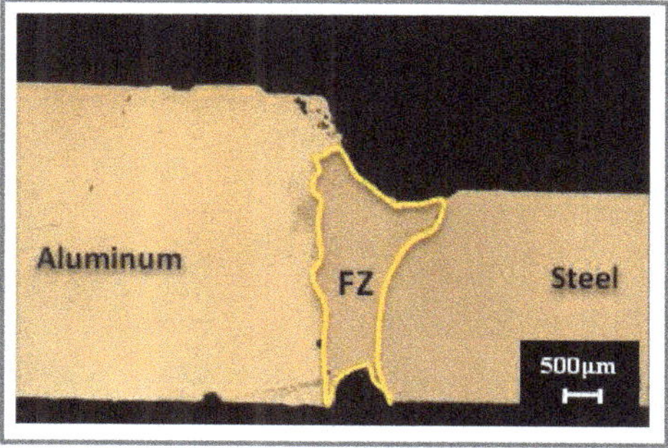

Figure 8. Optical macrograph of cross section obtained by LOW technique.

Full penetration was achieved. The undercut at the top surface was prevented, while the bottom part of the weld exhibited a slight sagging. Such a geometric defect mainly derived from the contraction of liquid walls during the solidification and recoil pressure inside the keyhole, which made the molten material fill the volume of the cavity as the source advanced.

Figure 9a,b shows the IMC layer formed by the Fe–Al reaction. The interface appeared curvilinear and Fe-based isles were distributed non-uniformly within the thickness. The low thermal input promoted a slight interaction between liquid-state materials. A magnification of zone A is shown in Figure 9b.

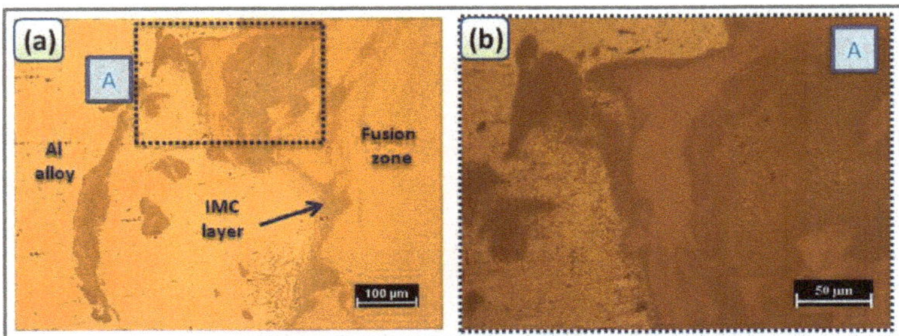

Figure 9. Optical micrographs of the intermetallic compounds (IMC) layer of the joint at: (**a**) 200×; (**b**) high-magnification image of area A in (**a**) at 500×.

The IMCs formation is diffusion controlled and dependent on time and thermal cycles. The key challenge in joining dissimilar materials is the accurate control of the fusion behaviour and the mixing of interfacing materials. Focusing the beam on the steel side confined the interaction between liquid phases into a narrow area. The keyhole was kept stable within the steel side, without affecting the interface zone. Firstly, the aluminum side was not subjected to the direct exposure to laser emission. Therefore, neither vaporization of alloying elements, nor liquid viscous flows towards the interface was observed. Secondly, since the beam was focused far enough from the interface, large liquid viscous forces were prevented. Thus, the growth of IMCs was limited and liquid flows were confined by the interface boundary, without creating any excessively large mixed zone. Laser offset welding significantly reduced the content of cracks and promoted a narrower intermetallic layer, which was limited to roughly 6 µm. (Figure 10). Moreover, the rapid process speed lead to a high cooling rate, enabling a narrower fusion zone [37]. Consequently, the shorter interaction time and narrower fusion area promoted a thinner IMC layer. Such a result is hugely beneficial for the joint strength [30].

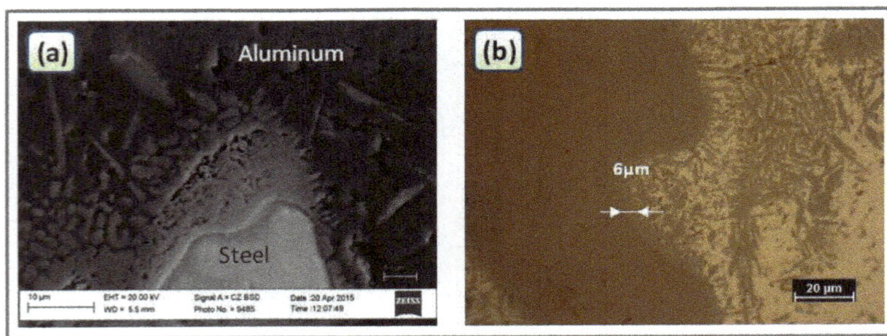

Figure 10. (**a**) Scanning electron microscopy (SEM) and (**b**) optical micrographs showing the IMC layer.

Firstly, the composition of the IMC layer was analyzed by the energy dispersive spectroscopy (EDS) analysis during the SEM investigation (Figures 11 and 12). Figure 11 shows the maps of elements at the interface (Figure 11b,c), meanwhile Table 4 gives the composition (at %) of Fe and Al in the points P1 (layer_1) and P2 (layer_2), which were located in Figure 12. The analysis of the maps in Figure 11b,c indicated that the light gray areas in Figure 11a contained mainly iron and aluminum, while the fused zones contained iron and aluminum together with their alloy elements. Particularly, the effect of the diffusion of Al alloy elements towards the steel fusion zone and Fe alloy towards the aluminum side can be observed. The local chemical analysis in Figure 12 revealed that the chemical composition of the IMC layer in the investigated points could be Fe_2Al_5- or $FeAl_2$-type according to Fe–Al phase diagram [5].

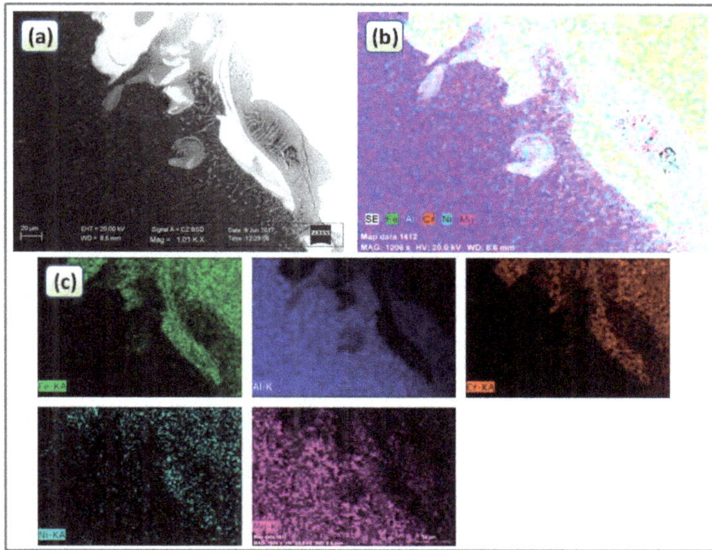

Figure 11. (**a**) SEM micrograph (Back scattered electrons) showing the IMC layer (**b**) alloy elements mapping and (**c**) maps of single elements at the joint interface.

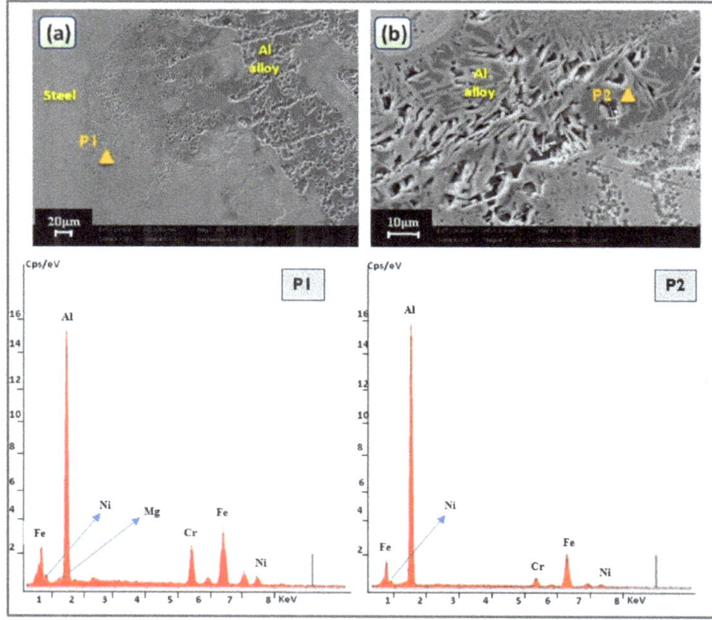

Figure 12. SEM micrographs of the IMC zone of the cross section with energy dispersive spectroscopy (EDS) analysis results in the highlighted points P1, P2 at (**a**) 5000× and (**b**) 10,000× magnifications.

Table 4. Chemical compositions at different points near aluminum/fusion zone interface (at %).

Point No.	Al	Fe	Cr	Ni	Mg
P1	63.9	26.2	5.8	3.3	0.8
P2	71.37	23.43	3.6	1.6	-

As shown in the Fe–Al binary phase diagram [5], primarily five types of Fe–Al IMCs (i.e., Fe_3Al, FeAl, $FeAl_2$, Fe_2Al_5, and $FeAl_3$ phases) are produced during the Fe/Al reaction process [38]. The sequence of the formation of Fe–Al IMCs based on the thermodynamic data of the free energy indicates that $\Delta G°$ (Fe_2Al_5) < $\Delta G°$ ($FeAl_3$) < $\Delta G°$ ($FeAl_2$) < $\Delta G°$ (FeAl) < 0 < $\Delta G°$ (Fe_3Al) [38–41], suggesting that the first phase to be formed is Fe_2Al_5.

The nature of the IMCs compounds in the joint was assessed by XRD analysis (Figure 13a). Because the stoichiometry of the Fe_2Al_5, $FeAl_3$, and $FeAl_2$ compounds is very close each other, it is not possible to distinguish them by EDS chemical analysis. Particularly, in the intermetallic area, a precision X-ray diffraction analysis was employed to try and identify the compounds in the thinner intermetallic layer. Both the fusion zone of aluminum and steel was analyzed by XRD (Figure 13b,c). In the steel fusion zone were detected the diffraction peaks of Aluminum and Iron, meanwhile in the Al zone only the aluminum peak were detected. The XRD spectrum in the intermetallic zone identified only the $FeAl_2$ compound together with the Al matrix, while no trace of more stable compounds was found. The reason could be the low amount of the more stable compounds (i.e., Fe_2Al_5). In fact, the joint solidification is a non-equilibrium process, which is characterized by high welding speed (2 m/min) and cooling rate. Therefore, the compounds' crystallization could be not strictly in accordance with thermodynamic condition. So, if thermodynamically more stable compounds (such as Fe_2Al_5), being diffusion controlled, do not have enough time to grow during the cooling of the joint, some other less stable compounds (such as $FeAl_2$) could nucleate and grow preferentially [42]. Vickers micro-hardness measurements on the intermetallic interface layer (Figure 14) were in accordance with the values reported in the literatures for FeAl or Fe_2Al_5 (Figure 14a) compounds [1,3]. In such a case, the values of hardness could be underestimated because of the Al matrix (Figure 14b).

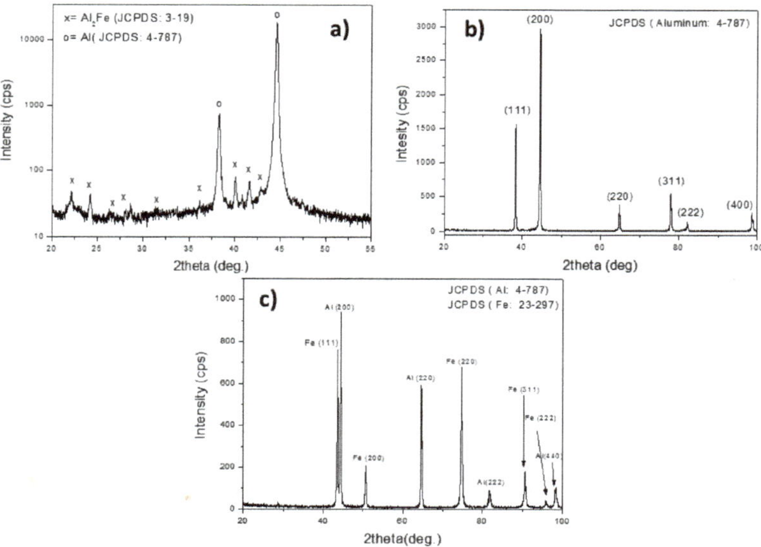

Figure 13. (**a**) Precision X-ray diffraction analysis in the thinner intermetallic layer. X-ray diffraction analysis in the (**b**) aluminum and (**c**) steel fusion zone.

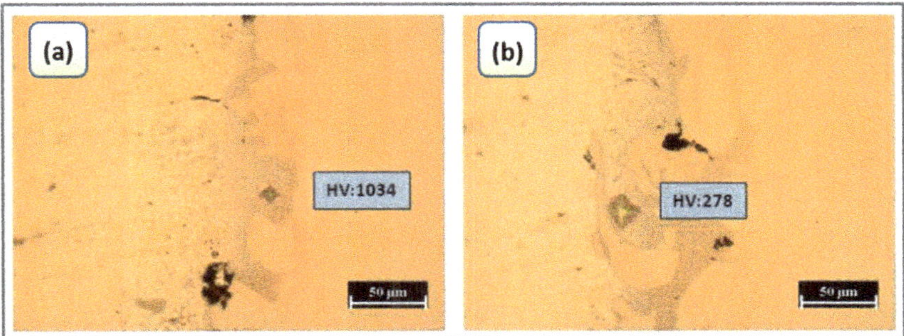

Figure 14. Optical micrographs of joint s1 shows the result of Vickers micro-hardness measurements on the interface aluminum/steel: (**a**) hardness value due to IMC (**b**) hardness value due to effects of the IMC and Al matrices.

Figure 15 represents the microhardness at the half thickness of the cross section. The hardness of the Al FZ was higher than that in the heat affected zone (HAZ) and the base material (BM). Rapid solidification, and therefore both grain refining and low grain boundaries' precipitation, increased the hardness in the FZ. The hardness of the Al HAZ was slightly larger than that in the base material. This result may be due to the dissolution of soluble compounds and the consequent strengthening by solid solution. Figure 15 shows that an increase of hardness in the steel was due to grain refinement promoted during the welding process [43,44].

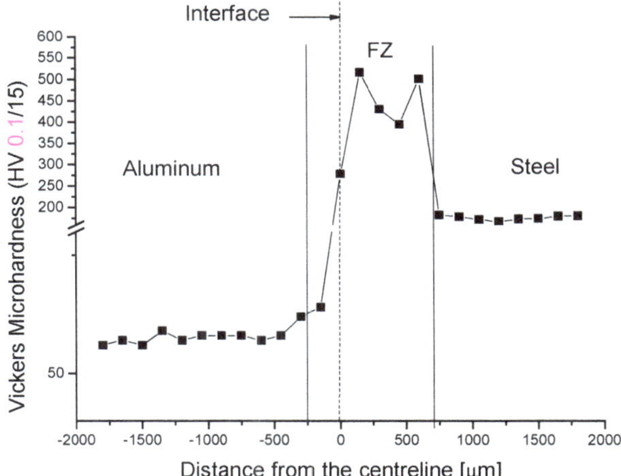

Figure 15. Cross weld section microhardness profile for the laser offset welding.

Microhardness was high in the fusion zone due to the finer grain size (Figure 16) The average size of the grains in the BM steel was 30 ± 5.7 µm, meanwhile in the FZ (at both the interface with steel HAZ and Al FZ) it was equal to 6 ± 1.2 µm. In the literature, there are several works on the laser welding steel/aluminum which showed an increased hardness due to the refinement of the grain size [44,45].

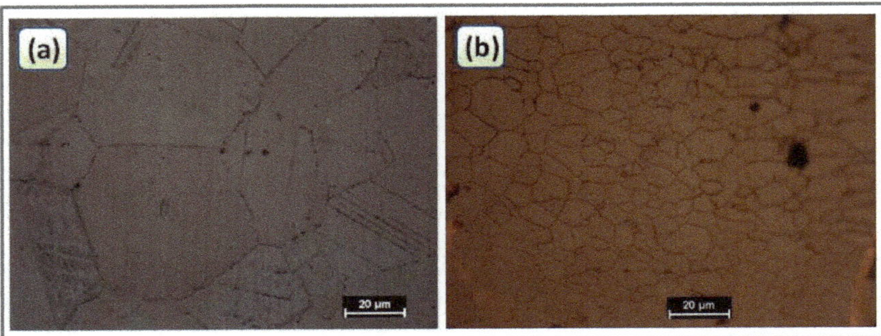

Figure 16. Optical micrographs showing (**a**) base material of 316L stainless steel and (**b**) fine structure in the fusion zone due to the fast solidification.

Grain boundary precipitation on the aluminum side did not occur during the joint solidification because of the rapid cooling. If grain boundary precipitation occurred, the hardness of AA5754 would strongly decrease. The precipitation of Al–Mg particles inside the grains leads to softening [27,36].

5. Laser Hybrid Welding Results

Figure 17 shows the cross section of the joint produced by hybrid laser-MIG welding. Excessive weld metal was observed at the top surface, while the bottom part presented a slight lack of penetration. The reason why these geometric defects occurred can be explained by assessing the process dynamics. Excessive weld metal resulted from the high wire deposition rate. Lower wire feeding speeds were adopted to enhance the geometric outcome and reduce the defectiveness. Anyhow, reducing the deposition rate of filler metal must correspond to a reduction of the heat input to keep the process energy balance and avoid wire overheating. Then, several experiments were performed with lower values of total power and wire feeding speed. The laser power was kept constant, because it is mainly responsible for penetration. However, even if the laser power was kept constant, the reduction of the MIG power had a detrimental effect on the geometry of the joint, since the amount of total energy was not enough to fully penetrate the sheets' thickness and generate a consistent bond. Thus, the most satisfactory outcome was evaluated for the present analysis (see Table 3). As shown in Figure 17, a good compromise between penetration and excessive weld metal was found. A slight lack of weld penetration was observed at the bottom part (less than 0.3 mm depth), while the excessive weld metal was limited to 0.6 mm.

As stated above, the stability of the process is dependent on the keyhole dynamics. The balance over different force contribution is needed to sustain the plasma inside the cavity in stable conditions. Either higher pressure of vapor gases or excessive viscous action of liquid walls might lead to keyhole collapse. Since the beam was focused at the interfaces between two dissimilar materials, such a condition was critical for keyhole stability in hybrid welding, which suffered from additional forces gradients. The risk of a collapsing keyhole is less significant in the fiber laser since the absorptivity by metal gas vapors is much less.

However, the significant difference in liquid metal viscosity, thermal properties, and surface tension compromised the process dynamics, leading to the collapse of the keyhole and consequent entrapment of gas bubbles. Because of the rapid contraction of liquid walls and the rapid solidification rate, gas inclusions did not have enough time and energy to escape from the weld pool [46]. Thus, macro voids formed in the fusion zone, and their direction was determined by the viscous metal flows during keyhole collapse.

Because of the significant difference in thermal expansion coefficient between the two metals and the brittleness of IMCs structures, solidification cracks formed at the interface.

A huge amount of thermal energy was directly provided at the interface between the sheets. Thus, large viscous forces of molten metals were generated. The boundary between the two metals was highly irregular and non-homogeneous. The behavior of the metal at the interface was not governed by controllable thermal gradients. Liquid flows were uncontrolled, since the distribution of viscous forces within the thickness was not scientifically predictable. The hydrodynamic pressure of the molten steel at the top part of the weld had enough energy to penetrate the aluminum substrate (see Figures 17 and 18), while a large volume of liquid aluminum was pulled down by gravity at the bottom side.

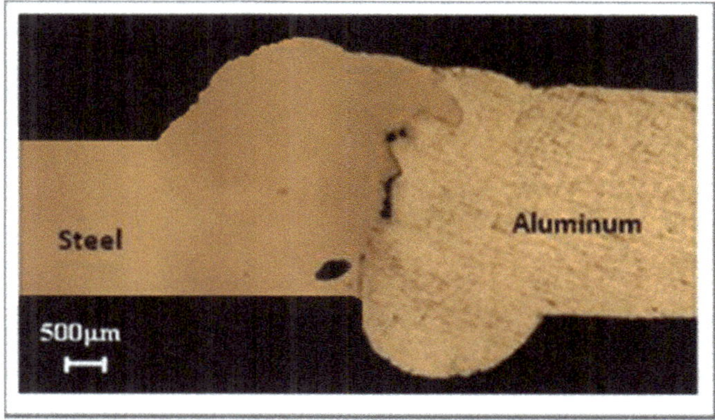

Figure 17. Optical macrograph of cross section by hybrid laser-MIG welding.

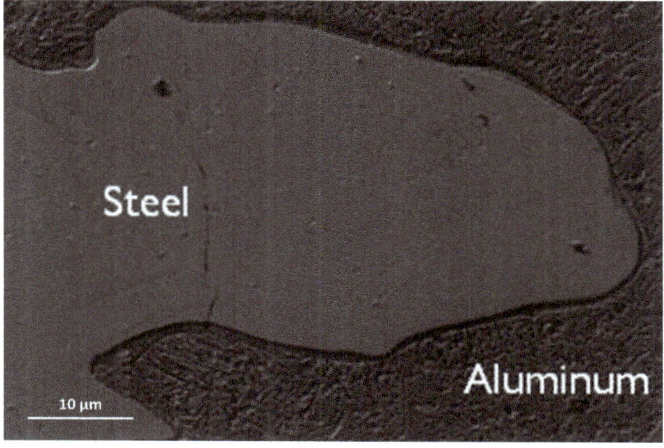

Figure 18. SEM picture showing the highly irregular and non-homogeneous Steel/Al interface.

The geometry of the IMC layer was inspected by electron microscopy. Figure 19a,b shows that the width exceeded the value of 14 μm. The process control did not benefit from the hybrid laser-arc combination. The arc promoted a lower cooling rate and non-uniform energy distribution within the thickness. Longer reaction time for IMC growth occurred. As mentioned above, such a result is undesirable for tensile properties [17]. Micro-cracks were detected at both the upper and lower side of the interface. IMC structures at the interface are hugely brittle and no plastic behavior was detected, because of the nearly-zero dislocation mobility. During solidification, local stresses exceeded the elastic limit, leading to small fractures both perpendicular and parallel to the layer width.

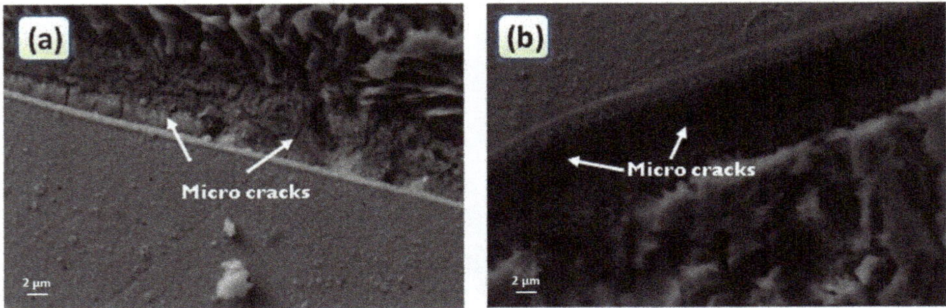

Figure 19. SEM pictures show: (**a**) Close-up of the upper side of the interface of the sample (5000×); (**b**) Close-up of the lower side of the interface of the sample (6000×).

Figure 20 shows the micro-hardness profile detected from the cross section of the sample. Adding arc increased the thermal energy and volume of molten metal compared to single laser welding. Therefore, even if the welding speed was the same as the autogenous laser process, the thermal inertia of the fusion zone was higher and the cooling rate was attenuated, leading to grain size widening and lower hardness values in the FZ.

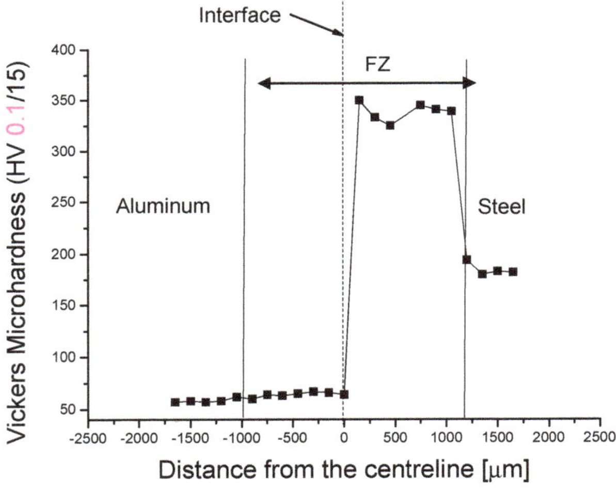

Figure 20. Cross weld section microhardness profile for the hybrid laser-MIG.

The indentation presented in Figure 21 refers to the aluminum/steel interface and reports on a micro-hardness value of 850 Vickers. The IMC area resulted much harder than single-alloy fusion zones and presented a high gradient of thermal expansion coefficient. Therefore, this area is locally subjected to compressive and tensile stresses during the non-uniform solidification, which potentially enables crack propagation.

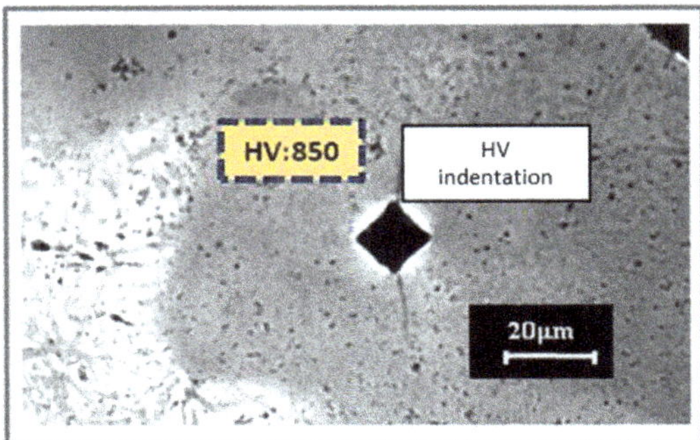

Figure 21. Indentation of the intermetallic layer.

6. Conclusions

The present work reported the characterization of the microstructure and the intermetallic compounds (IMCs) in Fe–Al dissimilar welds obtained by fiber off-set and hybrid laser welding. The following considerations were pointed out:

- The EDS/XRD analysis revealed the presence of $FeAl_2$ in the laser welded joint.
- Full penetration and low defectiveness were obtained by laser offset welding. The interaction between liquid phases was restricted. Viscous forces were attenuated by optimizing the process energy balance. Moreover, the high cooling rate and low mix between the two metals enabled IMC layer growth, which was as thick as 6 μm. As stated in the literature, a thin IMC layer improves the mechanical properties of the weld. Brittle phases were detected but hot cracks were avoided.
- Hybrid laser-arc welding resulted less effective. In fact, an excessive weld crown was observed and the weld presented a lack of penetration. The process was instable because of the significant difference in thermal- and fluid-dynamic properties of the two metals, which compromised the keyhole stability. The interface was highly irregular and non-homogeneous, due to the action of the viscous forces.

It can be concluded that LOW effects on Al–Fe dissimilar metallurgy and IMC can be controlled better than those of the hybrid laser-arc welding. Therefore, further investigation on the weldability of Al–Fe dissimilar welds will focus on the LOW welding technique.

Author Contributions: Giuseppe Casalino and Michelangelo Mortello dealt with the fabrication of the welds, Paola Leo and Patrizias Perulli handled the microstructure analisys and performed the X-ray analysis, Alessandra Varone advised on the interpretation of the X-ray spectra.

References

1. Yang, J.; Long, L.Y.; Zhang, H. Microstructure and mechanical properties of pulsed laser welded Al/steel dissimilar joint. *Trans. Nonferrous Met. Soc. China* **2016**, *26*, 994–1002. [CrossRef]
2. Fridlyander, I.N.; Sister, V.G.; Grushko, O.E.; Berstenev, V.; Sheveleva, L.M.; Ivanova, L.A. Aluminium alloys: Promising materials in the automotive industry. *Met. Sci. Heat Treat.* **2002**, *44*, 365–370. [CrossRef]
3. Torkamany, M.J.; Tahamtan, S.; Sabbaghzadeh, J. Dissimilar welding of carbon steel to 5754 aluminum alloy by Nd:YAG pulsed laser. *Mater. Des.* **2010**, *31*, 458–465. [CrossRef]
4. Chen, S.; Zhai, Z.; Huang, J.; Zhao, X.; Xiong, J. Interface microstructure and fracture behavior of single/dual-beam laser welded steel-Al dissimilar joint produced with copper interlayer. *Int. J. Adv. Manuf. Technol.* **2016**, *82*, 631–643. [CrossRef]

5. Hiroaki, O.; Schlesinger, M.E.; Mueller, E.M. *Alloy Phase Diagrams*, 3rd ed.; ASM International: Geauga County, OH, USA, 1992; ISBN 978-1-62708-070-5.

6. Çam, G.; Erim, S.; Yeni, Ç.; Koçak, M. Determination of mechanical and fracture properties of laser beam welded steel joints. *Weld. J.* **1999**, *78*, 193–201.

7. Çam, G.; Yeni, Ç.; Erim, S.; Ventzke, V.; Koçak, M. Investigation into properties of laser welded similar and dissimilar steel joints. *Sci. Technol. Weld. Join.* **1998**, *3*, 177–189. [CrossRef]

8. Mvola, B.; Kah, P.; Martikainen, J.; Suoranta, R. Dissimilar high-strength steels: Fusion welded joints, mismatches, and challenges. *Rev. Adv. Mater. Sci.* **2016**, *44*, 146–159.

9. Liu, X.; Lan, S.; Ni, J. Analysis of process parameters effects on friction stir welding of dissimilar aluminum alloy to advanced high strength steel. *Mater. Des.* **2014**, *59*, 50–62. [CrossRef]

10. Movahedi, M.; Kokabi, A.H.; Reihani, S.S.; Cheng, W.-J.; Wang, C.J. Effect of annealing treatment on joint strength of aluminum/steel friction stir lap weld. *Mater. Des.* **2013**, *44*, 487–492. [CrossRef]

11. Habibnia, M.; Shakeri, M.; Nourouzi, S.; Givi, M.B. Microstructural and mechanical properties of friction stir welded 5050 Al alloy and 304 stainless steel plates. *Int. J. Adv. Manuf. Technol.* **2015**, *76*, 819–829. [CrossRef]

12. Sahin, M. Joining of stainless-steel and aluminium materials by friction welding. *Int. J. Adv. Manuf. Technol.* **2009**, *41*, 487–497. [CrossRef]

13. Campanelli, S.L.; Casalino, G.; Casavola, C.; Moramarco, V. Analysis and comparison of friction stir welding and laser assisted friction stir welding of aluminum alloy. *Materials* **2013**, *6*, 5923–5941. [CrossRef]

14. Ratanathavorn, W.; Melander, A.; Magnusson, H. Intermetallic compounds in friction stirred lap joints between AA5754/galvanised ultra-high strength steel. *Sci. Technol. Weld. Join.* **2016**, *21*, 653–659. [CrossRef]

15. Oikawa, H.; Ohimiya, S.; Yoshimura, T.; Saitoh, T. Resistance spot welding of steel and aluminum sheet using insert metal sheet. *Sci. Technol. Weld. Join.* **1999**, *4*, 80–88. [CrossRef]

16. Qin, G.; Su, Y.H.; Meng, X.M.; Fu, B.L. Numerical simulation on MIG arc brazing-fusion welding of aluminum alloy to galvanized steel plate. *Int. J. Adv. Manuf. Technol.* **2015**, *78*, 1917–1925. [CrossRef]

17. Gao, M.; Chen, C.; Mei, S.W.; Wang, L.; Zeng, X.Y. Parameter optimization and mechanism of laser–arc hybrid welding of dissimilar Al alloy and stainless steel. *Int. J. Adv. Manuf. Technol.* **2014**, *74*, 199–208. [CrossRef]

18. Çam, G. Friction stir welded structural materials: Beyond Al alloys. *Int. Mater. Rev.* **2011**, *56*, 1–48. [CrossRef]

19. Chen, S.; Huang, J.; Ma, K.; Zhang, H.; Zhao, X. Influence of a Ni-foil interlayer on Fe/Al dissimilar joint by laser penetration welding. *Mater. Lett.* **2012**, *79*, 296–299. [CrossRef]

20. Mathieu, A.; Pontevicci, S.; Viala, J.; Cicala, E.; Matteï, S.; Grevey, D. Laser brazing of a steel/aluminum assembly with hot filler wire (88% Al, 12% Si). *Mater. Sci. Eng. A* **2006**, *435*, 19–28. [CrossRef]

21. Dharmendra, C.; Rao, K.; Wilden, J.; Reich, S. Study on laser welding-brazing of zinc coated steel to aluminum alloy with a zinc based filler. *Mater. Sci. Eng.* **2011**, *528*, 1497–1503. [CrossRef]

22. Song, J.; Lin, S.; Yang, C.; Ma, G.; Liu, H. Spreading behavior and microstructure characteristics of dissimilar metals TIG welding brazing of aluminum alloy to stainless steel. *Mater. Sci. Eng.* **2009**, *509*, 31–40. [CrossRef]

23. Choi, C.; Kim, D.; Nam, D.; Kim, Y.; Park, Y. A hybrid joining technology for aluminum/zinc coated steels in vehicles. *J. Mater. Sci. Technol.* **2010**, *26*, 858–864. [CrossRef]

24. Zhang, K.; Zhenglong, L.; Chen, Y.; Liu, M.; Liu, Y. Microstructure characteristics and mechanical properties of laser-TIG hybrid welded dissimilar joints of Ti–22Al–27Nb and TA15. *Opt. Laser Technol.* **2015**, *73*, 139–145. [CrossRef]

25. Thomy, C.; Vollertsen, F. Laser-Mig Hybrid Welding of Aluminium to Steel Effect of Process Parameters on Joint Properties. *Weld. World* **2012**, *56*, 124–132. [CrossRef]

26. Leo, P.; Renna, G.; Casalino, G.; Olabi, A.G. Effect of power distribution on the weld quality during hybrid laser welding of an Al–Mg alloy. *Opt. Laser Technol.* **2015**, *73*, 118–126. [CrossRef]

27. Casalino, G.; Campanelli, S.L.; Ludovico, A.D. Laser-arc hybrid welding of wrought to selective laser molten stainless steel. *Int. J. Adv. Manuf. Technol.* **2013**, *68*, 209–216. [CrossRef]

28. Zeng, Z.; Li, X.; Miao, Y.; Wu, G.; Zhao, Z. Numerical and experiment analysis of residual stress on magnesium alloy and steel butt joint by hybrid laser-TIG welding. *Comput. Mater. Sci.* **2011**, *50*, 1763–1769. [CrossRef]

29. Casalino, G.; Mortello, M.; Peyre, P. Yb–YAG laser offset welding of AA5754 and T40 butt joint. *J. Mater. Process. Technol.* **2015**, *223*, 139–149. [CrossRef]

30. Rathod, M.J.; Kustuna, M. Joining of aluminum alloy 5052 and low carbon steel by laser roll welding. *Weld. J. Res. Suppl.* **2004**, *83*, 16–26.

31. Casalino, G.; Mortello, M. Modeling and experimental analysis of fiber laser offset welding of Al-Ti butt joints. *Int. J. Adv. Manuf. Technol.* **2016**, *83*, 89–98. [CrossRef]

32. Casalino, G.; Guglielmi, P.; Lorusso, V.D.; Mortello, M.; Peyre, P.; Sorgente, D. Laser offset welding of AZ31B magnesium alloy to 316 stainless steel. *J. Mater. Process. Technol.* **2017**, *242*, 49–59. [CrossRef]

33. Miranda, R.M.; Assunção, E.; Silva, R.J.C.; Oliveira, J.P.; Quintino, L. Fiber laser welding of NiTi to Ti-6Al-4V. *Int. J. Adv. Manuf. Technol.* **2015**, *81*, 1533–1538. [CrossRef]

34. Pardal, G.; Meco, S.; Ganguly, S.; Williams, S.; Prangnell, P. Dissimilar metal laser spot joining of steel to aluminium in conduction mode. *Int. J. Adv. Manuf. Technol.* **2014**, *73*, 365–373. [CrossRef]

35. Kumar, S.; Nadendla, H.B.; Scamans, G.M.; Eskin, D.G.; Fan, Z. Solidification behavior of an AA5754 alloy ingot cast with high impurity content. *Int. J. Mater. Res.* **2012**, *103*, 1228–1234. [CrossRef]

36. Smith, W.F. *Structure and Properties of Engineering Alloys*, 1st ed.; McGraw-Hill: New York, NY, USA, 1992; pp. 186–191. ISBN 0-07-59172-5.

37. Davis, J.R. *Aluminum and Aluminum Alloys*, 1st ed.; ASM International: Geauga County, OH, USA, 1993; ISBN 978-0-87170-496-2.

38. Miao, Y.G.; Han, D.F.; Yao, J.Z.; Li, F. Microstructure and interface characteristics of laser penetration brazed magnesium alloy and steel. *Sci. Technol. Weld. Join.* **2010**, *15*, 97–103. [CrossRef]

39. Kobayashi, S.; Yakou, T. Control of intermetallic compound layers at interface between steel and aluminum by diffusion-treatment. *Mater. Sci. Eng. A* **2002**, *338*, 44–53. [CrossRef]

40. Sun, J.; Yan, Q.; Gao, W.; Huang, J. Investigation of laser welding on butt joints of Al/steel dissimilar materials. *Mater. Des.* **2015**, *83*, 120–128. [CrossRef]

41. Qiu, R.; Shi, H.; Zhang, K.; Tu, Y.; Iwamoto, C.; Satonaka, S. Interfacial characterization of joint between mild steel and aluminum alloy welded by resistance spot welding. *Mater. Charact.* **2010**, *61*, 684–688. [CrossRef]

42. Messler, R.W. *Principles of Welding: Processes, Physics, Chemistry, and Metallurgy*, 1st ed.; Wiley-VCH: New York, NY, USA, 2007; pp. 406–423. ISBN 0-471-25376-6.

43. Brooks, J.; Lippold, J. *Welding, Brazing, and Soldering*, 1st ed.; ASM International: Geauga County, OH, USA, 1993; ISBN 978-0-87170-382-8.

44. Kumar, N.; Mukherjee, M.; Bandyopadhyay, A. Comparative study of pulsed Nd:YAG laser welding of AISI 304 and AISI 316 stainless steels. *Opt. Laser Technol.* **2017**, *88*, 24–39. [CrossRef]

45. Sathiya, P.; Mishra, M.K.; Shanmugarajan, B. Effect of shielding gases on microstructure and mechanical properties of super austenitic stainless steel by hybrid welding. *Mater. Des.* **2012**, *33*, 203–212. [CrossRef]

46. Cao, X.; Wallace, W.; Immarigeon, J.P.; Poon, C. Research and Progress in Laser Welding of Wrought Aluminum Alloys. II. Metallurgical Microstructures, Defects, and Mechanical Properties. *Mater. Manuf. Process.* **2003**, *18*, 23–49. [CrossRef]

Article

Microstructure and Mechanical Properties of Ti5553 Butt Welds Performed by LBW under Conduction Regime

Jose Maria Sánchez-Amaya [1,*], Timotius Pasang [2], Margarita Raquel Amaya-Vazquez [1], Juan De Dios Lopez-Castro [1], Cristina Churiaque [1], Yuan Tao [2] and Francisco Javier Botana Pedemonte [1]

[1]   Departmento de Ciencia de los Materiales e Ingeniería Metalúrgica y Química Inorgánica, LABCYP, Universidad de Cádiz. Escuela Superior de Ingeniería, Av. Universidad de Cádiz, 11519 Puerto Real, Cádiz, Spain; margarita.amaya@uca.es (M.R.A.-V.); juan.lopezcastro@uca.es (J.D.D.L.-C.); cristina.churiaque@uca.es (C.C.); javier.botana@uca.es (F.J.B.P.)

[2]   Department of Mechanical Engineering, AUT University, Auckland 1020, New Zealand; timotius.pasang@aut.ac.nz (T.P.); yuan.tao@aut.ac.nz (Y.T.)

*   Correspondence: josemaria.sanchez@uca.es; Tel.: +34-956-483-339

Received: 20 June 2017; Accepted: 10 July 2017; Published: 13 July 2017

Abstract: Ti-5Al-5V-5Mo-3Cr (Ti5553) is a metastable β titanium alloy with a high potential use in the aeronautic industry due to its high strength, excellent hardenability, fracture toughness and high fatigue resistance. However, recent research shows this alloy has a limited weldability. Different welding technologies have been applied in the literature to weld this alloy, such as electron beam welding (EBW), gas tungsten arc welding (GTAW) or laser beam welding (LBW) under keyhole regime. Thus, in tensile tests, joints normally break at the weld zones, the strength of the welds being always lower than that of the base metal. In the present work, a novel approach, based on the application of LBW under conduction regime (with a High-Power Diode Laser, HPDL), has been employed for the first time to weld this alloy. Microstructure, microhardness and strength of obtained welds were analyzed and reported in this paper. LBW under conduction regime (LBW-CR) leads to welds with slightly higher values of Ultimate Tensile Strength (UTS) than those previously obtained with other joining processes, probably due to the higher hardness of the fusion zone and to lower porosity of the weld.

Keywords: laser beam welding; conduction regime; Ti-5Al-5V-5Mo-3Cr

1. Introduction

Titanium and its alloys have been used commercially in various industries since the 1950s. The industries that have taken the benefit of titanium range from automotive, jewelry, sporting equipment, petrochemical, marine, aerospace and, more recently, biomedical industry. The reasons behind the successful applications of titanium alloys in many industries include their (i) relatively low density (vs. steels and superalloys), (ii) excellent corrosion resistance (vs. steels), (iii) moderate-to-high strength (vs. aluminum), (iv) ability to withstand up to 600 °C (better than aluminum), and (v) compatibility with the human body. Titanium alloys are generally divided into three main families, the so-called α, α + β and β alloys. A great advantage of β alloys is that they can be processed at lower temperatures than α + β alloys, and some heavily stabilized β alloys are even cold deformable.

Depending on their classification, weldability of titanium may range from poor to excellent. Although commercially pure titanium, α-titanium and α + β titanium generally have excellent weldability, metastable β titanium alloys may have limited weldability due to the high content of β stabilizing elements [1]. Metastable β titanium alloys have good weldability in the annealed or solution heat-treated conditions [1]. In general, in the as-welded (AW) condition, the weld fusion zone (FZ) is comprised of coarse columnar β grains from solidification while the heat-affected zone (HAZ) adjacent to the fusion lines is characterized by retained β structure. In this condition, the strength and hardness of these welds are low, but their ductility is reasonable. Becker and Baeslack [2] conducted weldability studies on three different types of metastable β titanium alloys (Ti-15V-3Cr-3Al-3Sn, Ti-8V-7Cr-3Al-4Sn-1Zr and Ti-8V-4Cr-2Mo-2Fe-3Al) and showed that the alloys are readily weldable. For all three alloys, the strengths were increased with post-weld heat treatment (PWHT) at, however, the expense of ductility. Weldability of Beta-21S sheet using laser welding technique was investigated by Liu et al. [3]. Both the FZ and HAZ were narrow with fine retained β grain structure. The FZ was "crown-shaped" (also referred in the literature as "hour glass-shaped") with wider top and bottom surfaces compared with the mid-thickness area. Epitaxial grain growth was observed to form from the HAZ into the FZ. The FZ had transitioned from a solidification mode of a cellular-type along the fusion boundary to a complete cellular-dendritic (or columnar-dendritic) solidification mode at the weld centerline [3]. Welding investigations on Beta-CTM (metastable β titanium alloy Ti-3Al-8V-6Cr-4Mo-4Zr, Ti 38-644) using Gas Tungsten Arc Welding (GTAW or TIG) showed epitaxial growth from the near-HAZ into the FZ, solidified with a cellular mode and progressively formed a complete columnar-dendritic grain structure at the weld centerline [4].

In the early 2000s, a new metastable β titanium alloy known as Ti-5Al-5V-5Mo-3Cr, designated as Ti5553, was introduced. It is a variation of the alloy VT22, being an alternative to the Ti-10-2-3 alloy. The advantage of this alloy as compared to other β titanium alloys is the sluggish precipitation kinetics of the α phase [5,6]. Ti5553 offers high strength, excellent hardenability, fracture toughness as well as high fatigue resistance [7]. The potential applications of this alloy are in the high-strength related areas such as landing gear and pylon/nacelle areas. Note that the landing gear beam truck of a Boeing 787 has been successfully manufactured using this alloy [8]. To find more applications in different areas, a number of factors have to be investigated, and one of them is the weldability.

The most common welding techniques to joint titanium and its alloys are Gas Metal Arc Welding (GMAW), such as Metal Inert Gas (MIG); Plasma Arc Welding (PAW); Laser Beam Welding (LBW); and Electron Beam Welding (EBW) [1,9–12]. The first three methods fall in the arc welding category with high heat input and low power density of heat source, while the last two techniques belong to the high-energy beam group.

LBW can be performed by two mechanisms: keyhole and conduction, the basic difference between two modes depending on the characteristics of the generated weld pool [13]. The key difference between these two operational regimes is the power density applied to the welding area. Conduction takes place when the energy intensity is not enough to provoke boiling, while in Keyhole, the intensity is sufficiently high to cause evaporation in the weld zone, creating a cavity (keyhole) in the melt pool [14]. A basic scheme of both operational modes is included in Figure 1. The energy density of the laser beam is higher in keyhole welding than in conduction welding, therefore leading to narrower weld beads. LBW under Keyhole Regime (LBW-KR) generates a cavity in the weld pool as a consequence of the metal evaporation. This cavity is stabilized by the pressure of the own vapor generated [15]. The keyhole is associated with violent plasma generation, which consists of metal vapor, ionized ions and electrons. The plasma resides both outside and inside the keyhole, known as the plasma plume and keyhole plasma, respectively [16]. The stability of the keyhole depends on a balance between surface tension pressure and vapor pressure: surface tension pressure tends to close the keyhole, while vaporization tends to keep it open [17,18]. Keyhole is reported to oscillate throughout the welding process and can lead to porosity defects in the final welds, especially in the case of partial penetration laser welding. Keyhole-induced porosity has been identified through experiments on the welding processes and is considered to be one of the major causes of porosity defects in laser welds. Large porosity defects can significantly deteriorate some of the mechanical properties of the welds, such as the tensile and fatigue strength [18]. Thus, Pang et al. [18] have developed a quantitative model able to predict the porosity defects induced by keyhole instability of titanium laser welding.

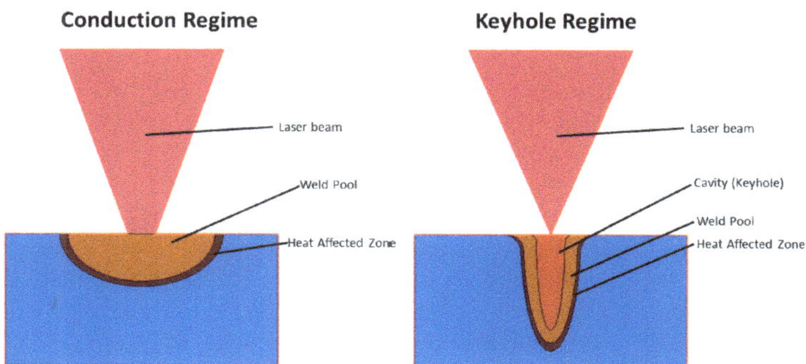

Figure 1. Schematic description of conduction and keyhole laser beam welding (LBW) regimes.

LBW under Keyhole Regime (LBW-KR) is more widely employed than LBW under Conduction Regime (LBW-CR), because it produces welds with high aspect ratios and narrow heat-affected zones [19,20]. On the other hand, LBW-CR is a more stable process, since metal vaporization occurs at a lower level than in LBW-KR [16,17,21]. In LBW-CR, the energy density of the laser beam locally heats the material up to a temperature above its melting point and below its boiling point [22]. Therefore, conduction welding offers an alternative joining mode for difficult-to-weld materials [13,23], such as the Ti5553 alloy. In fact, LBW-CR is being nowadays investigated for different purposes, such as to confirm the simulation results. Thus, simulation studies were performed to understand the weld pool flow patterns of titanium, showing that laser weld pool flow dynamics play a key role during the transition from conduction mode to keyhole [24]. A model is presented by Du et al. [25] for flow simulation of full penetration laser beam welding of titanium alloy, concluding that the molten pool becomes shorter and wider under the conduction regime, as a consequence of the Marangoni effect. More recently, a Finite Element Method (FEM) has been successfully developed to simulate the LBW-CR of Ti6Al4V alloy, employing for the first time a heat source with a double ellipsoid shape [26]. An excellent agreement between these simulations and experimental welding results was found, allowing the use of the refined FEM model to predict phase transformations, distortion and stresses of LBW of industrial titanium elements [26]. On the other hand, LBW-CR have been also employed to study the influence of some experimental variables on the welding process, as the superficial pre-treatments, concluding that chemical cleaning pre-treatments led to deeper and stronger welds than those treatments based on coatings [27]. Conduction regime also offers the possibility to perform laser heat treatments at localized zones of pieces, in order to provoke improvements of some properties, such as hardness [28,29], corrosion resistance [28] or tribocorrosion behavior [30,31].

In the present work, LBW-CR (with a High-Power Diode Laser, HPDL) has been applied for the first time to weld the Ti5553 alloy. Microstructure, microhardness and strength of obtained welds have been reported. The obtained results have been compared with the microstructure and properties of Ti5553 welds obtained in the literature with other joining techniques, such as GTAW, EBW and LBW under keyhole regime.

2. Materials and Methods

In this investigation, a metastable β titanium alloy, Ti-5Al-5V-5Mo-3Cr (Ti5553), was used. This alloy was provided by the Boeing Aircraft Company (Chicago, IL, USA). The alloy was in the annealed condition with a typical α/β microstructure. The chemical composition of the material is shown in Table 1.

Table 1. Chemical composition (wt %) of the Ti5553 alloy used in this study.

Material	O	N	Al	V	Mo	Cr	Fe	Ti
Ti5553	0.14	<0.01	5.03	5.10	5.06	2.64	0.38	Bal.

Autogenous (no filler), butt welding joints of Ti5553 were obtained by LBW-CR. The conduction welds were performed using a High-Power Diode Laser (HPDL) of LABCYP, UCA (Research Group of Corrosion and Protection of University of Cádiz, Spain), model DL028S, with 2.8 KW of maximum power acquired to ROFIN-SINAR (Plymouth, Mi, USA), shown in Figure 2. Joints were performed with treatments consisted of a linear laser scan of 35 mm, performed at a laser power of 2.75 kW and a welding speed of 5 mm/s, placing the surface of the samples at the focal distance (42 mm from the laser head). The spatial distribution of the laser beam source at the focal distance is included in Figure 3. The size of the spot of this laser source in this condition is 1.79 mm^2 (1.19 mm in X axe $\times$ 1.50 mm in Y axe). A laboratory-made conditioning chamber, shown in Figure 4, was employed to shield the welds, allowing the application of a continuous flow of argon of 10 L/min. The laser fluence (F) of these treatments was 46.21 kJ/cm^2, F being estimated as Equation (1):

$$F = \frac{P}{vD} \tag{1}$$

where F is the accumulated fluence (laser energy per area) of the laser treatment (46.21 kJ/cm^2), P is the Laser Power (2.75 kW), v is the processing rate (5 mm/s), and D is the spot width (1.19 mm, X axe).

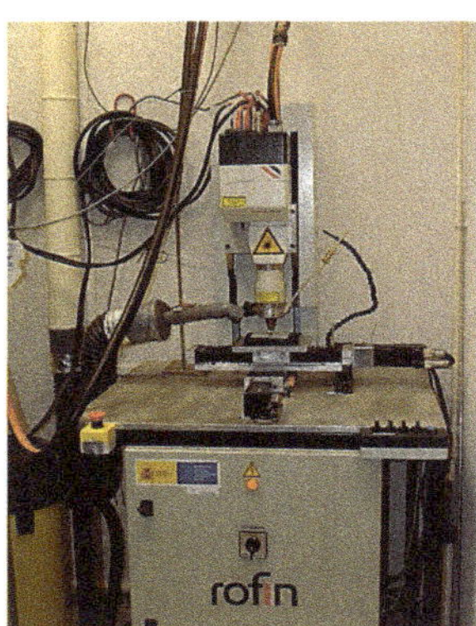

Figure 2. Equipment of High-Power Diode Laser (HPDL) employed to perform the welds under conduction regime.

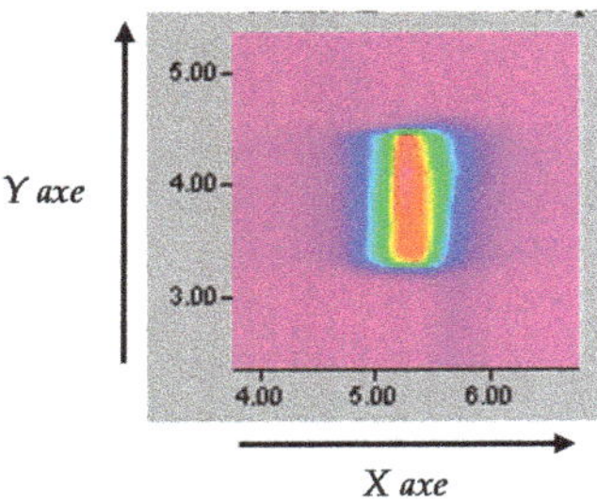

Figure 3. Spatial distribution of the laser beam source (HPDL) at the focal distance.

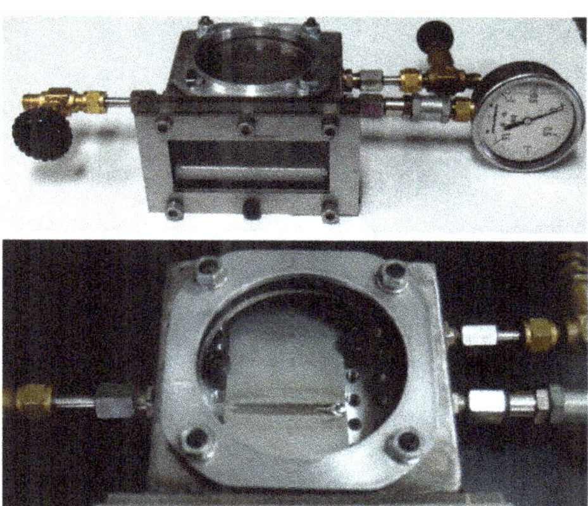

Figure 4. Laboratory-made conditioning chamber used to shield the welding samples.

The sizes of samples were $50 \times 50 \times 1.6$ mm^3 as shown in Figure 5. Before welding, the samples were cleaned according to ASTM B 600-91 standard. The welding direction was perpendicular to rolling direction. Metallographic samples for grain structure investigation and for hardness tests were prepared from the welded sheets (as shown in Figure 3). The metallographic preparation steps consisted of grinding from 120 to 2400 grit SiC paper, polishing to 0.3 μm colloidal alumina, followed by final polishing with 0.05 μm colloidal silica suspension. The samples were etched with Kroll's reagent with a composition of 100 mL water + 2 mL HF + 5 mL HNO$_3$. An optical microscope was used to characterize the microstructure of the welds.

Microhardness tests were carried out on the metallographically-prepared samples using a load of 300 g (HV300g) to produce the hardness profiles. Dog-bone flat samples (with a constant thickness of 1.6 mm) were taken from the welded sheets for tensile testing, as shown in Figure 5a, in accordance with ASTM E 8M-04, with the weld located perpendicular to the tensile axis. The tensile tests were conducted at room temperature, fixing a deformation speed of 0.005 mm/min at the elastic deformation regime, and 1.6 mm/min at the plastic deformation regime. Finally, the fractured surfaces of tensile samples were cleaned with an ultrasonic cleaner, and subsequently examined using a Scanning Electron Microscope (SEM) at different magnifications.

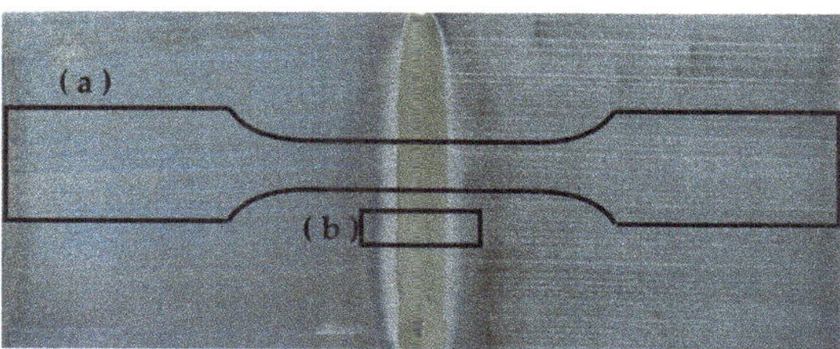

Figure 5. Schematic diagram indicating the weld joint and locations of samples for tensile tests (**a**) and metallographic and microhardness analysis (**b**).

The authors would like to clarify that welds were only made on 1.6 mm thick material. Future work needs to be performed to analyze the influence of the different material thickness and size on the developed microstructure and the laser fluence required to weld Ti5553 samples. This is a very important issue in LBW-CR, as the heat conduction can change when scaling the size of the welding samples.

3. Results and Discussion

Macrographic images of front and back views of Ti5553-Ti5553 butt weld performed with LBW-CR are shown in Figure 6a,b, respectively. The absence of colored tones confirms the appropriate protection provided by the conditioning shielding system. In general, LBW allows the minimization of thermal distortion. Note that the welds seem to present an apparent deformation, related to the relative disorientation of the mechanical machining lines of samples generated before welding. In fact, the obtained welds were flat, not showing evidences of thermal distortion.

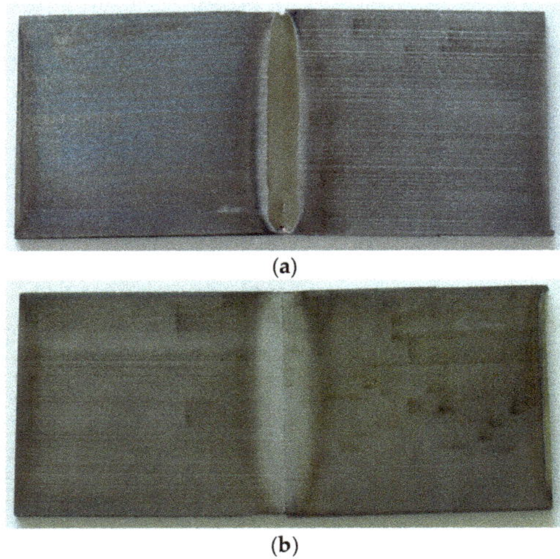

(a)

(b)

Figure 6. Front view (**a**) and back view (**b**) of Ti5553-Ti5553 butt joint performed with LBW under Conduction Regime (LBW-CR) using a HPDL.

3.1. Microstructure and Shape of Welds

For reference purposes, metallographic images of Ti5553 base metal (BM) are given in Figure 7a at 50× and Figure 5b at 1000×, showing α/β microstructure in which α particles with an average size of less than 5 μm are distributed within the β matrix. Cross-sections of Ti5553-Ti5553 LBW-CR butt welds were studied in detail, allowing the identification of the different zones of the weld (FZ, HAZ and BM), the analysis of the FZ morphology (shape) and the measurements of grain size at HAZ. Some metallographic images at low magnification (50×) of the welds are reported in Figure 8. The welds presented semicircle shape with relatively low depth/width ratio, allowing one to confirm that joints were performed under conduction regime. The shape of this LBW-CR weld can be compared with other Ti5553-Ti5553 butt welds previously obtained in the literature [31]. Thus, GTAW welds of this alloy presented V-shaped FZ, being also conduction-dominated. In contrast, the FZ for EBW and LBW performed with Nd:YAG are reported to provide hour glass-shaped welds, a morphology directly related in the literature with the keyhole regime.

Table 2 summarizes the width and depth of the FZ and HAZ of the LBW-CR welds obtained. For comparison purposes, the results obtained in recent research for Ti5553-Ti5553 butt welds performed with other technologies (GTAW, EBW and LBW-KR with Nd:YAG) [32] are also reported in the table. The widths of the FZ and the HAZ of the GTAW and LBW-CR welds were much wider compared with those of EBW and LBW-KR. The FZ generated by GTAW and LBW-CR were 5.4 and 4.9 mm, compared with 2.6 and 1.7 mm made by LBW-KR and EBW, respectively. The HAZ of the GTAW and LBW-CR samples reached up to 3 and 1.3 mm, respectively, being 0.8 mm for both

LBW-KR and EBW samples. The narrower weld zones in EBW and LBW-KR samples compared with the LBW-CR and GTAW samples are associated with the higher heat source localization (smaller beam size at the focus position) of the former technologies. The microstructural characteristics of the LBW-CR welds were similar to those previously reported for other welding methods [32]: FZ with a columnar-dendritic-typed grain morphology, HAZ decorated with retained equiaxed β grains (with larger grains at the fusion boundary and smaller grains towards BM) and epitaxial grain growth from the HAZ into the FZ.

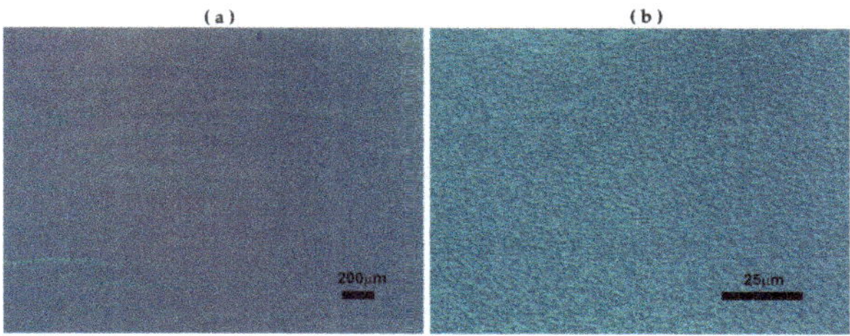

Figure 7. Metallographic images at 50× (**a**) and 1000× (**b**) of Ti5553 base metal (BM) showing the α/β microstructure (α particles are the white grains).

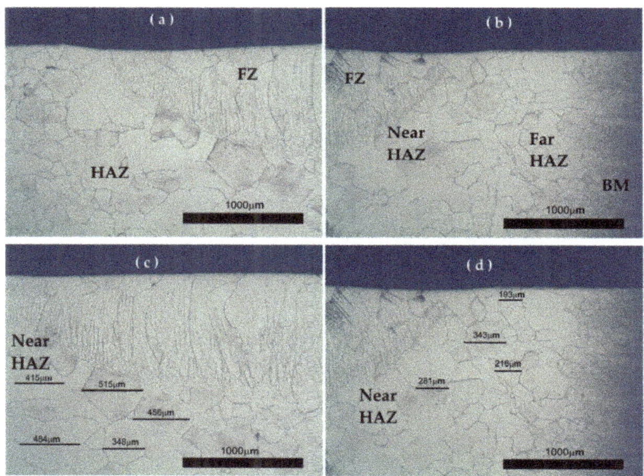

Figure 8. Metallographic images at 50× of Ti5553-Ti5553 LBW-CR butt welds showing the different zones of the welds (FZ, near HAZ, far HAZ and BM) in (**a**,**b**); and the grain size at HAZ (around 350 µm) in (**c**,**d**).

Table 2 also includes the averaged grain size of the welds zones. The grain sizes in the FZ for all welds were a few hundred microns. In the HAZ, large grains were observed at the near HAZ (along the fusion boundary) of up to 200 µm in the EBW and LBW-Nd:YAG samples, and up to 600 µm and 350 µm for GTAW and LBW-HPDL samples, respectively. The grains became gradually smaller towards the BM. The larger grain sizes in the near HAZ are associated with intermediate peak temperature during welding that facilitates grain growth. Therefore, taking into account the width and grain size of the welds, it can be stated that LBW-CR is a compensated and interesting welding technology, as it employs an energy density intermediate between arc welding, such as GTAW (high heat input and low power density) and EBW/LBW-KR (relatively low heat input and high power density).

Table 2. Width of fusion and heat-affected zone (FZ and HAZ) and grain size of HAZ of the different welds.

Conditions	Width of FZ (mm)	Width of HAZ (mm)	Grain Size of HAZ (µm)
LBW-CR (HPDL)	4.9	1.3	350
GTAW [32]	5.4	3.0	600
LBW-KR (Nd:YAG) [32]	2.6	0.8	200
EBW [32]	1.7	0.8	200

3.2. Microhardness

Hardness profile of the Ti5553-Ti5553 LBW-CR butt weld is shown in Figure 9. The hardness values (HV) follow a similar tendency than those previously obtained with other welding methods [32], i.e., 290–320 HV in the FZ, 300–360 HV in the HAZ and 370–390 HV for BM. The weld zones (FZ and HAZ) had lower hardness values compared with that of the base metal. In metastable β titanium alloys, the hardness of the FZ is usually lower than that of the BM, as the formation of the strengthening precipitates is suppressed because of the overwhelming amount of β stabilizing elements leading to a $[Mo]_{eq}$ around 12. According to Bania [33], α' (alpha prime) or martensite precipitates will not form if the $[Mo]_{eq}$ is greater than 10. The dissolution of the α phase in the heat-affected zone and the presence of entirely metastable/retained β phase in the fusion zone seem to be responsible for the significant hardness decrease in the HAZ and FZ [34]. It is interesting to note that the decrease of hardness in the FZ of LBW-CR is slightly lower than that in the FZ of welds performed with other technologies [32]. Thus, the minimum hardness value obtained in the FZ of LBW-CR weld was 310 HV, while the minimum values measured for LBW, EBW and GTAW were always below 300 HV. These results suggest that LBW-CR methodology can partially limit the dissolution of α phase in the FZ, keeping a slightly higher concentration of strengthening precipitates in the weld.

It is noteworthy that the hardness profiles of metastable β titanium alloys are different than those of α or α/β alloys. Hardness values of α or α/β alloys in the weld zones are generally comparable or higher than that of the BM, due to the formation of martensite (α' microstructure) [10,28,35].

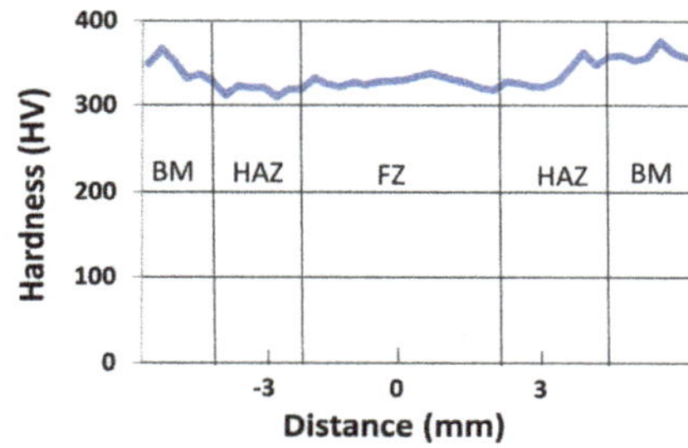

Figure 9. Hardness profile across the Ti5553-Ti5553 butt weld performed with LBW-CR with HPDL.

3.3. Tensile Test and Fracture Surface Analysis

Table 3 summarizes the results obtained from tensile testing of LBW-CR Ti5553-Ti5553 butt weld. It also includes the data reported in previous researches [32,36,37] for the base metal (BM), GTAW, EBW and LBW-Nd:YAG welds. Note that in [36], very low ductility (strain to rupture) values were obtained in EBW welds in comparison to [37]. The results published in [36] report the first trials to perform the EBW weld and therefore the experimental parameters were not optimised. Those results published in [37] were performed/tested with better measurement capability. All welds present a significant reduction in strengths and elongation values in comparison with the BM. The elongation of EBW and LBW-Nd:YAG welds were relatively comparable to that of the unwelded samples, while

LBW-HPDL and GTAW welds presented lower ductility. Concerning the strength, it is interesting to highlight that the highest Ultimate Tensile Strength (UTS) values were found in LBW-HPDL welds (78% of the BM), followed by LBW-Nd:YAG (72% of the BM), EBW (65% of the BM) and GTAW (56% of the BM). In general terms, tensile properties of all these welds are in good agreement with the results previously reported by Shariff et al. [34], in which Ti5553 welds show a maximum joint efficiency of around 75% and a variable reduction of ductility.

The slightly higher UTS value obtained in the present work for LBW-CR with HPDL is thought to be due to the higher hardness of the FZ and to the lower porosity generated in the melting pool within conduction welding. In fact, conduction modes are widely claimed to provide less porous joints than keyhole as a consequence of the lower evaporation [16–21]. However, unexpected low elongation values were obtained in LBW-CR. One would expect larger ductility in LBW-CR than in LBW-KR, which provided the lower porosity and larger grains at FZ obtained in LBW-CR. The reason for this behavior is not clear for the authors, although the higher concentration of strengthening precipitates in the LBW-CR or/and the possible lower weld depth of the LBW-CR weld can be possible explanations of these low ductility results. Further investigation needs to be performed to solve this issue.

Fracture surfaces from the tensile test of LBW-CR samples are given in Figure 10. Figure 10a,b were obtained with optical microscope (OM) at 30× and 72×, respectively; and Figure 10c–f were taken by Scanning Electron Microscope (SEM) at 300× (Figure 10c) and 2000× (Figure 10d–f), respectively. Samples fractured in the weld zones, and exhibited transgranular fracture modes with microvoid coalescence (dimples) mechanism. SEM images at 2000× (Figure 10d–f) in different zones of the fractured samples show the dimples formation at different zones.

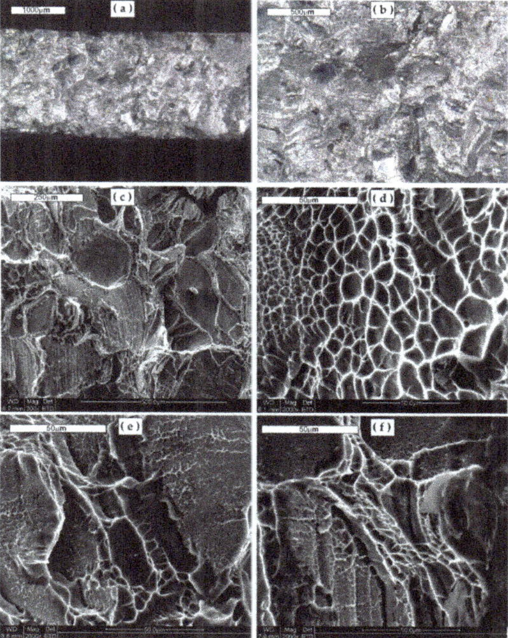

Figure 10. (**a,b**) Optical microscope (OM) images and (**c–f**) scanning electron microscope SEM images showing transgranular fracture with microvoid-coalescence mechanism of LBW-CW (HPDL) Ti5553-Ti5553 butt weld.

Table 3. Comparison of mechanical properties of as-welded Ti5553-Ti5553 butt joints performed with LBW-CR and with other different technologies [32,36,37].

Conditions	Yield Strength (MPa)	Ultimate Tensile Strength (MPa)	Elongation (%) (Strain to Rupture)
LBW-CR with HPDL	816	818	3.9
Base Metal [36]	1028	1053	12
EBW [36]	N/A	778	0.83
EBW [37]	680	680	11
LBW-KR with Nd:YAG [32]	N/A	757	9
GTAW [32]	N/A	591	N/A

4. Conclusions

Ti5553 butt welds have been obtained for the first time by means of Laser Beam Welding under Conduction Regime (LBW-CR) with a High-Power Diode Laser (HPDL). The microstructure, microhardness and strength of obtained welds have been analyzed. The welds presented a characteristic semicircle shape with relatively low depth/width ratio, a key feature of conduction-limited welding. The microstructure of LBW-CR welds was seen to be similar to those previously reported in the literature for other welding methods. Thus, the FZ presented a columnar-dendritic-typed grain morphology; the HAZ is decorated with retained equiaxed β grains (with larger grains at the fusion boundary and smaller grains towards BM); and an epitaxial grain growth is observed from the HAZ into the FZ. Interestingly, the width and grain size values of the LBW-CR welds were intermediate between the results previously obtained for GTAW and EBW. In addition, the FZ of LBW-CR presented slightly higher values of microhardness than those reported for other technologies. From the tensile tests, it is shown that the Ultimate Tensile Strength (UTS) of the welded specimens is around 78% of the BM, but still, it is slightly higher than the UTS values obtained with other joining processes, probably due to higher hardness of the FZ and to the lower porosity of the LBW-CR weld. Therefore, conduction welding offers an interesting joining mode for difficult-to-weld materials, such as metastable β titanium alloys.

Acknowledgments: The authors (Jose Maria Sánchez-Amaya, Margarita Raquel Amaya-Vazquez, Juan De Dios Lopez-Castro, Cristina Churiaque and Francisco Javier Botana Pedemonte) would like to thank the financial support of the project SOLDATIA, Ref. TEP 6180 (Junta de Andalucía, Spain). Timotius Pasang would like to thank The Loewy Family Foundation for sponsoring his research leave at Lehigh University.

Author Contributions: Jose Maria Sánchez-Amaya conceived and designed the experiments, and wrote the paper; Margarita Raquel Amaya-Vazquez, Juan De Dios Lopez-Castro and Cristina Churiaque performed the experiments; Timotius Pasang and Yuan Tao contributed the titanium alloy and revised the paper; Francisco Javier Botana Pedemonte revised the paper.

Conflicts of Interest: The authors declare no conflict of interest.

References

1. Donachie, M.J. *Titanium—A Technical Guide*, 2nd ed.; ASM International: Geauga County, OH, USA, 2000; p. 70.
2. Becker, D.W.; Baeslack, W.A., III. Property-microstructure relationships in metastable-beta titanium alloy weldments. *Weld. J. Res. Suppl.* **1980**, *59*, 85–92.
3. Liu, P.S.; Hou, K.H.; Baeslack, W.A., III; Hurley, J. Laser Welding of an Oxidation Resistant Metastable-Beta Titanium Alloy—Beta-21S. In *The Minerals, Metals and Materials Society in Titanium—World Conference, Titanium '92*; Froes, F.H., Caplan, I., Eds.; TMS: Pittsburgh, PA, USA, 1993; p. 1477.
4. Baeslack, W.A., III; Liu, P.S.; Barbis, D.P.; Schley, J.R.; Wood, J.R. Postweld Heat Treatment of GTA Welds in a High-Strength Metastable Titanium Alloy—Beta-CTM. In *The Minerals, Metals and Materials Society in Titanium—World Conference, Titanium '92: Science and Technology*; Froes, F.H., Caplan, I., Eds.; TMS: Pittsburgh, PA, USA, 1993; pp. 1469–1476.

5. Bettaieb, M.B.; Lenain, A.; Habraken, A.M. Static and fatigue characterization of the Ti5553 titanium alloy. *Fatigue Fract. Eng. Mater. Struct.* **2012**, *36*, 401–415. [CrossRef]

6. Lütjering, G.; Williams, J.C. *Titanium*, 2nd ed.; Springer: New York, NY, USA, 2003; p. 330.

7. Huang, J.; Wang, Z.; Zhou, J. Cyclic Deformation Response of β-Annealed Ti-5Al-5V-5Mo-3Cr Alloy under Compressive Loading Conditions. *Metall. Mater. Trans. A* **2011**, *42*, 2868–2880. [CrossRef]

8. Fanning, J.C.; Boyer, R.R. Properties of TIMETAL 555—A New Near- Beta Alloy for Airframe Components. In Proceedings of the 10th World Conference on Titanium, Hamburg, Germany, 13–18 July 2003; Volume 4, pp. 2643–2650.

9. Leyens, C.; Peters, M. (Eds.) *Titanium and Titanium Alloys: Fundamentals and Applications*; Willey-VCH Verlag GmbH & Co. KGaA: Weinheim, Germany, 2003. Available online: http://onlinelibrary.wiley.com/book/10.1002/3527602119 (accessed on 16 June 2017).

10. Cao, X.; Kabir, A.S.H.; Wanjara, P.; Gholipour, J.; Birur, A.; Cuddy, J.; Medraj, M. Global and Local Mechanical Properties of Autogenously Laser Welded Ti-6Al-4V. *Metall. Mater. Trans. A* **2014**, *45A*, 1258–1272. [CrossRef]

11. Gouret, N.; Dour, G.; Miguet, B.; Ollivier, E.; Fortunier, R. Assessment of the Origin of Porosity in Electron-Beam-Welded TA6V Plates. *Metall. Mater. Trans. A* **2004**, *35A*, 879–889. [CrossRef]

12. Liu, J.; Ventzke, V.; Staron, P.; Schell, N.; Kashaev, N.; Huber, N. Effect of Post-weld Heat Treatment on Microstructure and Mechanical Properties of Laser Beam Welded TiAl-based Alloy. *Metall. Mater. Trans. A* **2014**, *45A*, 16–28. [CrossRef]

13. Sánchez-Amaya, J.M.; Amaya-Vázquez, M.R.; Botana, F.J. Laser welding of light metal alloys. In *Handbook of Laser Welding Technologies*; Katayama, S., Ed.; Elsevier: Cambridge, UK, 2013; Chapter 8; pp. 215–254.

14. Quintino, L.; Assuncao, E. Conduction Laser Welding. In *Handbook of Laser Welding Technologies*; Katayama, S., Ed.; Elsevier: Cambridge, UK, 2013; Chapter 6; pp. 139–162.

15. Duley, W.W. *Laser Welding*; John Wiley and Sons: New York, NY, USA, 1999; 264p.

16. Jin, X.; Zeng, L.; Cheng, Y. Direct observation of keyhole plasma characteristics in deep penetration laser welding of aluminum alloy 6016. *J. Phys. D Appl. Phys.* **2012**, *45*, 245205. [CrossRef]

17. Sibillano, T.; Ancona, A.; Berardi, V.; Schingaro, E.; Basile, G.; Lugarà, P.M. Optical detection of conduction/ keyhole mode transition in laser welding. *J. Mater. Process. Technol.* **2007**, *191*, 364–367. [CrossRef]

18. Pang, S.; Chen, W.; Wang, W. A Quantitative Model of Keyhole Instability Induced Porosity in Laser Welding of Titanium Alloy. *Metall. Mater. Trans. A* **2014**, *45A*, 2808–2818. [CrossRef]

19. Okon, P.; Dearden, G.; Watkins, K.; Sharp, M.; French, P. Laser welding of aluminium alloy 5083. In Proceedings of the 21st International Congress on Applications of Lasers and Electro-Optics, Scottsdale, AZ, USA, 14–17 October 2002.

20. Kabir, A.S.H.; Cao, X.; Gholipour, J.; Wanjara, P.; Cuddy, J.; Birur, A.; Medraj, M. Effect of postweld heat treatment on microstructure, hardness, and tensile properties of laser-welded Ti-6Al-4V. *Metall. Mater. Trans. A* **2012**, *43A*, 4171–4184. [CrossRef]

21. Sánchez-Amaya, J.M.; Delgado, T.; De Damborenea, J.J.; López, V.; Botana, F.J. Laser welding of AA 5083 samples by high power diode laser. *Sci. Technol. Weld. Join.* **2009**, *14*, 78–86. [CrossRef]

22. Cao, X.; Wallace, W.; Poon, C.; Immarigeon, J.P. Research and progress in laser welding of wrought aluminum alloys. I. Laser welding processes. *Mater. Manuf. Process.* **2003**, *18*, 1–22. [CrossRef]

23. Sánchez-Amaya, J.M.; Boukha, Z.; Amaya-Vázquez, M.R.; Botana, F.J. Weldability of aluminum alloys with high-power diode laser. *Weld. J.* **2012**, *91*, 155–161.

24. Cho, J.H.; Farson, D.F.; Milewski, J.O.; Hollis, K.J. Weld pool flows during initial stages of keyhole formation in laser welding. *J. Phys. D Appl. Phys.* **2009**, *42*, 175502. [CrossRef]

25. Du, H.; Hu, L.; Liu, J.; Hu, X. A study on the metal flow in full penetration laser beam welding for titanium alloy. *Comput. Mater. Sci.* **2004**, *29*, 419–427. [CrossRef]

26. Churiaque, C.; Amaya-Vazquez, M.R.; Botana, F.J.; Sánchez-Amaya, J.M. FEM Simulation and Experimental Validation of LBW under Conduction Regime of Ti6Al4V Alloy. *J. Mater. Eng. Perform.* **2016**, *25*, 3260–3269. [CrossRef]

27. Sánchez-Amaya, J.M.; Amaya-Vázquez, M.R.; Gonzalez-Rovira, L.; Botana-Galvin, M.; Botana, F.J. Influence of Surface pre-treatments on laser welding of Ti6Al4V alloy. *J. Mater. Eng. Perform.* **2014**, *23*, 1568–1575. [CrossRef]

28. Amaya-Vázquez, M.R.; Sánchez-Amaya, J.M.; Boukha, Z.; Botana, F.J. Microstructure, microhardness and corrosion resistance of remelted TIG2 and Ti6Al4V by a high power diode laser. *Corros. Sci.* **2012**, *56*, 36–48. [CrossRef]

29. Lisiecki, A. Titanium Matrix Composite Ti/TiN Produced by Diode Laser Gas Nitriding. *Metals* **2015**, *5*, 54–69. [CrossRef]

30. Silva, D.P.; Churiaque, C.; Bastos, I.N.; Sánchez-Amaya, J.M. Tribocorrosion Study of Ordinary and Laser-Melted Ti6Al4V Alloy. *Metals* **2016**, *6*, 253. [CrossRef]

31. Astarita, A.; Rubino, F.; Carlone, P.; Ruggiero, A.; Leone, C.; Genna, S.; Merola, M.; Squillace, A. On the Improvement of AA2024 Wear Properties through the Deposition of a Cold-Sprayed Titanium Coating. *Metals* **2016**, *6*, 185. [CrossRef]

32. Pasang, T.; Sánchez-Amaya, J.M.; Tao, Y.; Amaya-Vazquez, M.R.; Botana, F.J.; Sabol, J.C.; Misiolek, W.Z.; Kamiya, O. Comparison of Ti-5Al-5V-5Mo-3Cr Welds Performed by Laser Beam, Electron Beam and Gas Tungsten Arc Welding. *Procedia Eng.* **2013**, *63*, 397–404. [CrossRef]

33. Bania, P.J. *Beta Titanium Alloys in the 1990's*; Eylon, D., Boyer, R.R., Koss, D.A., Eds.; TMS: Warrendale, PA, USA, 1993; pp. 3–14.

34. Shariff, T.; Cao, X.; Chromik, R.R.; Baradari, J.G.; Wanjara, P.; Cuddy, J.; Birur, A. Laser welding of Ti-5Al-5V-5Mo-3Cr. *Can. Metall. Q.* **2011**, *50*, 263–272. [CrossRef]

35. Caiazzo, F.; Alfieri, V.; Corrado, G.; Argenio, P.; Barbieri, G.; Acerra, F.; Innaro, V. Laser Beam Welding of a Ti-6Al-4V Support Flange for Buy-to-Fly Reduction. *Metals* **2017**, *7*, 183. [CrossRef]

36. Mitchell, R.; Short, A.; Pasang, T.; Littlefair, G. Characteristics of Electron Beam Welded Ti and Ti Alloys. *Adv. Mater. Res.* **2011**, *275*, 81–84. [CrossRef]

37. Sabol, J.C.; Pasang, T.; Misiolek, W.Z.; Williams, J.C. Localized tensile strain distribution and metallurgy of electron beam welded Ti-5Al-5V-5Mo-3Cr titanium alloys. *J. Mater. Process. Technol.* **2012**, *212*, 2380–2385. [CrossRef]

metals

Article

The Influence of Laser Welding on the Mechanical Properties of Dual Phase and Trip Steels

Emil Evin [1] and Miroslav Tomáš [2,*]

[1] Department of Automotive Production, Faculty of Mechanical Engineering,
Technical University of Košice, Mäsiarska 74, 040 01 Košice, Slovakia; emil.evin@tuke.sk
[2] Department of Computer Support of Technology, Faculty of Mechanical Engineering,
Technical University of Košice, Mäsiarska 74, 040 01 Košice, Slovakia
* Correspondence: miroslav.tomas@tuke.sk; Tel.: +421-55-602-3524

Received: 6 June 2017; Accepted: 26 June 2017; Published: 29 June 2017

Abstract: Nowadays, a wide range of materials is used for car body structures in order to improve both the passengers' safety and fuel consumption. These are joined by laser welding and solid state fiber lasers being used more and more in present. The article is focused on the research of laser welding influence on the mechanical and deformation properties, microstructure and microhardness of advanced high-strength steels: high-strength low-alloyed steel HC340LA, dual phase steel HCT600X and multi-phase residual austenite steel RAK40/70. The proper welding parameters have been found based on weld quality evaluation. The specimens for tensile test with longitudinal laser weld were used to measure mechanical and deformation properties. Microstructure and microhardness of laser welds were evaluated in the base metal, heat affected zone and fusion zone. The higher values of strength and lower ones for deformation properties of laser-welded materials have been found for dual and multi-phase steel. The microhardness strongly depends on the carbon equivalent of steel. Deformation properties are more sensitive than strength properties to the change of microstructure in the fusion zone and heat affected zone.

Keywords: laser welds; mechanical properties; low alloyed steel; dual phase steel; trip steel

1. Introduction

In the automotive industry, an emphasis on inventions, innovations and information is reflected in the constant increase of car models in the portfolio of individual automakers. On the one hand, this is due to intensification of individualism with current absolute orientation of enterprises on the customer, and, on the other hand, this is due to the models' diversification of vehicle manufacturers in terms of marketing of various technological innovations and new vehicle concepts on the market, with the aim to occupy the greatest number of market segments. Customers also demand cars that will respect their individual needs, and it affects not only the actual increase in the number of models on offer, but also increases the complexity of the car and a separate system.

Today, it is expected in the automotive industry that, based on the material selection strategies, such products are introduced to the market, involving more intellect and less material. Materials also have a close relationship to the passive safety of vehicles. While the vehicle's car body is the important passive safety component, the lives of the passengers and the other transport users depend on strength and deformation properties of the material. The car body designers have to choose materials with exact properties to control the deformation of each car body structural component. In the frontal deformation zones, materials able to absorb as much impact energy as possible on a larger path should be used. Otherwise, in the middle zone of the car-body, strong materials able to absorb impact energy on a shorter path (minimum deformation) should be used. The innovative potential of the thin-walled lightweight (shell) structures from aluminum alloys, magnesium alloys, composites, High Strength

Steels (HSS) and Ultra High Strength Steels (UHSS) is a big challenge when looking for the right material for each deformation zone [1,2].

Light-weight alloy materials used in the car body structural components are undoubtedly of great potential for lightening many structural parts of the cars. However, the overall share of steel still exceeds 50% of the total weight of the mass-produced cars. The uniqueness of the position of steel in the construction of cars is determined by the combination of its properties such as strength, stiffness, deformation work, formability and recyclability [2,3].

Standard low alloyed high-strength steels (IF, BH, and HSLA), advanced high-strength steels (AHSS) and ultra high-strength steels (UHSS—DP, CP, TRIP, TWIP and MS steels) commonly used in the automotive industry differ in their microstructure. Properties of steels are controlled by chemical composition and the strengthening mechanisms, such as: hardening by solid solutions, dislocation hardening, grain boundaries hardening, precipitation or transformation hardening [4].

The research in the paper is focused on the high strength low alloyed steel, dual phase steel and multiphase steel. In the case of high-strength low-alloy (HSLA) steels, the increase of the ferritic matrix strength is achieved especially by microalloying elements Ti, V, and Nb, which increase strength by grain size control, precipitation hardening, or substitution hardening of structure [4,5]. Dual phase steels consist of a ferritic matrix with 5% to 30% martensite content. Ferrite is a soft phase with good formability. Martensite is a hard phase that gives the material its strength and hardness [4,6]. Transformation induced plasticity (TRIP) steels consist of retained austenite, martensite, bainite and ferrite. Residual austenite transforms to martensite during deformation, and it is situated in the polygonal ferritic matrix, which also includes bainite, preserving very good formability [4,7]. Both dual phase and multi-phase steels show more work hardening (WH) and bake hardening (BH) (Figure 1a,b) than high strength low alloyed steel [3].

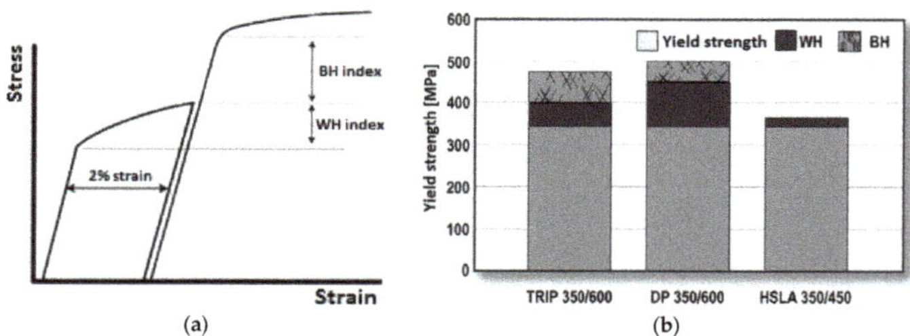

Figure 1. Contribution of work and bake hardening for some advanced high strength steels: (**a**) the effect of both, work hardening (WH) and bake hardening (BH) on the stress–strain curve; (**b**) comparison of the WH and BH for transformation induced plasticity—TRIP, dual phase—DP and high strength low alloyed steel—HSLA steel at a 2% level of strain [3].

Structural thin-walled components of security zones of cars must often meet conflicting requirements—on the one hand, the growing tendency for the reduction of the weight of stamped body parts and, on the other hand, the tendency of increasing the security (strength, stiffness, absorption capacity and programmable deformability under crash). The desired compromise between the requirements to reduce the weight of cars on the one hand and requirements for improving the safety parameters (strength, stiffness, both deformation and fatigue works), reduction of fuel consumption and emissions during operation of cars on the other hand, can be achieved by proper application of the combined tailored blanks: tailor-welded blanks (TWB), engineering blanks (EB), tailor-rolled blanks (TRB), tailor heat treated blanks (THTB) (with the same or different thickness, with different strength and deformation properties, especially in the deformation zones of thin-walled components of cars) [8–11]. The combined TWBs prepared by laser beam or electron beam welding have found the widest application.

In the areas with smaller load, the sheet metal with reduced thickness or less expensive material of less strength is applied. In the areas where greater loads occur, the material of greater thickness and higher strength is applied. Because of these advantages, combined blanks are widespread in the design of car body components and also improve their utility properties (e.g., stainless austenitic and ferritic steels for improving the corrosion resistance in the exhaust systems, the fuel tanks, etc.). According to the data of suppliers of materials for the automotive industry, their share in the structure of the car body should be increased in the coming years up to 30% [12,13].

Application of solid-state lasers in the industry currently comes with a significantly increasing perspective, especially due to the fact that new types of solid-state lasers such as fiber and disk lasers with their technical parameters surpass the parameters of gas CO_2 lasers [14,15].

The most important advantages of solid-state lasers include:

1. Conversion efficiency of electric energy to the energy of the laser beam being higher by 30% compared to the CO_2 lasers,
2. Higher focusability that expands the field of industrial applicability,
3. Improved flexibility and stability of output parameters,
4. Increased dynamics of beam parameters with the possibility of shaping the rise and decay, and the tip of the laser pulse.

Research results of weldability and formability indicate that influence of welded joints on the material plastic flow in the joint area is examined and analyzed by the same kind of tests as are used for evaluation of weldability and formability of base materials, when used for TWBs: tensile tests; braking tests; technological tests (cup drawing tests, Erichsen tests, the Swift method, hydraulic bulge tests, hole expanding test, limit dome height tests and three- or four-point bending tests); macro and microstructure analyses of individual zones of welded joints; hardness tests and Vickers micro-hardness in zones of welded joints and the base material [16,17].

The aim of the paper is to assess the impact of laser welding on the mechanical properties of selected AHSS materials for TWBs, used in the automotive industry for structural components of the car body. The study is focused on the measurement of strength and deformation properties for base material and laser welded by a solid-state fiber laser. Many authors have investigated the influence of laser welding on the tensile properties on specimens with transverse weld [5,6,18–20]. Thus, the research was focused on the specimens with longitudinal butt laser weld, as it is used for tailored TWBs. It is supposed that the longitudinal weld is more sensitive to the stress–strain distribution when a tensile test is performed. In addition, the microstructural changes and microhardness in the heat affected zone and fusion zone were examined.

2. Materials and Methods

The experimental research has been focused on the high strength low alloyed steel HC340LA, dual phase steel HCT600X and multi-phase residual austenite steel RAK40/70. Steels were electrolytically coated with zinc (100 g/m^2) and had special requirements for surface (M—normal surface finish, B—best quality, O—oiled). These are commonly used to enhance the safety characteristics—strength, stiffness, and deformation work—and reduce both the fuel consumption and car emissions. The chemical composition of materials, shown in Table 1, was measured by atomic emission spectrometry according to standard ASTM E415-14.

The first stage of the experimental research was focused on analyzing the influence of laser welding on mechanical and deformation properties of laser-welded base material HC340LA-HC340LA, HCT600X-HCT600X and RAK40/70-RAK40/70 when welded by a solid state fiber laser YLS-5000 (IPG Photonics, Oxford, MA, USA). The welding was done by continual mode without protective gas with parameters shown in Table 2. Samples for microstructure analysis taken from laser-welded sheets have been analyzed by an optical microscope Axio Observer 1M (Carl Zeiss Jena GmbH, Jena, Germay). The welding parameters have been optimized by quality control based on weld porosity and weld root quality in the heat affected zone and fusion zone. Optimized weld parameters were used to prepare samples for the research.

Table 1. Chemical composition of materials used for research.

Material	Chemical Composition wt %								
HC340LA	C	Si	Mn	P	S	Cu	Al	Cr	Mo
	0.083	0.039	0.181	0.038	<0.002	0.081	0.062	0.044	0.034
	Ni	V	Ti	Nb	Co	W	CE_Y acc. to (1)		
	0.048	0.018	<0.002	0.071	0.155	-	0.122		
HCT600X	C	Si	Mn	P	S	Cu	Al	Cr	Mo
	0.111	0.279	1.963	0.026	<0.002	0.019	0.031	0.206	<0.002
	Ni	V	Ti	Nb	Co	W	CE_Y acc. to (1)		
	<0.002	0.012	<0.002	0.02	0.017	<0.005	0.397		
RAK40/70	C	Si	Mn	P	S	Cu	Al	Cr	Mo
	0.197	0.165	1.576	0.015	<0.002	0.022	1.352	0.0455	0.022
	Ni	V	Ti	Nb	Co	W	CE_Y acc. to (1)		
	0.016	0.011	-	0.002	-	-	0.486		

Table 2. Parameters of laser welding by solid state fiber laser YLS-5000.

Material	Sheet Thickness [mm]	Laser Power [W]	Focal Point [mm]	Welding Speed [mm·s^{-1}]	Width of FZ [1] [mm]	Width of HAZ [2] [mm]	Note
HC340LA	0.78	1700	10	50	0.9	0.35	Optimal
		2100		70	0.92	0.36	-
		2300		70	0.96	0.38	-
HCT600X	0.77	2000	10	50	2.15	0.45	Optimal
		2700		70	2.21	0.47	-
RAK40/70	0.77	2000	10	50	1.05	0.40	Optimal
		2700		70	1.17	0.48	-

[1] FZ—fusion zone, [2] HAZ—heat affected zone.

The second stage of the experimental research was focused to analyze microstructure hardness in the base metal (BM), heat affected zone (HAZ) and fusion zone (FZ). The hardness measurement was done by the Vickers method according to STN EN ISO 6507-1 on the device Duramin A300 (Struers GmbH, Willich, Germany).

The mechanical properties—yield strength, ultimate tensile strength and extension—were measured on specimens from base material and specimens with longitudinal laser weld (Figure 2) using optimized parameters of laser welding. The tests were done according to STN EN ISO 6892-1, the normal anisotropy ratio r was measured according to STN EN ISO 10113 and strain-hardening exponent n was measured according to STN EN 10275.

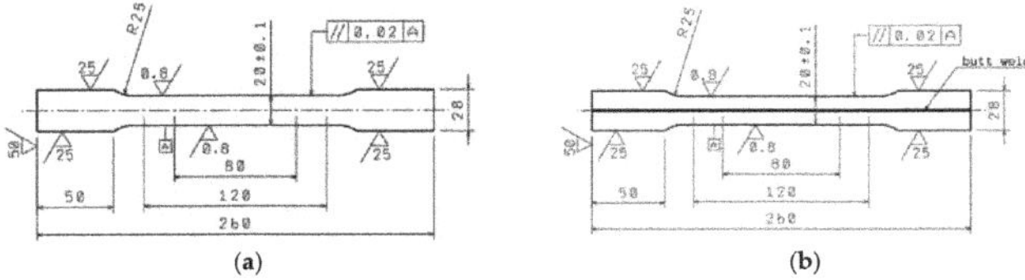

Figure 2. Specimens for tensile test: (**a**) base material; (**b**) laser-welded specimen (unit: mm).

3. Results and Discussion

3.1. Analysis of Microstructure

The microstructure of laser-welded materials in the base metal, heat affected zone and fusion zone have been analyzed for HC340LA-HC340LA, HCT600X-HCT600X and RAK40/70-RAK40/70 samples.

An overall photo of the microstructure of laser-welded steel HC340LA-HC340LA is seen in Figure 3. The overall width of the weld is approx. 1.6 mm, width of the fusion zone approx. 0.9 mm and width of the heat affected zone is approx. 0.35 mm.

The microstructure of base metal HC340LA (Figure 4) is uniform, polyhedral and fine-grained. It is composed of fine equiaxed ferrite grains, little pearlite units and tertiary cementite.

In the heat affected zone, a mostly acicular microstructure of the fusion zone comes to the polyhedral microstructure. It is created by a mixture of very fine grains of transformed ferrite, possibly bainite. Towards the base metal, in the heat affected zone, original equiaxed thicker ferrite grains are detected and their proportion is increasing (Figure 5).

A microstructure of the fusion zone is composed of ferrite, bainite and martensite in the shape of grains and needles of various sizes and shapes (Figure 6).

Figure 3. An overall photo of laser-welded steel HC340LA-HC340LA.

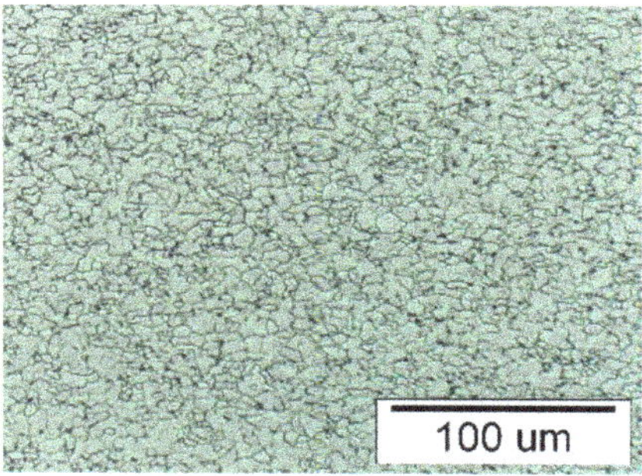

Figure 4. Microstructure of the base metal of laser-welded steel HC340LA-HC340LA.

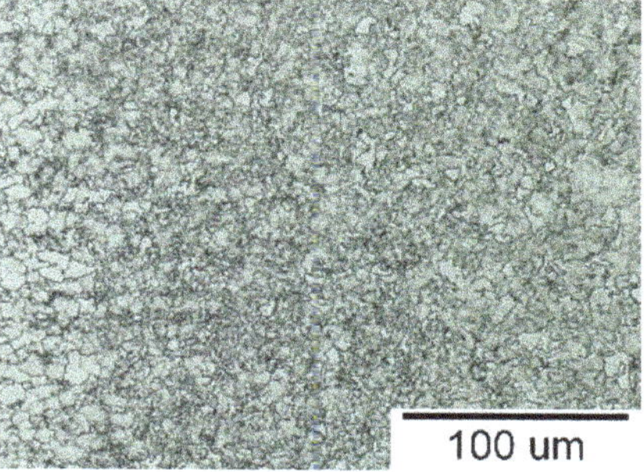

Figure 5. Microstructure of the heat affected zone of laser-welded steel HC340LA-HC340LA.

Figure 6. Microstructure of the fusion zone of laser-welded steel HC340LA-HC340LA.

An overall photo of the microstructure of laser-welded steel HCT600X-HCT600X is seen in Figure 7. The overall width of the weld is approx. 2.15 mm, width of the fusion zone is approx. 1.25 mm and width of the heat affected zone is approx. 0.45 mm.

Microstructure of base metal HCT600X is a fine-grained polyhedral made of equiaxic ferrite grains and very fine martensite grains (Figure 8). However, there are visible lines in the distribution of phases in structure and impurity particles. Tertiary cementite also occurs in the structure. The volume fraction of the martensite was $18.2 \pm 2\%$ determined by lattice method using image analyzer Image J (open source software) at a scale of $1000\times$.

Subtilisation of martensite is visible in the heat affected zone (Figure 9). In subtle martensite, which exists in the heat affected zone, there are subtle ferrite grains, a proportion of which is increasing towards the primary material. In proximity to the primary material, the portion of ferrite grains is higher than the portion of martensite.

Due to morphology of the fusion zone microstructure, a bilateral orientation of its solidification from primary material to the welding midline is clearly seen. Microstructure of the fusion zone is composed of martensite, which was created in extended austenitic grains (Figure 10). Towards the heat affected zone, orientation of martensite laths decreases, which is a result of the decrease of a size and reduction of orientation of austenite grains, where martensite was being created.

Figure 7. Overall photo of laser-welded steel HCT600X-HCT600X.

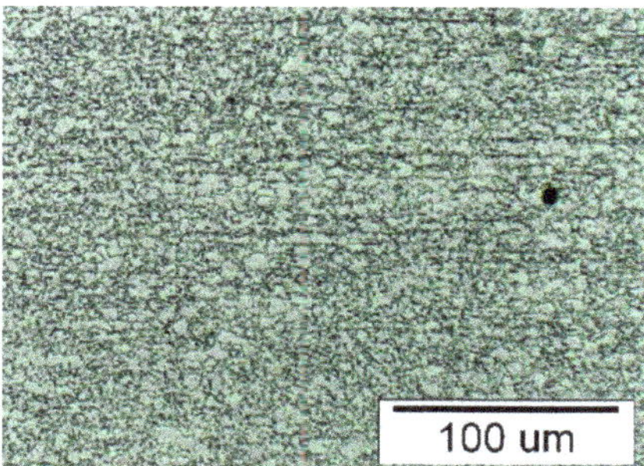

Figure 8. Microstructure of the base metal of laser-welded steel HCT600X-HCT600X.

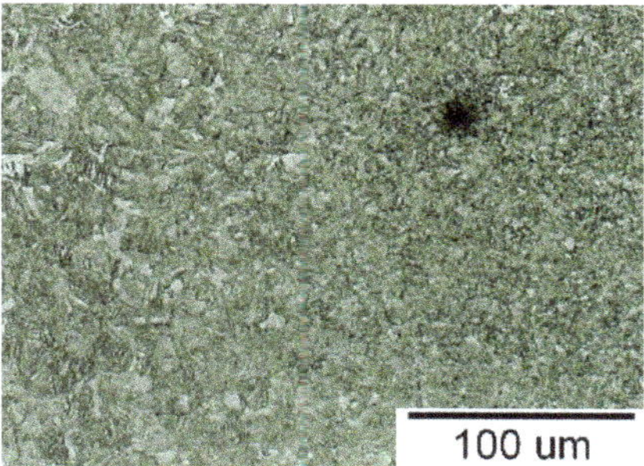

Figure 9. Microstructure of the heat affected zone of laser-welded steel HCT600X-HCT600X.

Figure 10. Microstructure of the fusion zone of laser-welded steel HCT600X-HCT600X.

An overall photo of the microstructure of laser-welded steel RAK40/70 is in Figure 11. Overall weld width is approx. 1.85 mm, width of fusion zone area is approx. 1.05 mm and width of heat affected zone is approx. 0.4 mm.

Microstructure of the base metal (Figure 12) is fine-grained, homogeneous with signs of line organization of the structure elements. It is mainly composed of fine ferrite grains and has a minority of very fine units of residual austenite, bainite and martensite. The volume fraction of residual austenite measured by RTG analysis was 22.3%. The fraction of ferrite, bainite and martensite is 77.7%. Phase components of bainite and martensite were impossible to identify due to a small distortion of lattice.

In the heat affected zone of the weld joints (Figure 13), a mixed martensite-ferrite structure occurs. The proportion of ferrite grains in this mixed microstructure increases with the distance towards primary material. In the close proximity to the primary material, in the microstructure of heat affected zone, ferrite preponderates in the form of polyhedral grains. Martensite, in the form of fine units of different shapes is in this area a minority phase.

Results of microstructural changes received when similar materials were welded comply with [5–7,20]: the fusion zone of dual phase steel consisted of fully martensitic structure; the fusion zone of high strength low alloyed steel consisted of ferrite, bainite and martensite; and the fusion zone of multi-phase steel consisted of martensite with laths oriented randomly.

Figure 11. Overall photo of laser-welded steel RAK40/70-RAK40/70.

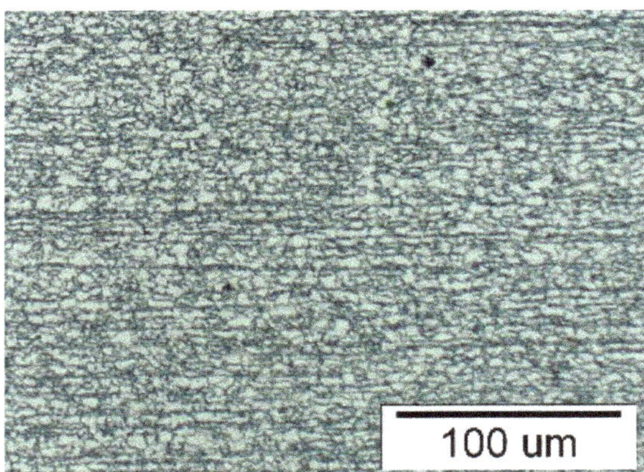

Figure 12. Microstructure of the base metal of laser-welded steel RAK40/70-RAK40/70.

Figure 13. Microstructure of the heat affected zone of laser-welded steel RAK40/70-RAK40/70.

Figure 14. Microstructure of the fusion zone of laser-welded steel RAK40/70-RAK40/70.

3.2. Analysis of Microhardness and Its Distribution

The microstructure hardness was measured in the base metal (BM), heat affected zone (HAZ) and fusion zone (FZ). The hardness measurement was done by the Vickers method according to STN EN ISO 6507-1 on the device Duramin A300. The device was calibrated before measurement of microhardness HV05 on samples prepared for microstructure observation. Microhardness with load 0.5 within 10 s was automatically evaluated by software Ecos. In addition, 15 points in a row were measured on samples through the weld joint—from left to right (three points for each zone): the base metal, heat affected zone, weld joint, and heat affected zone to the base metal. The average values of HV05 microhardness and standard deviation (Stdev) for each material are shown in Table 3. The microhardness distributions in base metal, heat affected zone and fusion zone are shown in Figure 15.

Table 3. Measured values of microhardness HV05.

Material	Base Metal	Heat Affected Zone		Fusion Zone	
	HV05	HV05	HV_{HAZ}/HV_{BM}	HV05	HV_{FZ}/HV_{BM}
HC340LA	152 ± 2	183 ± 7	1.2	213 ± 3	1.4
HCT600X	205 ± 3	314 ± 49	1.5	390 ± 3	1.9
RAK40/70	219 ± 3	314 ± 56	1.4	474 ± 3	2.2

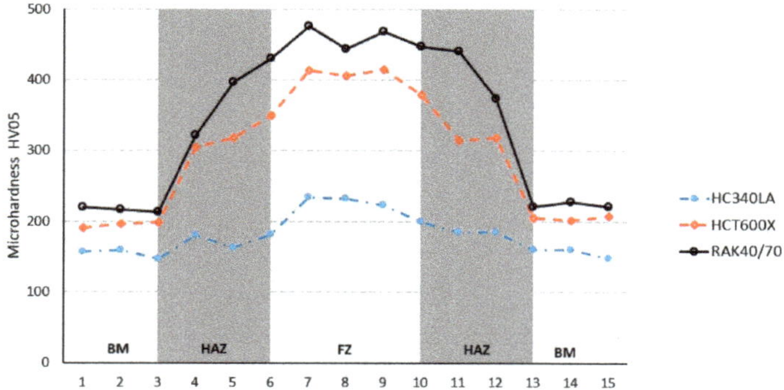

Figure 15. Microhardness distribution in the laser-welded joint.

The microhardness HV05 of base metal for HC340LA was 152 ± 2, for HCT600X, it was 205 ± 3, and for RAK40/70, it was 219 ± 3. In the transition zone between the base metal and heat affected zone, microhardness has not declined markedly, i.e., any substantial soft zone hasn't been found for each material. Otherwise, other researchers identified soft zones when thicker materials with high power were welded [5–7,20].

The microhardness HV05 of the heat affected zone laid for HC340LA-HC340LA within the interval 183 ± 7, for HCT600X-HCT600X, it was within interval 314 ± 49, and for RAK40/70-RAK40/70, it was within interval 314 ± 56. Thus, the microhardness HV05 of the heat affected zone increased for high strength low alloyed steel HC340LA 1.2 times, for dual phase steel HCT600X 1.5 times and for multi-phase steel RAK40/70 1.4 times when compared to the microhardness of base metal. Higher values of microstructure were identified for dual phase and multi-phase steels in the direction towards the fusion zone.

The highest values of microhardness for each material were measured in the fusion zone—it was for HC340LA-HC340LA within interval 213 ± 3, for HCT600X-HCT600X, it was within interval 390 ± 3, and, for RAK40/70-RAK40/70, it was within interval 474 ± 3. Hence, the microhardness HV05 of fusion zone increased for high strength low alloyed steel HC340LA 1.4 times, for dual phase steel HCT600X 1.9 times, and for multi-phase steel RAK40/70 2.2 times, when compared to the microhardness of base metal. The microhardness HV05 measured in the fusion zone for HCT600X-HCT600X and RAK40/70-RAK40/70 differ markedly, and even the welding parameters for these steels were the same. It is given by different chemical composition, i.e., carbon equivalent when calculated according to Yurioki [21,22] from chemical composition shown in Table 1—$CE_{Y\ H220PD} = 0.122$; $CE_{Y\ HCT600X} = 0.397$; $CE_{Y\ TRIP40/70} = 0.486$:

$$CE_Y = C + 0.75 + 0.25\tanh[20(C - 0.12)]\left[\frac{Si}{24} + \frac{Mn}{6} + \frac{Cu}{15} + \frac{Ni}{20} + \frac{Cr + Mo + Nb + V}{5}\right], \quad (1)$$

The microhardness HV05 measured in the fusion zone is well correlated to the calculated carbon equivalent CE_Y, as is shown in Figure 16.

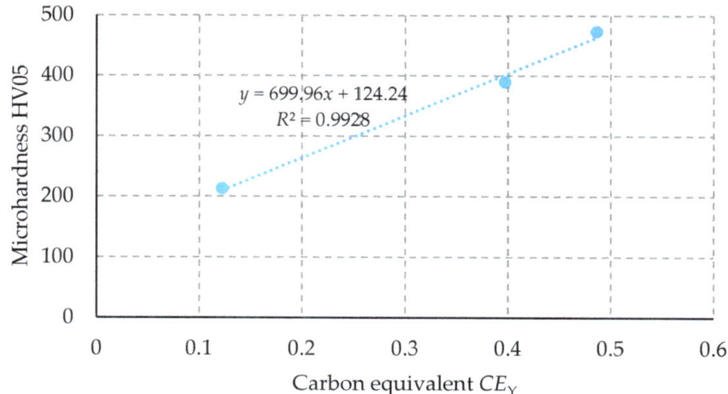

Figure 16. Microhardness distribution in the laser-welded joint.

3.3. Analysis of Mechanical Properties

Nowadays, the modern design of stampings' production is realized by numerical simulation, due to time shortening and optimization of the production process. Thus, material deformation properties have to be defined as equations describing the stress–strain relationship and material anisotropy. When simulating cold forming processes, Hill 48 yield criteria and models of stress–strain are commonly used as follows [17,23,24]:

Hill 48:

$$\sigma_{11}^2 + \sigma_{22}^2 - \frac{2 \cdot r}{1+r}\sigma_{11}\sigma_{22} + \frac{2(1+r)}{1+r}\sigma_{12}^2 = \sigma_Y, \tag{2}$$

Krupkowski:

$$\sigma = K(\varphi_0 + \varphi)^n, \tag{3}$$

Hollomon:

$$\sigma = K \cdot \varphi^n, \tag{4}$$

Power law:

$$\sigma = Re + \varphi^n, \tag{5}$$

where σ_{11}, σ_{22}, σ_{12} are stress tensor components, σ_Y is yield strength, r is average normal anisotropy ratio, K is material constant, φ_0 is pre-strain, φ is logarithmic strain and n is a strain-hardening exponent. These are used with basic mechanical properties—yield strength Re, ultimate tensile strength Rm, extension A, Young modulus E, Poisson's constant μ, and density ρ.

Lankford's coefficients or normal anisotropy ratio r measured at $0°$, $45°$ and $90°$ describe normal anisotropy of steel sheets as a consequence of deformation texture after rolling. These were measured according to the standard mentioned previously. Measured values of mechanical properties (Re or $R_{p0.2}$, Rm, Ag, A) and deformation properties (K, n, r) for base material and a laser-welded one are shown in Tables 4–6 for each material [25–27].

According to the stress–strain curves for base and laser-welded material (Figure 17), the considerably lowest work hardening has been found for low alloyed steel HC340LA with ferritic structure of base material (WH$_{BM}$ 148 MPa, WH$_{LW}$ 148 MPa), when compared to dual phase steel HTC600X with ferritic-martensitic structure of base material (WH$_{BM}$ 374 MPa, WH$_{LW}$ 360 MPa) or multi-phase steel RAK40/70 with ferritic-martensitic-bainitic structure of base material with residual austenite (WH$_{BM}$ 532 MPa, WH$_{LW}$ 483 MPa). The highest deformation hardening has been found for multi-phase steel due to phase transformation of austenite to martensite when plastically deformed. A similar tendency has been found for laser-welded samples. Comparing stress-strain curves for laser-welded samples HC340LA-HC340LA, HTC600X-HTC600X and RAK40/70-RAK40/70, these were similar to the stress–strain curves of base material. The fracture occurred in the base material for laser-welded samples, which imply good quality of laser-welded joints.

Table 4. Mechanical properties of low alloyed steel HC340LA.

Material	Dir.	Re ($R_{p0.2}$) [MPa]	Rm [MPa]	Ag [%]	A [%]	K	n	r
HC340LA *	0°	378	449	17.4	28.5	728	0.178	0.66
	45°	383	455	16.8	30.4	722	0.161	0.85
	90°	389	446	18.1	29.5	724	0.169	0.76
Average		383	450	17.4	29.5	725	0.169	0.78
Stdev		5.5	4.6	0.7	1	3	0.009	0.09
HC340LA **	0°	375	447	17	29	726	0.175	0.65
	45°	380	451	16.4	30.4	718	0.157	0.84
	90°	386	440	17.3	26.5	717	0.163	0.71
Average		380	444	16.7	26	718	0.163	0.76
Stdev		5.5	5.6	0.5	2	5	0.009	0.09

* Base material, ** Laser welded.

Table 5. Mechanical properties of dual phase steel HC340LA.

Material	Dir.	Re ($R_{p0.2}$) [MPa]	Rm [MPa]	Ag [%]	A [%]	K	n	r
HCT600X *	0°	376	632	19	28.4	1096	0.217	0.77
	45°	378	626	19.5	28	1084	0.215	0.84
	90°	371	627	19.3	30.7	1095	0.220	0.81
Average		375	628	19.3	29	1092	0.217	0.81
Stdev		4	3	0.3	1	7	0.003	0.04
HCT600X **	0°	435	677	17	20	1109	0.182	0.76
	45°	430	672	17.3	22.3	1117	0.185	0.81
	90°	427	674	17.1	23.3	1112	0.187	0.78
Average		431	674	17.1	21.9	1112	0.185	0.79
Stdev		4	3	0.2	2	4	0.003	0.02

* Base material, ** Laser welded.

Table 6. Mechanical properties of multi-phase steel RAK40/70.

Material	Dir.	Re ($R_{p0.2}$) [MPa]	Rm [MPa]	Ag [%]	A [%]	K	n	r
RAK40/70 *	0°	440	764	25.3	30.3	1497	0.295	0.658
	45°	462	761	22.8	26.6	1452	0.275	0.689
	90°	457	766	24.3	29.4	1474	0.281	0.623
Average		453	764	24.1	28.8	1474	0.284	0.66
Stdev		5	3	0.8	1.5	13	0.004	0.03
RAK40/70 **	0°	466	657	20	21	1459	0.249	0.66
	45°	483	659	19	20.1	1436	0.235	0.63
	90°	474	662	20.3	22.3	1458	0.244	0.60
Average		474	659	19.8	21.1	1451	0.243	0.63
Stdev		9	2	1	1	13	0.007	0.03

* Base material, ** Laser welded.

Comparison of measured values for base material and laser welded are shown in Figure 18 for strength properties and in Figure 19 for deformation properties.

For laser-welded samples from HC340LA-HC340LA, 1% lower values of strength properties has been found (yield strength, ultimate tensile strength, material constant K) and deformation properties (uniform extension *Ag*, total extension *A*, strain-hardening exponent *n*, normal anisotropy ratio *r*) as well as 5% lower value of r.n.1000 when compared to base material HC340LA.

For laser-welded samples from HTC600X-HTC600X, approx. 15% higher values of yield strength has been found, approx. 3% higher values of ultimate tensile strength and material constant *K* and lower values of deformation properties—uniform extension *Ag* approx. 12%, total extension *A* approx. 25%, strain-hardening n approx. 15%, normal anisotropy ratio *r* approx. 2% and 20% lower value of r.n.1000 when compared to base material HTC600X.

For laser-welded samples from RAK40/70-RAK40/70, approx. 5% higher values of yield strength has been found, approx. 14% lower values of ultimate tensile strength, approx. 2% lower values of material constant *K*, approx. 18% lower values of uniform extension Ag and approx. 27% lower values of total extension A, approx. 15% lower values of strain-hardening exponent *n*, approx. 5% lower values of normal anisotropy ratio r as well as 19% lower values of r.n.1000 when compared to base material RAK40/70.

It is supposed that the highest difference in strength and deformation properties for HTC600X-HTC600X and RAK40/70-RAK40/70 compared to low alloyed steel HC340LA-HC340LA is given by the volume of martensite, its morphology and distribution on each area of weld joints [23–25]. Similar results have been reached by other authors [5–7,18–20,28,29], although they used specimens with transverse laser weld and dissimilar materials were joined.

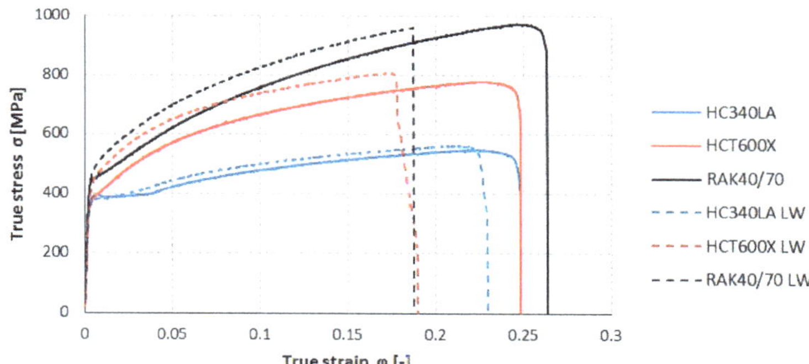

Figure 17. Stress-strain curves of base material and laser welded.

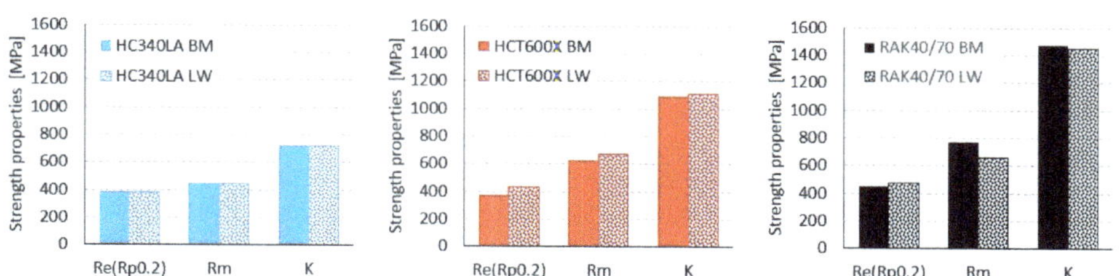

Figure 18. Strength properties of base material and laser welded.

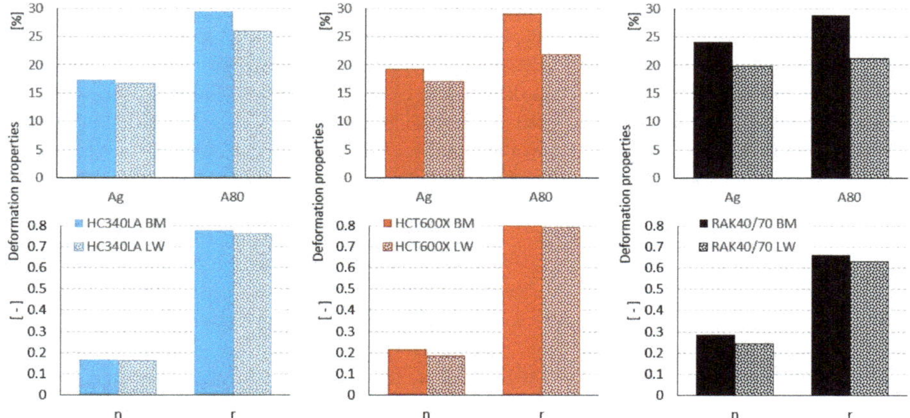

Figure 19. Deformation properties of base material and laser welded.

4. Conclusions

A wide range of metal sheets with the same or different thicknesses are joined by laser welding. However, there is a change in mechanical properties in the area of the fusion zone and the heat affected zone, due to its microstructure and width of weld joints. The experiments were focused on the research of laser welding impact on strength and deformation properties of metal sheets with thickness 0.78 mm from high strength low alloyed steel HC340LA, dual phase steel HTC600X and multi-phase steel RAK40/70, prepared as TWBs with longitudinal weld when tested.

Based on the research performed, the findings are as follows:

1. The proper quality of laser-welded joints has been reached for lower welding speed and power: 50 mm·s^{-1} and 1700 W for low alloyed steel HC340LA-HC340LA; 50 mm·s^{-1} and 2000 W for dual phase steel HTC600X-HTC600X; and 50 mm·s^{-1} and 2000 W for multi-phase steel RAK40/70-RAK40/70. This is well correlated with results presented in [18,19,28,29].
2. Laser-welded joints of experimental materials have shown proper quality of weld root and no porosity, and the good quality of joints has also been proven by tensile test, where fracture occurs in the base metal for each material.
3. Microhardness HV05 measured in samples for microstructural analysis has shown the highest values in the fusion zone for each material, and these gradually declined in heat affected zones towards the base material. Any substantial soft zone has not been found for each material. Microhardness of the fusion zone is well correlated to the carbon equivalent calculated by Yoriuki. Higher scattering of microhardness has been found for dual phase and multi-phase steel as a result of microstructure in the fusion zone and heat affected zone.
4. Stress strain curves for base material and laser-welded ones show the same tendency for each material. The highest work hardening has been found for multi-phase steel RAK40/70 due to phase transformation of austenite to martensite when plastically deformed.
5. Deformation properties are more sensitive than strength properties to the change of microstructure in the fusion zone. A greater change of uniform extension Ag and total extension A as well as a lower change of normal anisotropy ratio *r* and strain hardening exponent *n*—has been found for dual phase steel HTC600X-HTC600X and multi-phase steel RAK40/70-RAK40/70 when compared to low alloyed steel HC340LA-HC340LA. This is related to martensitic structure, its morphology and distribution in the fusion zone for dual and multi-phase steels, and ferrite, bainite (or martensite) structure in the fusion zone for high strength low-alloyed steel.

A possible solution to improve deformation properties of DP-DP, TRIP-TRIP or DP-TRIP steels can be found in their heat treatment or welding by two laser beams to reduce the cooling rate after laser welding.

Acknowledgments: The authors are grateful for the support of experimental works to Slovak Research and Development Agency, under project APVV-0273-12 "Supporting innovations of autobody components from the steel sheet blanks oriented to the safety, the ecology and the car weight reduction", as well as the grant agency for the support of the project VEGA 2/0113/16 "Influence of laser welding parameters on structure and properties of welded joints of advanced steels for the automotive industry".

Author Contributions: Emil Evin conceived and designed the experiments; Miroslav Tomáš performed the experiments of tensile tests; Emil Evin performed the experiments of microhardness tests; and Emil Evin and Miroslav Tomáš analyzed the data.

References

1. Drewes, E.J.; Prange, W. Innovative Halbzeuge für den Leichtbau mit Stahl. In *16th Steel Colloquium Aachen, "Umformtechnik, Stahl und NE-Werkstoffe-Innovative Halbzeuge—Basis für Hochleistungsprodukte"*; Verlag Mainz: Aachen, Germany, 2001; pp. 207–221.

2. Evin, E.; Tkáčová, J.; Tkáč, J. Aspects of steel sheets selection for car body components. *AI Mag. No. 2* **2012**, *5*, 88–91.

3. Evin, E.; Tkáčová, J.; Tkáč, J. Aspects of steel sheets selection for car body components. *AI Mag. No. 3* **2012**, *5*, 96–98.

4. Kvačkaj, T. Development of steels for automotive industry. In *MAT/TECH for Automotive Industry*; Technical University of Košice: Košice, Slovakia, 2015; pp. 58–67.

5. Yuce, C.; Tutar, M.; Karpat, F.; Yavuz, N. The optimization of process parameters and microstructural characterization of fiber laser welded dissimilar HSLA and MART steel joints. *Metals* **2016**, *6*, 245. [CrossRef]

6. Farabi, N.; Chen, D.L.; Zhou, Y. Microstructure and mechanical properties of laser welded dissimilar DP600/DP980 dual-phase steel joints. *J. Alloys Compd.* **2011**, *509*, 982–989. [CrossRef]

7. Mujica, L.; Weber, S.; Pinto, H.; Thomy, C.; Vollertsen, F. Microstructure and mechanical properties of laser-welded joints of TWIP and TRIP steels. *Mater. Sci. Eng. A* **2010**, *527*, 2071–2078. [CrossRef]

8. Chena, W.; Linb, G.S.; Hub, S.J. Comparison study on the effectiveness of stepped binder and weld line clamping pins on formability improvement for tailor-welded blanks. *J. Mater. Process. Technol.* **2008**, *207*, 204–210. [CrossRef]

9. Meyer, A.; Wietbrock, B.; Hirt, G. Increasing of the drawing depth using tailor rolled blanks—numerical and experimental analysis. *Int. J. Mach. Tools Manuf.* **2008**, *48*, 522–531. [CrossRef]

10. Chan, S.M.; Chan, L.C.; Lee, T.C. Tailored welded blanks of different thickness ratios effects on forming limiting diagrams. *J. Mater. Process. Technol.* **2003**, *132*, 95–101. [CrossRef]

11. Merklein, M.; Lechner, M. Manufacturing flexibilisation of metal forming components by tailored blanks. In Proceedings of the COMA 13th International Conference on Competitive Manufacturing, Stellenbosch, South Africa, 30 January–1 February 2013; pp. 1–6.

12. Monaco, A.; Sinke, J.; Benedictus, R. Experimental and numerical analysis of a beam made of adhesively bonded tailor-made blanks. *Int. J. Adv. Manuf. Technol.* **2009**, *44*, 766–780. [CrossRef]

13. Reisgen, U.; Schleser, M.; Mokrov, O.; Ahmed, E. Statistical modeling of laser welding of DP/TRIP steel sheets. *Opt. Laser Technol.* **2012**, *44*, 92–101. [CrossRef]

14. Sennaroglu, A. *Solid-State Lasers and Applications*, 1st ed.; CRC Press: Boca Raton, FL, USA, 2007.

15. Koechner, W.; Bass, M. *Solid-State Lasers: A Graduate Text*, 1st ed.; Springer: New York, NY, USA, 2003.

16. Hyrcza-Michalska, M.; Grosman, F. The evaluate of laser welded tailor and tubular blanks formability for automotive vehicle elements stamping. *Arch. Civ. Mech. Eng.* **2009**, *9*, 69–81. [CrossRef]

17. Piela, A.; Lisok, J.; Rojek, J. Experimental study and numerical simulation of tailor welded blanks. In Proceedings of the International Conference on Advanced Materials & Processing Technologies AMPT 2003, Dublin, Ireland, 8–11 July 2003.

18. Esquivel, A.S.; Nayak, S.S.; Xia, M.S.; Zhou, Y. Microstructure, hardness and tensile properties of fusion zone in laser welding of advanced high strength steels. *Can. Metall. Q.* **2012**, *51*, 328–335. [CrossRef]

19. Švec, P.; Schrek, A.; Hrnčiar, V.; Csicsó, T. Fibre laser welding of dual phase steels. *Acta. Metall. Slovaca* **2015**, *21*, 311–320. [CrossRef]

20. Saha, D.C.; Westerbaan, D.; Nayak, S.S.; Biro, E.; Gerlich, A.P.; Zhou, Y. Microstructure-properties correlation in fiber laser welding of dual-phase and HSLA steels. *Mater. Sci. Eng. A* **2014**, *607*, 445–453. [CrossRef]

21. Yurioka, N.; Okumura, M.; Kasuya, T.; Cotton, H. Prediction of HAZ hardness of transformable steels. *Met. Constr.* **1987**, *4*, 217R–223R.

22. Yurioka, N.; Kasuya, T. A chart method to determine necessary preheat temperature in steel welding. *Q. J. Jpn. Weld. Soc.* **1995**, *13*, 347–357.

23. ESI Group. *PAM-STAMP 2G 2011: Users's Guide*, 1st ed.; ESI Group: Paris, France, 2010.

24. Banabic, D. *Sheet Metal Forming Processes: Constitutive Modelling and Numerical Simulation*, 1st ed.; Springer: Berlin, Germany, 2010.

25. Evin, E.; Tomáš, M. *Annual Report of the Project APVV-0273-12*; Technical University of Kosice: Kosice, Slovakia, 2014.

26. Evin, E.; Tomáš, M. *Annual Report of the Project APVV-0273-12*; Technical University of Kosice: Kosice, Slovakia, 2015.

27. Evin, E.; Tomáš, M. *Annual Report of the Project APVV-0273-12*; Technical University of Kosice: Kosice, Slovakia, 2016.

28. Zhang, C.H.; Song, X.; Lu, P.; Hu, X. Effect of microstructure on mechanical properties in weld-repaired high strength low alloy steel. *Mater. Des.* **2012**, *36*, 233–242. [CrossRef]

29. Singh, S.; Nanda, T.; Ravi Kumar, B.; Singh, V. Controlled phase transformation simulations to design microstructure for tailored mechanical properties in steel. *Mater. Manuf. Process.* **2016**, *31*, 2064–2075. [CrossRef]

Article

Multiphysics Simulation and Experimental Investigation of Aluminum Wettability on a Titanium Substrate for Laser Welding-Brazing Process

Morgan Dal * and Patrice Peyre

Process and Engineering in Mechanics and Materials (PIMM) Laboratory, UMR 8006 CNRS–ENSAM, 75013 Paris, France; patrice.peyre@ensam.eu
* Correspondence: morgan.dal@ensam.eu; Tel.: +33-171-936-536

Received: 5 May 2017; Accepted: 9 June 2017; Published: 13 June 2017

Abstract: The control of metal wettability is a key-factor in the field of brazing or welding-brazing. The present paper deals with the numerical simulation of the whole phenomena occurring during the assembly of dissimilar alloys. The study is realized in the frame of potential applications for the aircraft industry, considering the case of the welding-brazing of aluminum Al5754 and quasi-pure titanium Ti40. The assembly configuration, presented here, is a simplification of the real experiment. We have reduced the three-dimensional overlap configuration to a bi-dimensional case. In the present case, an aluminum cylinder is fused onto a titanium substrate. The main physical phenomena which are considered here are: the heat transfers, the fluid flows with free boundaries and the mass transfer in terms of chemical species diffusion. The numerical problem is implemented with the commercial software Comsol Multiphysics™, by coupling heat equation, Navier-Stokes and continuity equations and the free boundary motion. The latter is treated with the Arbitrary Lagrangian Eulerian method, with a particular focus on the contact angle implementation. The comparison between numerical and experimental results shows a very satisfactory agreement in terms of droplet shape, thermal field and intermetallic layer thickness. The model validates our numerical approach.

Keywords: dissimilar joining; laser welding-brazing; finite element method; titanium; aluminium

---

1. Introduction

Assembling dissimilar alloys is widely used to reduce the weight of built structures while keeping satisfactory mechanical properties. For instance, in the aircraft industry, the material couple frequently encountered is titanium associated to aluminium. In this configuration, the titanium has better mechanical properties and is less sensitive to the corrosion attack, whereas the aluminium is 40% lighter.

The main difficulties to address when joining two dissimilar materials with conventional processes are frequently linked to the formation of intermetallic compounds. As shown by Gupta et al. [1], and confirmed by Al-Ti binary phase diagram, the aluminum has a solubility in titanium. For this reason, there is a diffusion phenomenon of the titanium into aluminium and that tends to generate intermetallic compounds, which are known to be very brittle ceramic materials that affect mechanical properties of assemblies. Moreover, the large differences in microstructures, melting temperatures and thermal properties favour the use of solid state joining methods: diffusion bonding, brazing or welding-brazing, whereas the bonding between the two materials is obtained by the diffusion of chemical species through the interface and the resulting formation of intermetallic (IM) compounds (Ti_3Al, $TiAl$, $TiAl_3$ . . .) [2,3].

A lot of works presented in literature are focused on the mechanical effect of the intermetallic layer on the interface strength. On one hand, Majumdar et al. [4] and Chen et al. [5] have studied the crack initiation and growth in Ti/Al assembly, whereas Rastkar et al. [6] have addressed the hardness changes in the intermetallic layer. On the other hand, Liu et al. [7] have worked on the numerical representation of the IM phase growth.

Chen et al. have also considered the process of welding-brazing with lasers [8] and they have focused on the improvement of the interfacial layer [9].

Previous works have been carried out to try to quantify the microstructural and mechanical properties of T40/A5754 joints in overlap configuration (Figure 1) [10], also called reactive wetting [11]. The adhesion strength was characterized by the laser shock adhesion test (LASAT) and an important sensitivity to the reaction layer thickness was shown. The best mechanical results were obtained with a small intermetallic layer thickness, around 1 or 2 micrometers. Lower thicknesses do not lead to sufficient bonding and higher values produce a brittle layer which reduces global material properties. It was also observed that, due to tensio-active effects, the aluminium oxide at the surface of the melt-pool strongly reduced its wettability on a titanium substrate. The interfacial contact surface was thus reduced resulting in lower tensile and adhesion strength for the assembly. This confirms that the mechanical resistance of a heterogeneous Ti/Al assembly was dependent on both the reaction layer area and the layer thickness. The latter is directly linked to the diffusion process, i.e., to the thermal field, and the laser parameters (velocity, intensity ...).

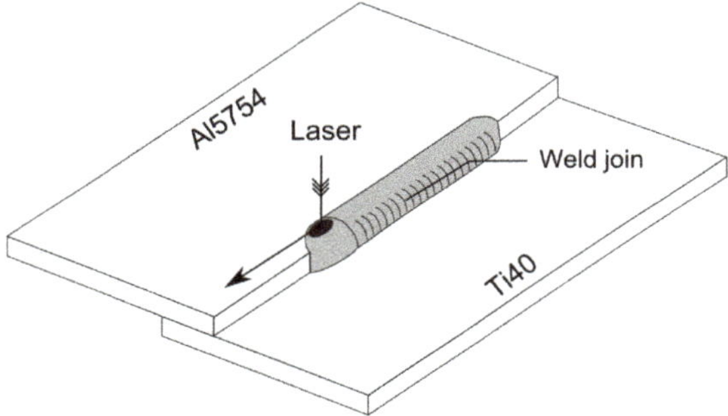

Figure 1. Scheme of the welding-brazing overlap configuration.

In this context, the aim of our numerical study was to compute the dynamics of the wetting process, to determine the temperature distribution $T = f(x, t)$ at the Ti-Al interface and then to correlate it to the TiAl$_3$ layer thickness obtained experimentally (Figure 2). In a next step, a validated model should allow for optimization of all of the Ti/Al assembly parameters.

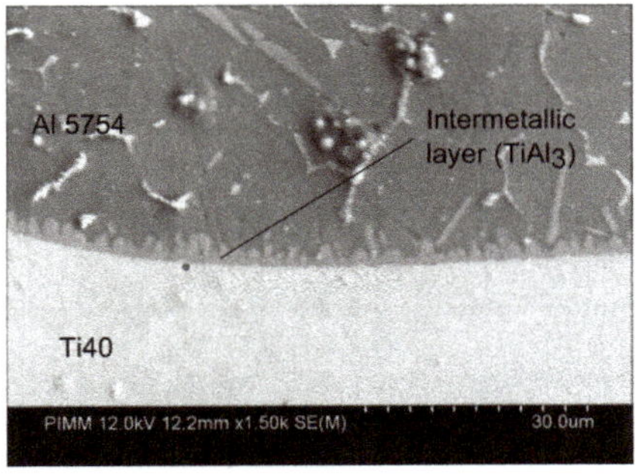

Figure 2. Scanning Electron Microscopy analysis of TiAl$_3$ serrated intermetallic layer.

It should be noted that the present study aims to propose a complete multiphysic simulation of complex phenomena occurring during the Ti/Al reactive-wetting process. In this paper, the physical problem was simplified and reduced to a bi-dimensional one, for which specific experimental measurements were implemented to validate numerical simulations. Nevertheless, the present numerical developments will be improved to further three dimensional studies.

This paper is divided into the following steps, after (1) a detailed description of the physical problem (heat transfer + fluid flow + mass transfer), the corresponding mathematical equations to be solved, and the main numerical conditions and assumptions, (2) the numerical results are presented and validated experimentally. An analysis of the process physical phenomena is then proposed (3) and future investigations are envisaged.

2. Experimental, Physical and Mathematical Description of the Ti/Al Reactive Wetting Assembly

2.1. Experimental Work

As previously indicated, the present paper is focused on a simplified Ti/Al reactive wetting assembly with a static laser irradiation. Indeed, before envisaging a more complex 3D simulation of the process, a preliminary study had to be made to ensure the validity of our thermo-hydraulic model. Moreover, in an overlap configuration, a specific focus on Al over Ti wetting parameters (surface tension and wetting angle) was expected to be an important point to address, because of a direct impact on joint shape and contact area. For this reason, the present work is dedicated to the numerical simulation of a real but simplified 2D axisymmetric case (Figure 3) where a top-hat laser beam irradiates during 1 s an aluminum cylinder that wets a titanium solid sheet.

Because aluminium is extremely sensitive to oxygen and to avoid a lack of wetting due to an alumina film between aluminium and titanium, the whole cylinder was coated by a fluor-based (Nocolok^{TM}, Solvay, Bruxelles, Belgium) anti-oxidation flux before melting. The effect of fluxing on surface tension is not well known but the influence of oxygen has already been studied, among other authors, by Molina et al. [12]. They indicated a decrease of surface tension coefficient when surfacic aluminum oxide occured, in the form of isolated Al_2O_3 islands. The corresponding surface tensions variation with temperature ($\gamma(T) = 0.883 - 0.185 \times 10^{-3}(T - 660)$) were used in the present paper. Such values were confirmed by [13].

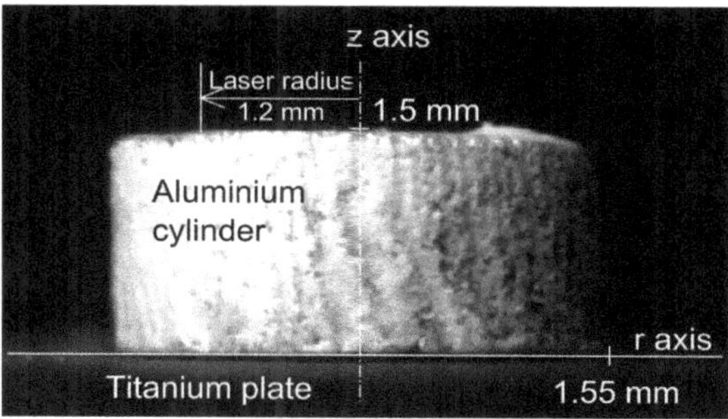

Figure 3. Axisymmetric experimental set-up of laser-assisted aluminum wetting on titanium.

2.2. Physical Description of the Reactive Wetting Phenomenon

In this sub-section, the physical phenomena are described and the corresponding mathematical models are written.

The phenomena occurring in the axisymmetric case are mostly like those appearing during the real 3D welding-brazing process.

In a first step, the laser beam heats the upper part of the cylinder in solid state. As soon as the metal is fused, a fluid flow is generated, and the aluminium droplet starts spreading on the Ti substrate. This flow is mainly driven by the thermocapillary forces but the buoyancy forces are also present. The resulting droplet has a shape which is theoretically (Young-Laplace principle [14]) dependent on the equilibrium between the surface tension and the local pressure (Figure 4). However, considering that surface tensions are highly sensitive to the temperature and because temperature is continuously varying with time during the wetting event, the process is dynamic. Similarly, the wetting angle (θ_σ) is a result of the surface tensions equilibrium (Young Duprés principle [14]) at the triple point (gas, liquid and solid) but the Al droplet is moving with time. As the substrate is not perfectly smooth, a viscous dissipation is observed. The wettability is thus highly time dependent because of the temperature dependence of the surface tension. Another important phenomenon to consider is the influence of the oxidation on surface tension values, which has already been studied by Laurent et al. [15]. A large increase of the wetting angle with the oxidation rate was experimentally observed at a constant temperature.

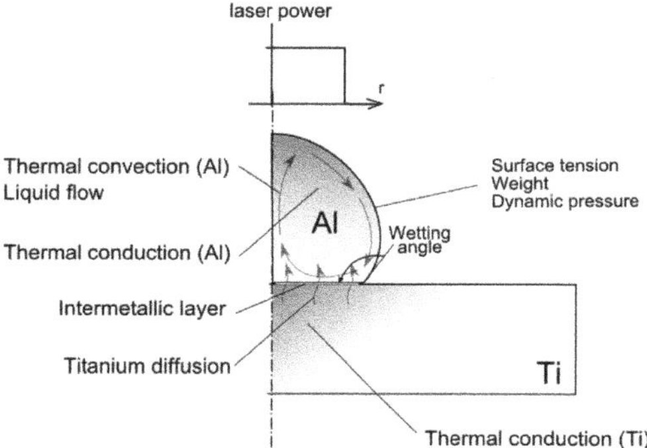

Figure 4. Simplified scheme of the physical phenomena involved during the fusion of an aluminium cylinder on a thin titanium substrate.

In a last step, once the cylinder is entirely fused and exhibits a sufficient temperature level, the titanium starts diffusing directly towards molten aluminium at the beginning of the wetting, and through a thin layer of intermetallic phase in a second step [16]. Additional contributions like fluid flow at the Ti/Al interface are also expected to favour Titanium mass transfer versus aluminium.

The inter-diffusion process occurring inside the binary Ti-Al, forms different compounds, Ti_3Al, $TiAl$, $TiAl_2$, Ti_5Al_{11}, Ti_9Al_{23} and $TiAl_3$. The phase diagram of Ti-Al, leads to two main remarks, firstly, some intermetallic phases ($TiAl_2$, Ti_5Al_{11}, Ti_9Al_{23} and $TiAl_3$) appear for a line concentration. Secondly, in case of titanium diffusion in aluminum the first intermetallic appearing is the $TiAl_3$. Moreover, as reported by Sujata et al. [2], in different works, $TiAl_3$ is the only phase formed in case of liquid aluminum reacting with solid titanium.

2.3. Mathematical Formulation

The physical problem detailed in the previous section has been addressed by solving partial differential equation (PDE) with adequate boundary conditions. The main details and assumptions of the corresponding model have been described below. The main assumption of the model is that a cylindrical coordinates system was used for the numerical calculations (simulated geometry in Figure 5), which seemed to be particularly appropriate for reproducing experiments.

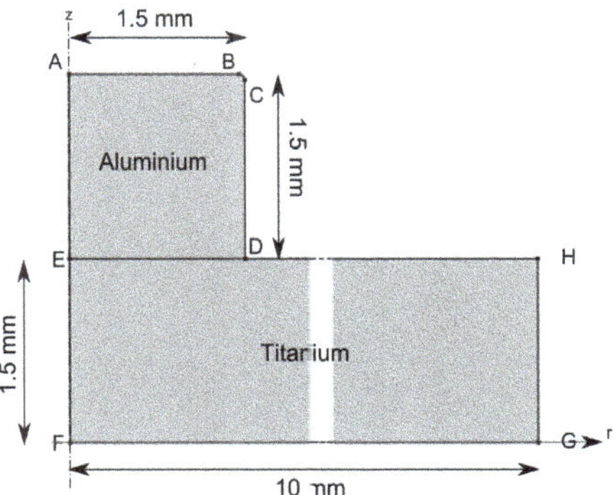

Figure 5. Geometrical model.

2.3.1. Thermal Problem

The energy conservation PDE was solved in the whole domains (Al $\cup$ Ti):

$$\rho(T)c_p^{eq}(T)\left(\frac{\partial T}{\partial t}\right) + \rho(T)c_p(T)\vec{V}\cdot\vec{\nabla}T = \vec{\nabla}\cdot\left(\lambda(T)\vec{\nabla}T\right) \tag{1}$$

In this equation, c_p^{eq} is an equivalent specific heat used to account for the latent heat of fusion (L_f) [17].

$$c_p^{eq}(T) = c_p^0(T) + L_f\frac{exp\left(\frac{-(T-T_{fus})^2}{\Delta T^2}\right)}{\sqrt{\pi\Delta T^2}} \tag{2}$$

The energy of phase change was thus distributed, through a Gaussian law, between the solidus and the liquidus temperatures (ΔT).

Thermal properties (ρ, c_p^0 and λ) were assumed to be temperature dependent for both materials and were taken from the literature [18]. Since physical properties of Al 5457 and Ti40 alloys are not easily available in literature data, the properties of equivalent materials (Al 5182 and pure titanium) were considered.

An axisymmetric condition was set at AE $\cup$ EF:

$$-\lambda(T)\frac{\partial T}{\partial n} = 0 \tag{3}$$

Convective and radiative heat losses were considered at DH $\cup$ HG $\cup$ GF:

$$-\lambda(T)\frac{\partial T}{\partial n} = h_{cv}(T - T_{inf}) + \varepsilon\sigma\left(T^4 - T_{inf}^4\right) \tag{4}$$

At AB $\cup$ BC $\cup$ CD, the aluminum workpiece was both irradiated by the top-hat (uniform) laser source and submitted to radiative or convective heat losses. Thus, the inward heat flux was written as:

$$\lambda(T)\frac{\partial T}{\partial n} = \frac{\alpha P}{\pi r_{laser}^2} - h_{cv}(T - T_{inf}) - \varepsilon\sigma\left(T^4 - T_{inf}^4\right) \tag{5}$$

It should be noted that, experimentally, aluminum and titanium workpieces were not perfectly in contact at the ED boundary (Figure 5). For this reason, the thermal contact between aluminum and titanium was addressed by a contact resistance (R_c) which is equivalent to a 10 μm layer of air before the fusion and 1 μm of intermetallic layer after the fusion (λ_{air} = 0.026 W·m^{-2}·K^{-1} et λ_{TiAl3} = 1.03

$W \cdot m^{-2} \cdot K^{-1}$ [19]). We assumed here, that the thermally-assisted diffusion of Ti to Al started just after the wetting. The interfacial condition becomes:

$$-\lambda(T)\frac{\partial T}{\partial n} = \frac{\Delta Ti}{R_c} \tag{6}$$

With ΔTi the temperature difference between the two materials at the interface.

2.3.2. Fluid Flow Problem

In order to reduce the number of degrees of freedom (DOF), the fluid flow problem was solved only in the aluminium volume. Moreover, as this material is submitted to high thermal expansion, the medium was assumed to be weakly compressible.

In the Al domain, the continuity Equation (7) and the momentum conservation (8) were solved.

$$\frac{\partial \rho(T)}{\partial t} - \vec{\nabla} \cdot \left(\rho(T)\vec{V} \right) = 0 \tag{7}$$

$$\rho(T)\frac{\partial \vec{V}}{\partial t} + \rho(T)\left(\vec{V} \cdot \vec{\nabla} \right)\vec{V} = \vec{\nabla} \cdot \left[-pI + \mu\left(\vec{\nabla}\vec{V} + \left(\vec{\nabla}\vec{V}\right)^{T} \right) - \frac{2}{3}\mu\left(\vec{\nabla} \cdot \vec{V} \right)I \right] + \vec{F_v} \tag{8}$$

where $\vec{F_v}$ is a volumic force which includes the gravity component $\rho(T)\vec{g}$ on the z direction.

It should be noted here that the previous Equation (8) is only suitable for Newtonian flows with Mach number smaller than 0.3; this condition is satisfied in the present case.

An axial symmetry i.e., a zero-normal velocity was set at the AE boundary.

$$\vec{V} \cdot \vec{n} = 0 \tag{9}$$

As titanium is expected to remain at a solid state during the whole process, the real boundary condition at ED was assumed to be a flow with no slip ($V = 0$). However, this condition nullifies the r component of the velocity at point D whereas the motion of the free boundary surface with the ALE method needs a non-zero value. For this reason, at this boundary, a condition of wall with slip was assumed:

$$\vec{V} \cdot \vec{n} = 0 \tag{10}$$

The no-slip condition was virtually set as follows, considering an artificial dynamic viscosity:

$$\mu = \left\{ \begin{array}{ll} \mu^0 & (1) \\ +10f_{sol}(T) + 1.10^5(T > T_s) & (2) \\ +100(1 - H(z - z_0, dz)) \} & (3) \end{array} \right. \tag{11}$$

With $H(...,..)$ the Heaviside function, z_0 the titanium thickness and dz a smoothing length set as small as possible (the size of the first element).

The Equation (11) is composed of three parts. The first (1) corresponds to the real property of the fluid. The second (2) allows increasing the viscosity in the mushy zone and nullifying the velocity in the solid phase. The third (3) was used to stop the flow at the titanium boundary.

The motion of boundaries $AB \cup BC \cup CD$ was computed with the Laplace law:

$$\vec{\sigma} \cdot \vec{n} = \gamma(T)\kappa \tag{12}$$

With κ is the local curvature of the free boundary and γ the surface tension assumed to be linearly dependent on the temperature:

$$\gamma(T) = \gamma^0 + \frac{\partial \gamma}{\partial T}\left(T - T_{fus} \right) \tag{13}$$

The slope of this function $\frac{\partial \gamma}{\partial T}$ is also known as the thermocapillary coefficient, responsible for the Marangoni effect.

The wetting angle was considered with a "driving" [11] or "wetting" [20] force located at the D point and written as follows:

$$f_w = \epsilon \gamma(T)\left(t_f \cdot t_s + \cos \theta_s\right) \cdot t_s \tag{14}$$

The wetting angle (θ_s) results from the equilibrium between the three surface tensions: liquid-vapour, liquid-solid and solid-vapour at the D point [14]. t_f and t_s, are the tangential unity vectors respectively to the liquid part and to the solid phase. While the interaction between the liquid and its vapour is well known in our case, the values of the two other surface tensions could not easily be found in the literature. Nevertheless, this parameter is very easy to observe experimentally and the few works found in the literature indicate a linear decrease of the angle with the temperature [21,22].

2.3.3. Free Boundary Motion

The domain mesh displacement and the free boundary motion were computed with the Arbitrary Lagrangian Eulerian (ALE) method [23]. This method is based upon a physical Lagrangian computation of the boundary shape (AB ∪ BC ∪ CD) and an arbitrary Eulerian calculation of the domain nodes displacement. The elements into the bulk are moved arbitrarily but with a smoothing method named "hyperelastic". This method searches the minimum of mesh deformation energy [24].

At boundaries AE, EF and GH the displacement was set free on the z axis and set to zero on the r axis. Conversely, the motion condition at boundaries ED, DH and FG was free on r and zero on z.

The free boundaries (AB ∪ BC ∪ CD) are then moved by the computed normal velocity (Equation (15)).

$$\vec{V} \cdot \vec{n} = u \times n_r + w \times n_z \tag{15}$$

With (u, w) and (n_r, n_z) the components of, respectively, the velocity vector and the normal vector to the boundary.

2.3.4. Mass Transfer Problem

This problem was not solved with the three others because of element sizes used for the diffusion problem (<100 nm) that leads to the resolution of a too large problem (>2,000,000 degrees of freedom). Nevertheless, as the temperature was rapidly homogeneous on all the interface aluminium/titanium, we assumed that the diffusion could be computed in 1D, along the z axis. Of course, this strong assumption will be justified in the results section.

The conservation equation that was solved here is the second Fick's law (Equation (16)):

$$\frac{\partial c_i}{\partial t} = \vec{\nabla} \cdot \left(D_{Ti/Al}(T)\vec{\nabla} c_i\right) \tag{16}$$

where c_i is the titanium concentration and $D_{Ti/Al}$ a temperature-dependent Arrhenian diffusion coefficient (Equation (17)):

$$D_{Ti/Al} = D^0_{Ti/Al} \exp\left(\frac{-\Delta H}{RT}\right). \tag{17}$$

With $D^0_{Ti/Al}$ a pre-exponential factor and ΔH the activation energy.

As the diffusion of aluminium in the titanium is very weak ($10^{-21} < D^0_{Al/Ti} < 10^{-12}$ m$^2 \cdot$s^{-1} and $\Delta H > 10^{23}$ J$\cdot$mol^{-1} [25]), we only considered the titanium diffusion towards aluminium ($D^0_{Ti/Al} = 4.29 \times 10^{-7}$ m$^2 \cdot$s^{-1} [26] and $\Delta H = 180 \times 10^3$ J$\cdot$mol^{-1} [3]). The value of activation energy here represents the diffusion of titanium in aluminium through an existing diffusion barrier, the TiAl$_3$ layer.

As the problem is mono-dimensional, only two boundary conditions were considered: (1) at the titanium/aluminum interface, the titanium concentration was assumed to be 100% (Equation (17)) and

(2) at the boundary representing the upper part of the aluminum, the titanium flux was assumed to be zero (Equation (18)).

$$c(z = 0) = 0 \tag{18}$$

$$\frac{-\partial c_i}{\partial n} = 0 \tag{19}$$

For the initial condition, we assumed that there is no titanium in the aluminum.

2.4. Material Properties

Table 1 summarizes the main model parameters used for the calculations. Some of these values were taken from the literature. Other ones were identified experimentally and the last ones are process parameters.

Table 1. Materials' properties.

Properties	Values/References
Thermal conductivity (Al-Ti)	$\lambda(T)$ [18]
Specific heat (Al-Ti)	$c_p^0(T)$ [18]
Density (Al-Ti)	$\rho(T)$ [18]
Convective coefficient	$h_{cv} = 15\ \mathrm{W \cdot m^{-2} \cdot K^{-1}}$
Aluminum Absorptivity	$\alpha = 0.18$
Aluminum Emissivity	$\varepsilon = 0.18$
Titanium Emissivity	$\varepsilon = 0.5$
Latent heat of fusion	$L_m = 2.6 \times 10^5\ \mathrm{J \cdot kg^{-1}}$
Stefan-Boltzmann constant	$\sigma = 5.67 \times 10^{-8}\ \mathrm{W \cdot m^{-2} \cdot K^{-4}}$
Aluminium melting range	815–913 K
Titanium melting range	1921–1941 K
Reference surface tension	γ^0 [25]
Surface tension variation	$d\gamma/dT = -1.44 \times 10^{-5}\ \mathrm{N \cdot m^{-1} \cdot K^{-1}}$
Aluminium Dynamic viscosity	$\mu^0 = 0.0011\ \mathrm{Pa \cdot s}$
Contact resistance	$R_c = 1 \times 10^{-4}\ \mathrm{K \cdot m^{-2} \cdot W^{-1}}$
Solid fraction	$f_{sol}(T)$ [18]
Laser power	1250 W
Beam diameter	2.4 mm

2.5. Numerical Considerations

The PDE equations were solved by the finite elements method, with the commercial software Comsol Multiphysics™ (V4.4, Comsol INC, Burlington, MA, USA, 2016).

Due to the deformation of a droplet and to the remeshing process (based on the mesh quality), triangular element shapes were chosen. The mesh (Figure 6) was built in three steps. The first meshed part is the aluminium boundaries. Then, the aluminium domain and finally, the titanium domain was freely meshed. In each domain, the elements growth factor was set at 1.3. A 40 µm maximal length of boundary elements was set in order to allow a compromise between computational times and solution accuracy. It has been performed by a convergence study. Decreasing the size of mesh has no significant effect on the results (thermal and velocity fields).

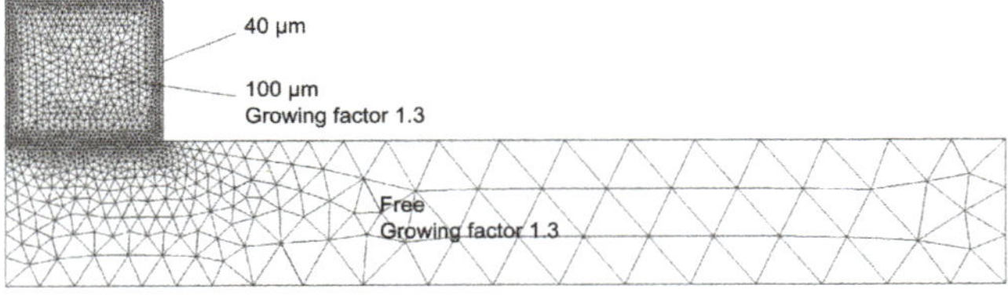

Figure 6. Mesh configuration.

Moreover, the thermal (T) and fluid flow (u, w and p) problems were solved with linear interpolations. For the ALE problem (r and z coordinates of the mesh), the geometry shape was represented with linear functions. To reduce the numerical instabilities due to linear interpolations, the calculation of velocity and pressure was stabilized by adding numerical crosswind diffusion.

This mesh leads to solve nearly 8000 degrees of freedom. A re-meshing process was used to avoid the loss of elements quality during the deformation, and each re-meshing was done when the mesh quality was below 0.2.

A fully coupled resolution was performed with the parallel sparse direct linear solver PARDISO. The numerical scheme used to solve the time-dependent problem was the implicit generalized-alpha method, and an adaptative time-stepping was used with a maximum time step of 0.0005 s. During the resolution, the use of a maximal time step value allowed for avoiding severe transition between two iterations and thus reduced time loss due to time step recalculation.

The error was controlled with an absolute tolerance of 0.01 (with the unit of each calculated variable) and a relative tolerance of 0.01%.

Such numerical conditions led to a calculation time lower than one hour for the global thermohydrodynamics process with 1.6 GHz CPU frequency on 4 threats and with 1600 MHz random access memory (on a laptop computer).

3. Results and Discussion

Before extensively using such a numerical model, we had to check the validity of its results. For this purpose, numerical solutions were compared to a dedicated Al/Ti reactive wetting experiment, where appropriate diagnostics were considered to validate or reject the numerical solution. As indicated in Section 1, the following diagnostics were used: (1) a C-Mos fast camera to monitor the droplet shape evolution with time, (2) thermocouples to measure the temperature of the titanium surface, and (3) optical micrographs to determine the reactive layer thickness.

As shown in Figure 3, a camera was positioned in front of the work-piece. This camera analysis, using a frame rate of 100 Hz, recorded the transient evolution of the droplet shape. In order to have a sufficient lighting of the work-piece, a halogen light is used.

A type-K thermocouple was spot-welded on the titanium surface (aluminium side) at a 5 mm distance from the laser beam axis (Figure 7). As the thermal measurement is quite far from the aluminium drop, the sensitivity is quite poor, and the TC data was mainly used to measure the initial temperature and to observe the diffusion through the two mediums.

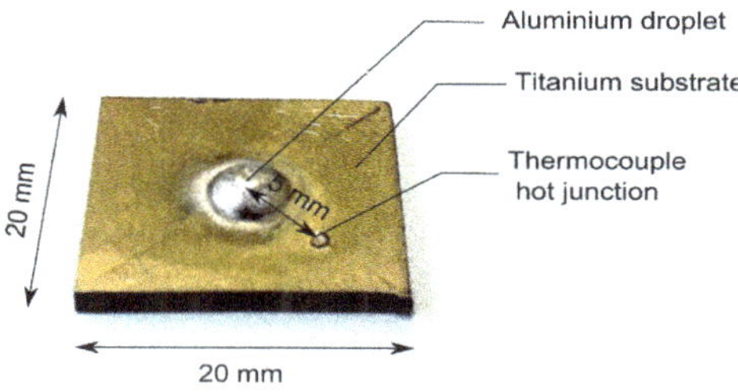

Figure 7. Experimental configuration.

Calculations were carried out using process parameters that were summarized in Table 1.

3.1. Droplet Shape Dynamics

Figure 8 shows a comparison between the experimental and numerical droplet shape variation with time. For each time step, the simulated droplet is placed besides the corresponding experimental picture. The numerical model is shown to provide a satisfactory agreement with fast droplet distortions.

Figure 6 shows three steps in the droplet behaviour. (1) the fusion process occurs until $t = 0.15$ s, then (2) a spheroidisation is observed between $t = 0.15$ s and $t = 0.25$ s and (3) an increase of wetting appears until the end of the shot.

At $t = 0.1$ s the camera observation allows distinguishing between the three classical zones: the liquid phase (on the top), an intermediate zone (probably not exactly the mushy zone) and the solid phase (at the bottom). Simulation results clearly show the isothermal lines corresponding to the solidus temperatures, which correspond quite well to camera observations. Similar conclusions can be found at 0.15 s time for the transition between liquidus and intermediate zone. In this picture, only two parts are shown: the liquid (which exhibits a mirror-like surface aspect) and a partially oxidized liquid zone (the bottom). Both wetting times globally validate the transient thermal field, and confirm macroscopically the thermal assumptions of the model.

In Figure 8, arrows are used to show the convection flow. The first step corresponds to the initiation of convection flow at the droplet surface by the Marangoni convection (thermocapillary effect). The second step is the creation of two convective cells, moving with time inside the droplet contour.

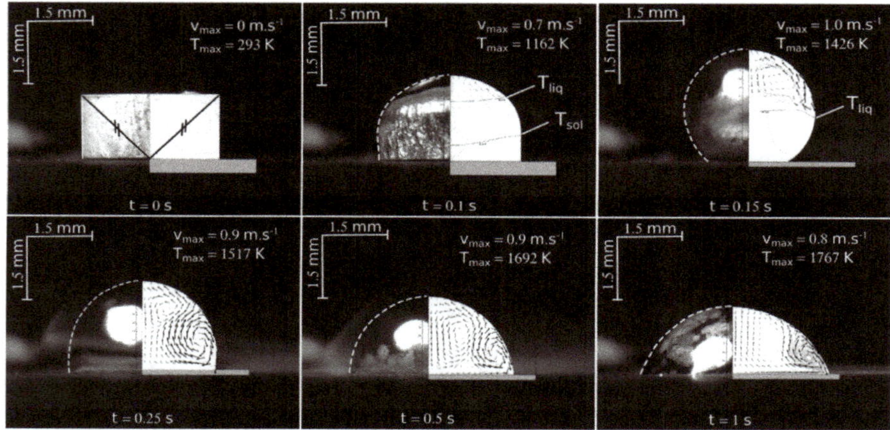

Figure 8. Experimental and numerical shapes comparison.

3.2. Thermal Validation

Figure 9 shows a comparison of experimental and numerical temperature versus time thermal cycles, considering one point of the titanium substrate, 5 mm distant from the laser beam axis (cf. Figure 7). A rather good agreement can be shown between numerical and experimental results. Both the dynamic evolution with time and the maximum amplitude are well reproduced by the model, except concerning the first temperature peak observed experimentally for the short times (between 0.1 and 0.3 s), which is attributed to a laser beam reflexion towards the thermocouple. Indeed, this time interval corresponds to the high droplet deformation phase (transition between no-wetting and wetting, Figure 8).

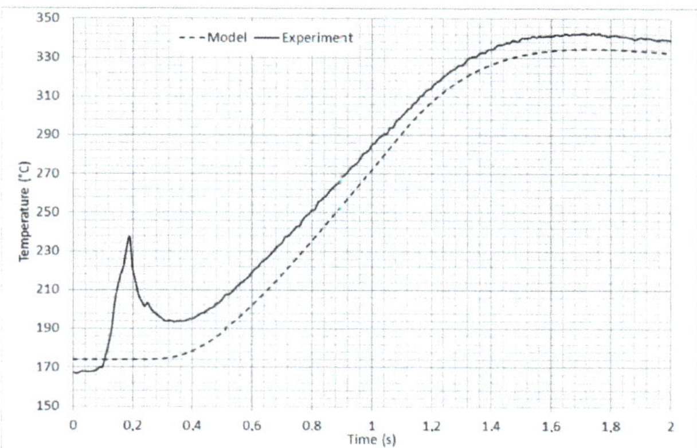

Figure 9. Experimental and numerical temperature versus time profile at a 5 mm distance from the droplet axis.

3.3. Intermetallic Layer

After validating the thermal and fluid flow numerical model, the mass transfer problem was addressed, with the objective of computing the intermetallic layer thickness.

Before considering numerically, the diffusion problem we have checked the assumption concerning the thermal homogeneity of the interface. Figure 10 shows the temperature distributions along the radius for different time steps. This graph exhibits a quite homogeneous temperature distribution along the r axis, which validates afterwards our initial assumption of interfacial thermal homogeneity.

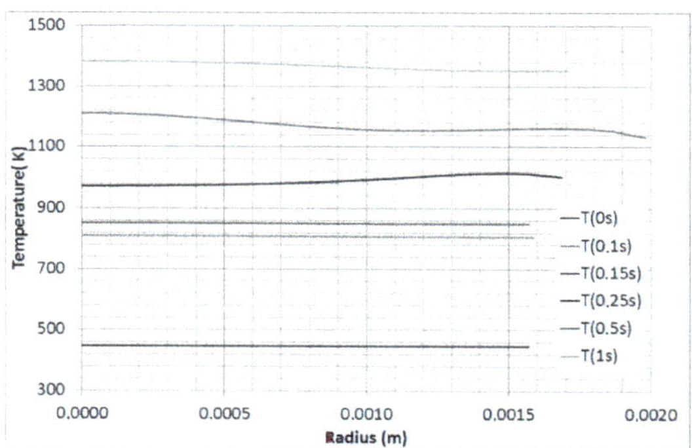

Figure 10. Interface temperature distribution with time along-r radial distance

With this graph, we observe a quite homogeneous temperature along the r axis, thus our assumption seems to be realistic.

The results of the diffusional problem are shown in Figure 11 where the titanium concentration is put along the symmetry axis for different time whereas Figure 12 shows the experimentally observed layer thickness. The horizontal line shown in Figure 11 is the assumed level of $TiAl_3$ formation (25%Ti and 75%Al). The numerical value obtained at the last time step is 2 µm. Compared to the experimental observation (1.5 µm), the numerical result was considered satisfactory, considering the assumptions made on input parameters, and especially on diffusion coefficient and activation energy.

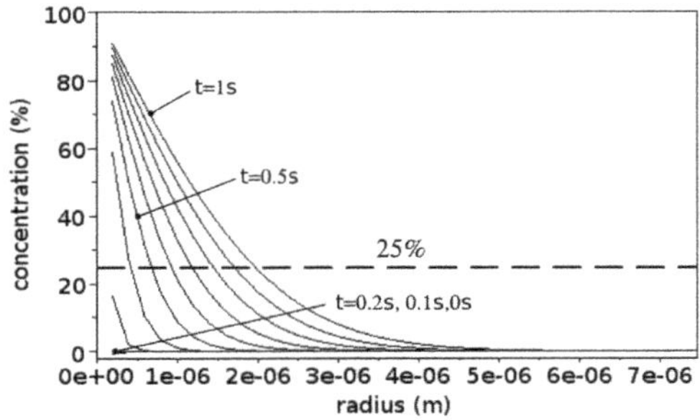

Figure 11. Titanium concentration in aluminum along the *z* axis.

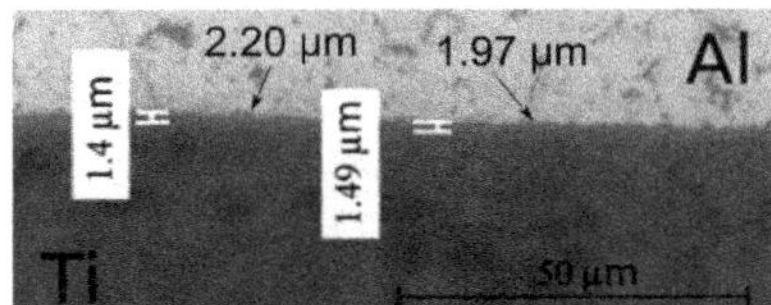

Figure 12. Experimental measurement of intermetallic thickness.

This comparison between simulation and experiment leads to 25% of relative error in terms of intermetallic thickness. This value seems to be very high; nevertheless, it is satisfactory considering the high level of uncertainty on diffusion parameters coming from the literature. Moreover, as shown in Figure 2, the layer is very serrated thus accurate experimental measurements are very difficult. For instance, two measurements made at local maximums (arrows in Figure 12) give values higher or equivalent to the 2 µm coming from simulation.

4. Conclusions

As a conclusion, this paper gives the different steps of the reactive wetting process simulation (welding-brazing process). The full assumption and mathematical models are described and justified. A thermo-hydrodynamic problem is firstly considered and then the diffusion of species is computed. The results, in terms of thermal field, velocity field and intermetallic thickness were compared to an experiment and results are very satisfactory.

The numerical layer thickness is a little bit larger than the experimental one (0.5 µm). Nevertheless, the simulated intermetallic layer is assumed to be homogeneous on the whole contact surface, which is not the case experimentally. While the aspect of intermetallic grain growth should be introduced in the model in order to obtain these level of details, the numerical resources available do not allow us to treat it. Indeed, the scale difference between the thermal problem (millimeter) and the diffusion problem (micrometer) and the resulting mesh lead to a large problem. That is why the uncoupled resolution looks to be the current best solution.

In future works, this method will be applied to the real welding-brazing case (overlap configuration) which means that the geometry will be three dimensional. The results will help the process user to set coherent process parameters.

Acknowledgments: The authors are very grateful to the Carnot institute for financial support.

Author Contributions: Patrice Peyre conceived, designed and performed the experiments; Morgan Dal analyzed the data, conceived the model, and performed the simulation; Morgan Dal wrote the paper.

References

1. Gupta, S.P. Intermetallic compounds in diffusion couples of Ti with an Al-Si eutectic alloy. *Mater. Charact.* **2003**, *49*, 321–330. [CrossRef]
2. Sujata, M; Bhargava, S.; Sangal, S. On the formation of TiAl$_3$ during reaction between solid Ti and liquid Al. *J. Mater. Sci. Lett.* **1997**, *16*, 1175–1178.
3. Wang, T.; Lu, Y.X.; Zhu, M.L.; Zhang, J.S. Identification of the comprehensive kinetics of thermal explosion synthesis Ti + 3Al $\rightarrow$ TiAl$_3$ using non-isothermal differential scanning calorimetry. *Mater. Lett.* **2002**, *54*, 284–290. [CrossRef]
4. Majumdar, B.; Galun, R.G.; Weisheit, A.; Mordike, B.L. Formation of a crack-free joint between Ti alloy and Al alloy by using a high-power CO$_2$ laser. *J. Mater Sci.* **1997**, *32*, 6191–6200. [CrossRef]
5. Chen, Y.; Chen, S.; Li, L. Influence of interfacial reaction layer morphologies on crack initiation and propagation in Ti/Al joint by laser welding–brazing. *Mater. Des.* **2010**, *31*, 227–233. [CrossRef]
6. Rastkar, A.R.; Parseh, P.; Darvishnia, N.; Hadavi, S.M.M. Microstructural evolution and hardness of TiAl$_3$ and TiAl$_2$ phases on Ti–45Al–2Nb–2Mn–1B by plasma pack aluminizing. *Appl. Surf. Sci.* **2013**, *276*, 112–119. [CrossRef]
7. Liu, J.P.; Luo, L.S.; Su, Y.Q.; Xu, Y.J.; Li, X.Z.; Chen, R.R.; Guo, J.J.; Fu, H.Z. Numerical simulation of intermediate phase growth in Ti/Al alternate foils. *Trans. Nonferr. Met. Soc. China* **2011**, *21*, 598–603. [CrossRef]
8. Chen, S.; Li, L.; Chen, Y.; Huang, J. Joining mechanism of Ti/Al dissimilar alloys during laser welding-brazing process. *J. Alloys Compd.* **2011**, *509*, 891–898. [CrossRef]
9. Chen, S.; Li, L.; Chen, Y.; Dai, J.; Huang, J. Improving interfacial reaction nonhomogeneity during laser welding–brazing aluminum to titanium. *Mater. Des.* **2011**, *32*, 4408–4416. [CrossRef]
10. Peyre, P.; Berthe, L.; Dal, M.; Pouzet, S.; Sallamand, P.; Tomashchuk, I. Generation and characterization of T40/A5754 interfaces with lasers. *J. Mater. Process. Technol.* **2014**, *214*, 1946–1953. [CrossRef]
11. Dezellus, O. Contribution à L'étude des Mécanismes du Mouillage Réactif. Ph.D. Thesis, Institut National Polytechnique de Grenoble, Grenoble, France, 12 May 2000.
12. Molina, J.M.; Voytovych, R.; Louis, E.; Eustathopoulos, N. The surface tension of liquid aluminium in high vacuum: The role of surface condition. *Int. J. Adhes. Adhes.* **2007**, *27*, 394–401. [CrossRef]
13. Garcia-Cordovilla, C.; Louis, E.; Pamies, A. The surface tension of liquid pure aluminium and aluminium-magnesium alloy. *J. Mater. Sci.* **1986**, *21*, 2787–2792. [CrossRef]
14. De Gennes, P.G.; Brochard-Wyart, F.; Quéré, D. *Gouttes, Bulles, Perles et Ondes*; Belin: Paris, France, 2005; p. 254.
15. Laurent, V.; Rado, C.; Eustathopoulos, N. Wetting kinetics and bonding of Al and A1 alloys on α-SiC. *Mater. Sci. Eng.* **1996**, *A205*, 1–8. [CrossRef]
16. Luo, J.G.; Acoff, V.L. Interfacial Reactions of Titanium and Aluminum during Diffusion Welding. *Weld. J.* **2000**, *79*, 239–243.
17. Bonaccina, C.; Comini, G.; Fasano, A.; Primicerio, M. Numerical solution of phase-change problems. *Int. J. Heat Mass Transf.* **1973**, *16*, 1825–1832. [CrossRef]
18. Mills, K.C. *Recommended Values of Thermophysical Properties for Selected Commercial Alloys*; Woodhead Publishing: Cambridge, UK, 2002; p. 246.
19. Duan, Y.H.; Sun, Y.; Lu, L. Thermodynamic properties and thermal conductivities of TiAl$_3$-type intermetallics in Al–Pt–Ti system. *Comput. Mater. Sci.* **2013**, *68*, 229–233. [CrossRef]
20. Teyssèdre, H.; Gillormini, P. Extension of the natural element method to surface tension and wettability for the simulation of polymer flows at the micro and nano scales. *J. Non-Newtonian Fluid Mech.* **2013**, *200*, 9–16. [CrossRef]
21. Champion, J.A.; Keene, B.J.; Sillwood, J.M. Wetting of Aluminium Oxide by Molten Aluminium and Other Metals. *J. Mater. Sci.* **1969**, *4*, 39–49. [CrossRef]
22. Han, D.S.; Jones, H.; Atkinson, H.V. The wettability of silicon carbide by liquid aluminium: The effect of free silicon in the carbide and of magnesium, silicon and copper alloy additions to the aluminium. *J. Mater. Sci.* **1993**, *28*, 2654–2658. [CrossRef]
23. Hirt, C.W.; Amsden, C.C.; Cook, J.L. An arbitrary lagrangian-eulerian computing method for all flow speeds. *J. Comput. Phys.* **1997**, *135*, 203–216. [CrossRef]

24.   Multiphysics, C. *Comsol Multiphysics Reference Manual*; COMSOL: Grenoble, France, 2013; p. 1084.
25.   Mishin, Y.; Herzig, C. Diffusion in the Ti-Al system. *Acta Mater.* **2000**, *48*, 589–623. [CrossRef]
26.   Du, Y.; Chang, Y.A.; Huang, B.; Gong, W.; Jin, Z.; Xu, H.; Yuan, Z.; Liu, Y.; He, Y.; Xie, F.Y. Diffusion coefficients of some solutes in FCC and liquid Al: Critical evaluation and correlation. *Mater. Sci. Eng.* **2003**, *A363*, 140–151. [CrossRef]

 *metals*

Article

Laser Beam Welding of a Ti–6Al–4V Support Flange for Buy-to-Fly Reduction

Fabrizia Caiazzo [1], **Vittorio Alfieri** [1,*], **Gaetano Corrado** [1], **Paolo Argenio** [1], **Giuseppe Barbieri** [2], **Francesco Acerra** [3] and **Vincenzo Innaro** [3]

[1] Department of Industrial Engineering, University of Salerno, Via Giovanni Paolo II 132, Fisciano 84084, Italy; f.caiazzo@unisa.it (F.C.); gcorrado@unisa.it (G.C.); pargenio@unisa.it (P.A.)

[2] Department of Sustainability, ENEA Casaccia, Via Anguillarese 131, Santa Maria di Galeria 00123, Italy; giuseppe.barbieri@enea.it

[3] Department of Aerostructure Technology and Development, Leonardo Company S.p.A., Viale dell'Aeronautica snc, Pomigliano d'Arco 80038, Italy; francesco.acerra@leonardocompany.com (F.A.); vincenzo.innaro@leonardocompany.com (V.I.)

*   Correspondence: valfieri@unisa.it; Tel.: +39-089-964-086

Academic Editors: João Pedro Oliveira and Zhi Zeng
Received: 12 April 2017; Accepted: 16 May 2017; Published: 20 May 2017

Abstract: Titanium and its alloys are increasingly being used in aerospace, although a number of issues must be addressed. Namely, in the framework of welding to produce complex parts, the same mechanical strength and a reduced buy-to-fly ratio are desired in comparison with the same components resulting from machining. To give grounds to actual application of autogenous laser beam welding, Ti–6Al–4V L- and T-joints have been investigated in this paper, as they are a common occurrence in general complex components. Discussions in terms of possible imperfections, microstructure, and microhardness have been conducted. Then, a real part consisting of a support flange for aerospace application has been chosen as a valuable test-article to be compared with its machined counterpart both in terms of strength and buy-to-fly. The feasibility and the effectiveness of the process are shown.

Keywords: laser beam welding; buy-to-fly; L-joint; T-joint; Ti–6Al–4V

---

1. Introduction

New materials and methods are being tested for aerospace in order to meet the challenges of innovation and reduce operating costs, but extensive studies are mandatory before introducing any change in industrial environments, as strict standards apply. In this regard, mechanical assembly is generally preferred because a reduction in waste material is achieved compared with its machined counterpart; this trend leads to shorter lead times and lower buy-to-fly ratio (i.e., the weight ratio between the amount of raw material to manufacture a component and the amount of the final part) [1]. Namely, improvements at the design stage are aimed to remove any mechanical fastening, such as screwing and riveting to introduce welded assemblies, thus preventing any increases in thickness.

Laser beam welding (LBW) is regarded as the logical solution to accomplish these different needs, as a number of benefits are provided in comparison with conventional welding technologies [2,3]. For instance, the process can be performed in remote locations and over three-dimensional components from single side access [4], autogenously and generally with no need for post-processing such as mechanical finishing; increased processing speed is achieved, and as a consequence, productivity is improved [5]. Advantages come from the primary feature of narrowly focusing the heat source; deep penetration is achieved when performing the process in key-hole mode, thus focusing the beam energy where required, preventing overheating of the base metal, which would suffer from thermal distortion and degradation of metallurgical properties otherwise [6].

A number of papers are available in the literature dealing with LBW of titanium alloys, which are widely used in aerospace thanks to high strength in combination with low density and good tensile properties; medical and surgical devices are even produced thanks to high biocompatibility [7]. Given these reasons, Ti–6Al–4V accounts for more than half of all titanium tonnage in the world and no other titanium alloy is deemed to threaten such a dominant position [8]: it is normally and extensively employed for turbine disks, compressor blades, airframe and space capsule structural components, rings for jet engines, pressure vessels, rocket engine cases, helicopter rotor hubs, fasteners, and engine exhausts [9]; thick titanium plates of up to 16 mm, for seagoing vessels and submarines, are even welded by means of LBW in the maritime industry [10].

In comparison with traditional technologies, tight laser beams have been proved to be effective in reducing both angular distortion and longitudinal bending on thin sheets [11]. Furthermore, a reduced mean grain size in the fusion zone is achieved; the overall mechanical quality is hence improved, considering that the grain growth is deemed to be one of the reasons for the reduction of tensile strength upon welding [12]. On the other hand, a significant reduction in ductility may occur due to aluminum oxides and micro pores. Provided that inert shielding is accomplished for the purpose of bead protection to produce sound joints and prevent oxidation [13], it has been shown that as few as 2% total porosity implies an 85% decrease in the ultimate tensile strength of the joint with respect to the base metal [12]; moreover, due to the specific evolution of the thermal fluid flow, porosity is found to occur mainly when the fusion zone is only partially penetrating through the thickness of the material [14]. In this way, uniform flow in the welding pool can be driven by the shielding gas, when the diameter of the key-hole is made larger and more stable [15].

A reduction of the mechanical properties may also result from non uniform fusion and geometry imperfections such as undercut and shrinkage groove [16] in the cross-sections, although a reliable predictive model to assess the extent of these as a function of the processing parameters has been proven to be unfeasible [13], as for other metal alloys [17]. Therefore, wire feeding [18] or hybrid welding with an assisting gun [19] are considered the usual practice to prevent imperfections.

Changes in the base microstructure and microhardness are produced as a consequence of welding. The parent metal is a two phase allotropic alloy [9]: the typical annealed microstructure consists of α hexagonal close-packed matrix with a body-centered cubic β phase at grain boundaries. A non–diffusional transformation into a martensitic α' microstructure is induced upon welding and cooling [8,20], hence the resulting hardness of the fusion zone is increased with respect to the parent metal [21]. Namely, a remarkable increase in the order of even 140 HV normally occurs in the fusion zone, although no clear trends in the mean value are reported as a function of the welding speed [8].

In order to improve ductility and fatigue properties in the fusion zone, post-welding heat treatment is required [22,23]. It has been reported in the literature [24] that full annealing of welds at 700 °C for 2 h results in improved fracture toughness in the operating range from room temperature to 500 °C, with a moderate decrease of hardness (i.e., from 361 to 356 HV on average) in the fusion zone.

To provide further understanding of LBW of Ti–6Al–4V and to provide valuable insight to be used in real parts, a number of tests are discussed in this paper relating to both L-joint (i.e., corner joint, fillet joint) and T-joint (i.e., right-angled joint), as both of them occur commonly in general complex components. The processing window has been set, then the outcome in terms of deviations from the intended geometry, the microstructure and the microhardness has been discussed. Furthermore, a proper test-article for aerospace application has been chosen, welded and tested to assess the effectiveness of LBW on actual components where multiple welds are required. Namely, welding, destructive and non destructive testing on a support flange have been conducted; a comparison has been drawn with the machined counterpart and the improvement in terms of buy-to-fly ratio has been eventually discussed.

2. Materials and Methods

2.1. Process Set-Up

To set the processing window, L- and T- joints (Figure 1) have been investigated; 3 and 5 mm thick plates have been considered. The welding set-up to access the joints with the laser beam has been chosen to comply with usual customer regulations in aerospace, since uniform complete penetration is mandatory and transparent welding, as investigated before [25], is not allowed. As a consequence, a proper approaching angle α of the laser beam to access the joint from inside must be considered. Furthermore, in view of shifting the results on a real component and given the need for combined, integrated clamping and effective inert shielding of the welding bead (i.e., top- and back-side shielding), autogenous welding has been considered; additional complex and expensive devices for wire feeding or hybrid welding would be required otherwise, whereas proper room for managing the laser head along the welding path on the test-article is crucial.

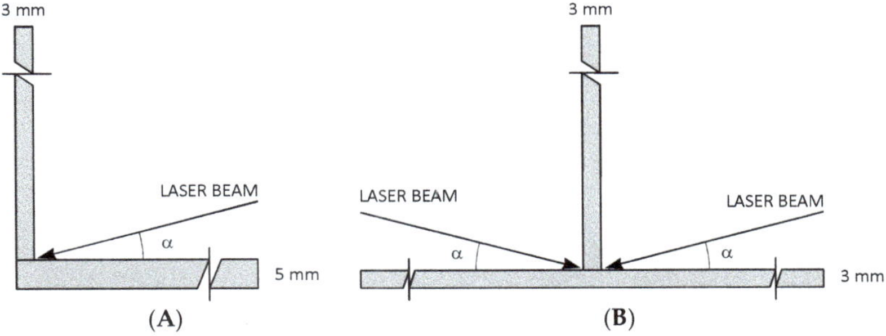

Figure 1. Welding joints to be investigated in the process set-up, (**A**) L-joint and (**B**) T-joint.

Although widespread research has been conducted with the same laser source on the same alloy on square butt joints [13,26], a convenient adjustment of the processing window has been required to comply with the current geometry. Irrespective of the joint type, the laser beam nominal power has been taken as a constant, 6 kW in continuous wave operation mode; a focused beam has been delivered to the theoretical interface. As regarding the L-joint, the effect of welding speed has been investigated; the beam angle to approach the joint has been taken as a constant, 25°, with respect to the 5 mm thick plate. In regards to the T-joint, a fully penetrative bead must be obtained by means of two welding passes, one pass at each side of the joint, each pass being partially penetrative (Figure 2); hence the welding speed has been conveniently adjusted (i.e., increased) with respect to the L-joint and the approaching angle α has been investigated for the purpose of uniform fusion along the interface. A recap of the processing conditions to be tested on sample joints is given in Table 1.

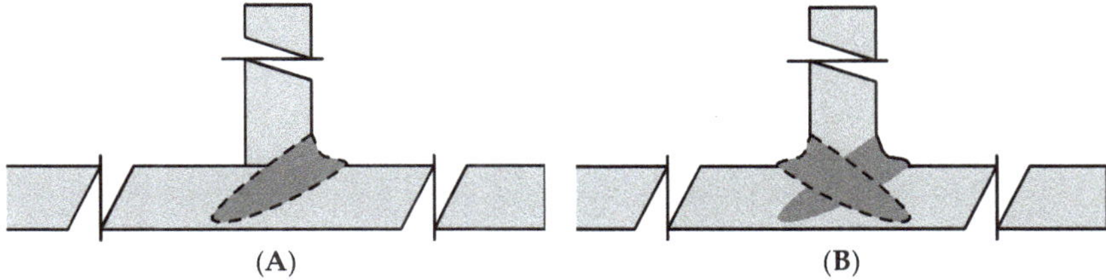

Figure 2. T-joint, scheme for welding by means of two passes: (**A**) first pass, (**B**), second pass, each one being partially penetrative.

Table 1. Processing conditions for sampling, based on the type of joint; focused beam.

Joint Type	Power (kW)	Speed (mm·min^{-1})	Approaching Angle (°)
L-joint	6.0	2400 3000 3600	25
T-joint	6.0	4800	20 25 30

Positions of the devices to perform laser beam welding (LBW) on the specimens is a carryover of a patent [27]. With respect to the direction of welding, a leading nozzle and helium supplying has been used for metal plasma blow. A trailing diffuser has been considered for argon shielding of the top surface of the welds instead. Shielding of the root in case of fully penetrative welding on L-joints has been accomplished by means of additional argon supply via a grooved box. Based on preliminary trials, a flow rate of 10 L/min has been set for assisting helium, whereas the argon flow for the surface and root has been set to 50 and 30 L/min, respectively.

Prior to welding, the samples have been manually polished with abrasive grinding paper, then degreased. LBW has been performed using a Yb:YAG fiber laser source (IPG Photonics, Oxford, MI, USA), whose main technical features are given in Table 2; three specimens have been processed for each welding condition. A 735 °C, 120 min heat treatment has been conducted in vacuum on the welding beads (TAV Vacuum Furnaces, Caravaggio, Italy) to the purpose of stress relieving. Inspections in three cross-sections coming from each weld have been conducted upon chemical etching with a solution of hydrofluoric acid (48%, 10 mL), nitric acid (65%, 15 mL), and water (75 mL) at room temperature. Vickers microhardness testing (Zwick Roell, Ulm, Germany) has been performed to investigate the response of the material in terms of microstructure evolution; namely, an indenting load of 0.300 kgf has been used for a dwell period of 10 s; a 150 μm step has been allowed between consecutive indentations, in agreement with the usual international standards on Vickers testing [28].

Table 2. Laser welding system, main technical data.

Parameter	Value
Maximum output power (kW)	10.0
Laser light wavelength (nm)	1030
Beam parameter product (mm × mrad)	6.0
Delivering fibre core diameter (μm)	200
Focus diameter (μm)	300

2.2. The Test-Article

A Ti–6Al–4V support flange test-article (Figure 3) has been processed. The part is designed to be used where metals and composites are in place; to this purpose, titanium alloys are suggested thanks to better corrosion resistance with respect to aluminum. The specimen is composed of three parts, resulting from water-jet cutting (Waterjet Corporation, Monza, Italy) and post-process milling (EMCO GmbH, Hallein, Austria) of the abutting surfaces: a 5 mm thick cap-plate, a 3 mm thick web-plate and a 3 mm thick supporting rib (Figure 4). Both L- and T-joint are involved, the former to weld the cap-plate to the web-plate, the latter to weld the rib, at its sides, with the web- and the cap-plate.

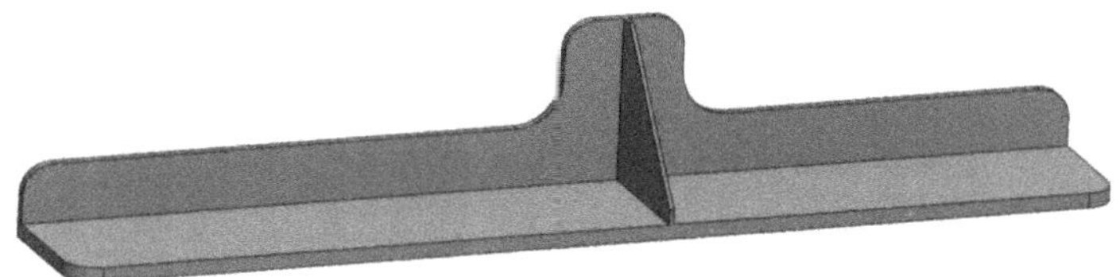

Figure 3. The test-article for the assessment of laser beam welding (LBW) on titanium alloy.

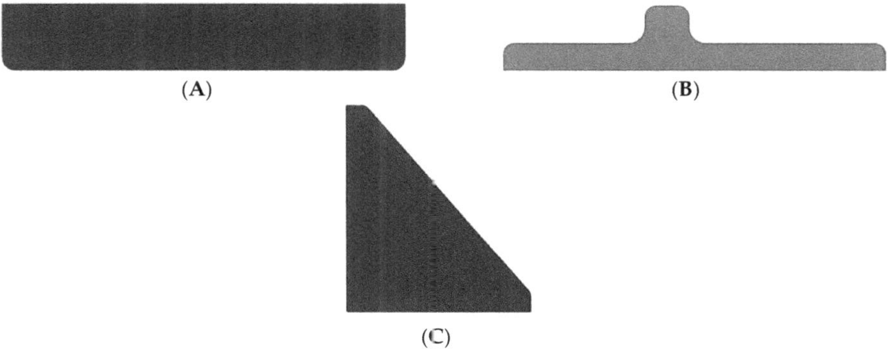

Figure 4. Components of the test-article: (**A**) cap-plate, (**B**) web-plate and (**C**) supporting rib; not to scale.

A proper method to prevent any possible deviation from the intended geometry must be taken into account to match specific customer regulations. Wire feeding or hybrid welding with an assisting gun are not fit to LBW of the test-article, as additional devices would be required in the system set-up. Therefore, for the purpose of easier automation and control of the actual process, increased thickness has been considered for the parts to be welded, 8 mm instead of 5 mm for the cap-plate, 5 mm instead of 3 mm for the web-plate and the rib: the driving idea is to perform machining by means of side and face milling upon welding in order to restore the nominal thickness of the test-article, thus removing any undercut and shrinkage groove concurrently. A significant favorable reduction of the buy-to-fly ratio is deemed to be achieved anyway with respect to machining from wrought metal and will be proven in the following.

Therefore, the welding parameters resulting from the experimental campaign (i.e., the process set-up) have been conveniently adjusted, so to effectively weld L- and T-joints of increased thickness (Table 3). Basically, the approaching angle and the power level have been taken, whilst the processing speed has been decreased. Nevertheless, in order to obtain deep penetrative welds, a 1 mm offset of the laser beam with respect to the theoretical welding line and towards the thinner plate has been required for L-joint configuration. Flow rates of assisting helium and argon supply have been taken as constant.

Table 3. Processing conditions for laser beam welding (LBW) of the test-article, based on the type of joint; focused beam.

Joint Type	Power (kW)	Speed (mm·min^{-1})	Approaching Angle (°)
L-joint (5 mm web-plate to 8 mm cap-plate)	6.0	1650	25
T-joint (5 mm rib to 8 mm cap; 5 mm to 5 mm web-plate)	4.5	1650	20

A non-commercial device to clamp the components of the test-article and concurrently provide inert shielding has been developed for the mere purpose of this research, consisting of two right-angled steel supports to accommodate the cap- and the web-plate, providing a hollow duct at their virtual intersection for the purpose of gas supply for back-side shielding. The cap- and the web-plate have been clamped using adjustable plain clamps (Figure 5); upon local spot welding with 100 mm step to better tighten the components to prevent unwanted slippage from the nominal position, LBW in L-joint configuration has been performed in a single pass with the suggested processing parameters and a 1 mm offset of the laser beam towards the web-plate with respect to the theoretical welding line (Figure 6) has been set. Assisting gases have been supplied: argon at back-side via the hollow duct on the clamping system, at top-side via the diffuser moving with the laser head; helium via the leading nozzle for metal plasma blow (Figure 7): a proper time delay has been allowed before welding to prepare inert atmosphere with stable inert flux; additional delay has been required to effectively shield the bead at laser switch-off.

Figure 5. Adjustable plain clamping of cap- and web-plates.

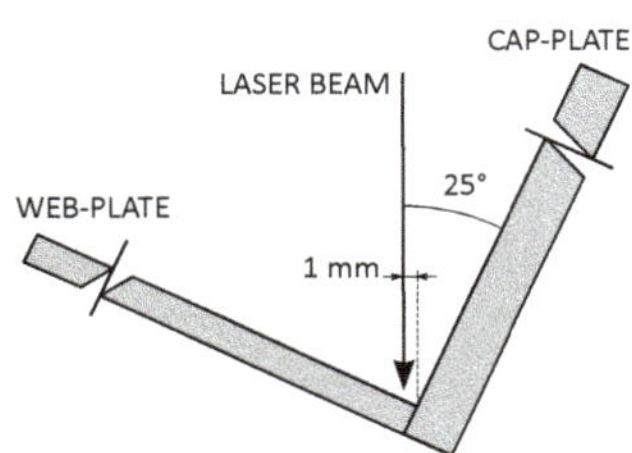

Figure 6. Beam offset towards the web-plate with respect to the theoretical welding line.

Figure 7. Positioning of the nozzle for plasma blow and the diffuser for inert shielding supply; back-side view in the inset.

The device to clamp the rib has been placed afterwards. An additional part is used to house, position, and concurrently shield the rib (Figure 8): a movable clamp allows tightening of the rib; shielding is supplied at both sides via separate ducts. Upon local spot welding to prevent unwanted slippage from the nominal position, LBW in T-joint configuration has been performed with two welding passes, one pass at each side of the joint, each pass being partially penetrative, with the suggested processing parameters. Helium has been supplied for top-side shielding; additional flow for metal plasma blow has not been arranged, the purpose being fulfilled by helium itself when leaving; argon has been supplied for back-side shielding against oxidation due to overheating, although full penetration is prevented.

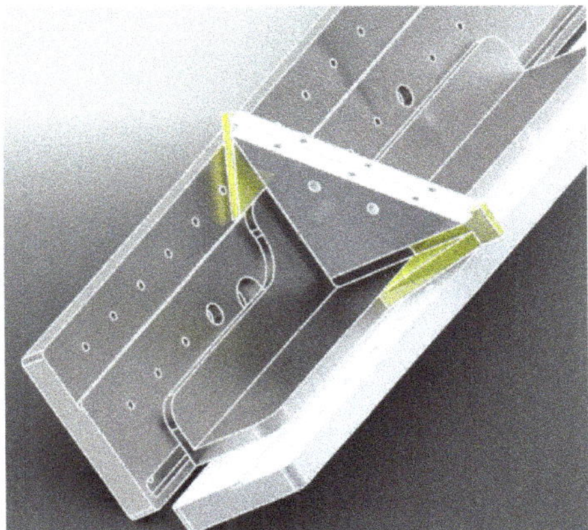

Figure 8. Detail of the device for housing, positioning and concurrent shielding of the rib.

Three test-articles have been welded (Figure 9), then heat treatment has been conducted, as for the samples in the process set-up. Machining by means of side and face milling has been performed to achieve the nominal thickness of the parts upon welding. In addition, ultrasonic non-destructive testing (GE Panametrics, Fairfield, United States, with custom interface developed by ENEA) have been performed to inspect the penetration depth.

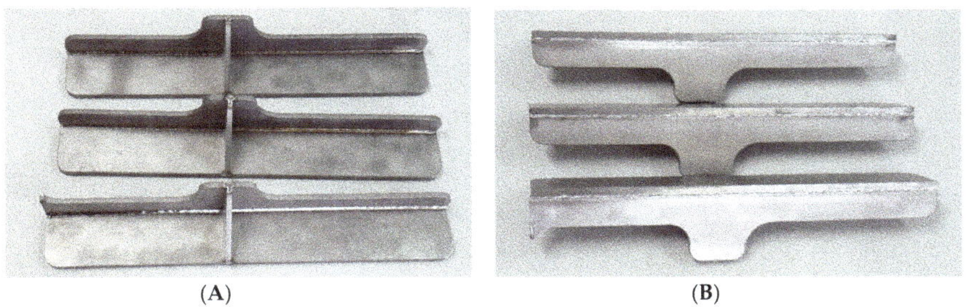

(A)   (B)

Figure 9. Test-articles as resulting from LBW: (**A**) inner view; (**B**) outer view.

3. Results and Discussion

3.1. Geometry

With respect to the L-joints, for a given approach angle of 25°, the effect of the welding speed has been discussed. Weldability is proven, although discontinuous joining at the interface of the abutting plates may occur for increasing welding speed (Figure 10). Therefore, as uniform fusion along the interface is mandatory for aerospace application of LBW, a welding speed of 2400 mm·min^{-1} is thought to be adequate to perform the process.

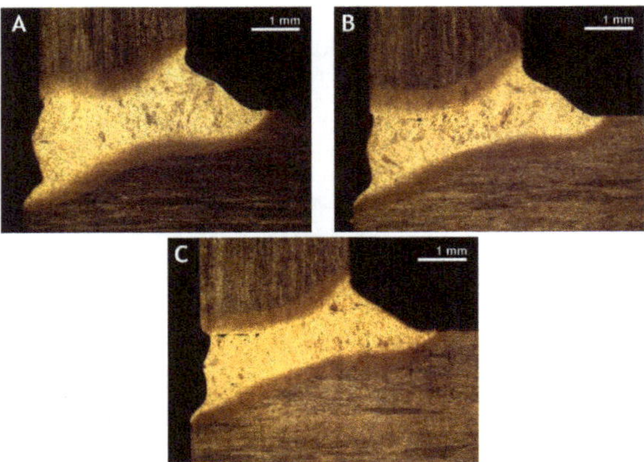

Figure 10. Welding beads in the cross-sections, L-joints, welding speed of (**A**) 2400 mm·min^{-1}, (**B**) 3000 mm·min^{-1}, (**C**) 3600 mm·min^{-1}.

With respect to the T-joints, uniform fusion of the interface between the abutting plates depends on the approach angle of the laser beam (Figure 11). A 20° angle is thought to be adequate to perform the process, irrespective of the thickness of the lower plate. For both the L- and the T-joints, imperfections in the cross-section for the suggested welding conditions have been found to comply with usual standards [29], although stricter customer regulations may not be matched, depending on the application. Hence, to address the issue of possible deviations from the intended geometry, a number of adjustments have been made when setting the experimental procedure for the test-article.

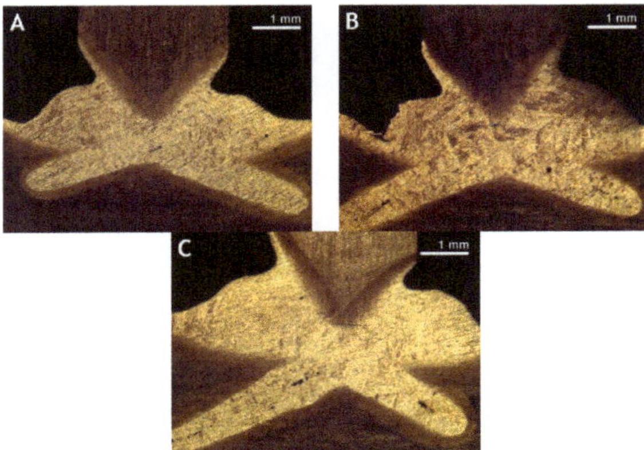

Figure 11. Welding beads in the cross-sections, T-joints, approaching angle of (**A**) 20°, (**B**) 25°, (**C**) 30°.

3.2. Microstructure and Microhardness

The evolution of the microstructure across the welding bead is worth investigating, as this would provide crucial information to discuss the response to loading. Prior to heat treatment, the original microstructure of the base metal has been lost in the fusion zone, a martensitic α' microstructure has been induced (Figure 12A). A similar structure results when quenching from the β phase region above the β-transus temperature. The heat-affected zone has been found to be a mixture of α' and primary α phase instead (Figure 12B), as matching a structure which is quenched from a region below the β-transus temperature. The grain size is affected by the welding speed, as shown via numerical models on the same alloy [13]: namely, any increase in the welding speed yields a decrease in the grain size, since the cooling rate is increased for the given power and defocusing.

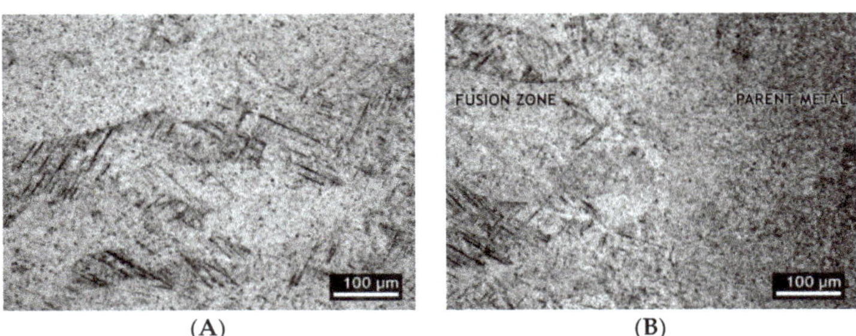

Figure 12. L-joint, 2400 mm·min^{-1}: (**A**) martensitic α' in the fusion zone; (**B**) microstructure of the heat-affected zone at the interface between fusion zone and unaffected parent metal.

To further discuss the subject, the trend of Vickers microhardness in the cross-section has been investigated, both prior to and upon heat treatment (Figure 13). An increase of hardness resulted in the fusion zone, from 340 HV$_{0.3}$ in the base metal to average 430 HV$_{0.3}$ with as high as 443 HV$_{0.3}$ peak values in the as-welded condition. In agreement with the literature [24], it is worth noting that a decrease of hardness to average 382 HV$_{0.3}$ in the fusion zone, hence a reduced mismatch with respect to the base metal, has been benefited upon heat treating. Furthermore, based on the step between consecutive indentations (i.e., 150 µm), the extent of the heat-affected zone is deemed to be shorter than 0.3 mm at both sides of the fusion zone and is not considered to be affected by heat treatment. Similar trends have been found, both in L- and T-joint. As regarding the latter, the microstructure in the fusion zone resulting from the first welding pass is unaffected by the second pass [30], as new fusion and cooling are experienced with the same rates. It is worth noting that no clear trends in the mean value have been found as a function of the welding speed, in agreement with the literature [8], irrespective of the grain size, which is expected to affect the mechanical strength instead [12].

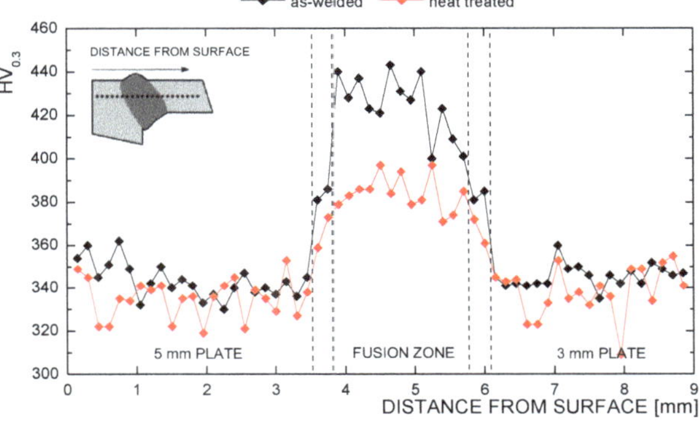

Figure 13. Vickers microhardness trend at half-thickness of the 3 mm plate; L-joint, 2400 mm·min^{-1}.

3.3. Testing of the Test-Article

Sound beads resulted along the L-joint between the cap- and web-plate; a number of indications to be ascribed to micro-pores have been detected when scanning T-joints instead (Figure 14). Imperfections are mainly located at the tip of the rib, as two welding beads cross; in terms of size and cumulated area, these are not deemed to affect the overall mechanical response.

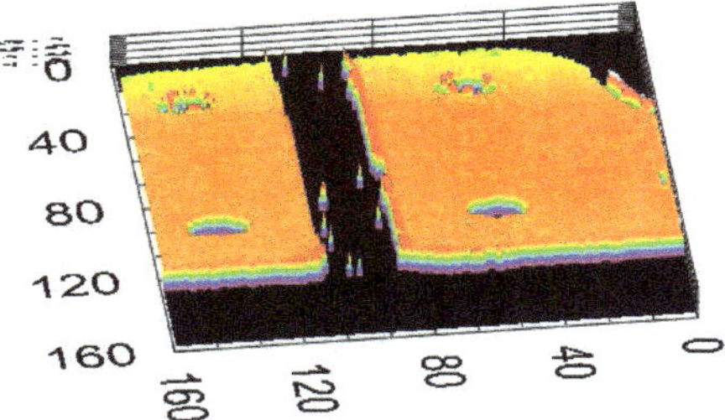

Figure 14. Ultrasonic non-destructive testing, indications due to micro-pores in scanning the T-joint.

Drilling of the web-plate of the test-article has been performed in order to allow clamping on the static testing device which has been specifically developed and manufactured in order to accurately reproduce the nominal condition of stress of the test-article (Figure 15); the device has then been employed on a conventional tensile testing machine. Two strain gauges (SG–1 and SG–2), one for each side of the supporting rib (Figure 16), have been monitored. Three test-articles resulting from welding and three test-articles resulting from machining of wrought titanium have been tested.

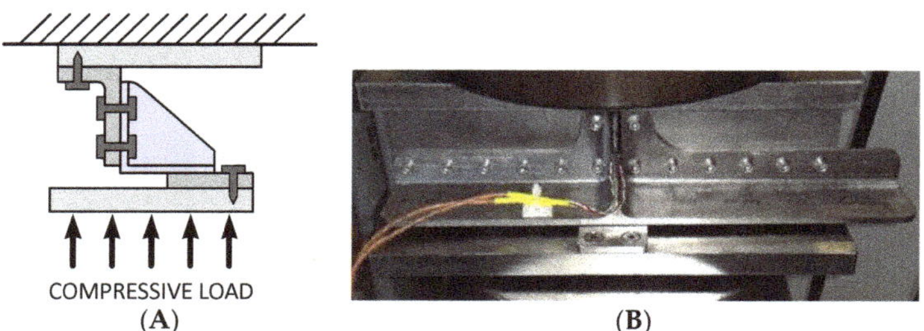

Figure 15. Testing device for the test-article: (**A**) scheme, (**B**) real view.

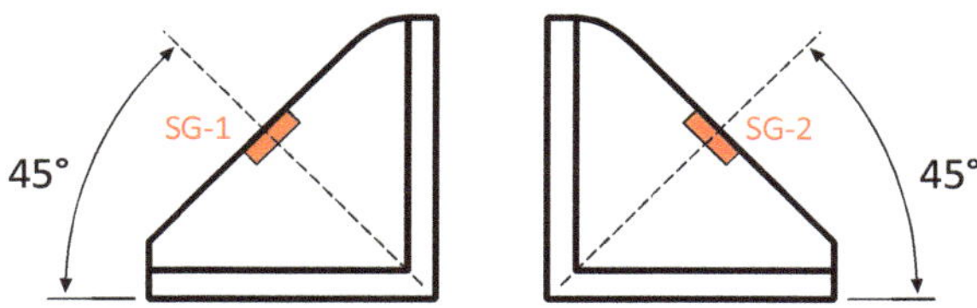

Figure 16. Positioning of the strain gauges on the supporting rib for static testing.

Based on the application of the real component, the threshold to be matched has been set to 15 kN, which is intended to be the limit load; nevertheless, as no evidence of failure were found in the form of cracks for all of the specimens when approaching the threshold, testing has been further progressed until there was contact between the test-article and the device for testing.

The average strain as monitored by SG–1 and SG–2 at 10, 15, and 20 kN are given in (Table 4). The load-strain diagram resulting from testing of welded and machined test-articles have been compared (Figure 17). Interestingly, the same behavior has been found for the test-articles in the operating nominal loading window and up to 50 kN, irrespective of the processing technology; a difference has been found above 60 kN instead, where higher strain is experienced by the samples resulting from machining. Namely, welding is deemed to provide improved strength thanks to microstructural evolution upon processing, as discussed in the process set-up, based on Vickers microhardness testing, although reduced percent elongation resulted.

Table 4. Average strain as monitored by SG–1 and SG–2 as a function of load for welded and machined test-articles.

Testing Condition	Welded Test-Articles		Machined Test-Articles	
Compressive load (kN)	Strain (SG–1) (10^{-3} mm/mm)	Strain (SG–2) (10^{-3} mm/mm)	Strain (SG–1) (10^{-3} mm/mm)	Strain (SG–2) (10^{-3} mm/mm)
10	1.17	1.11	1.32	1.19
15	1.79	1.71	1.97	1.83
20	2.43	2.32	2.59	2.45

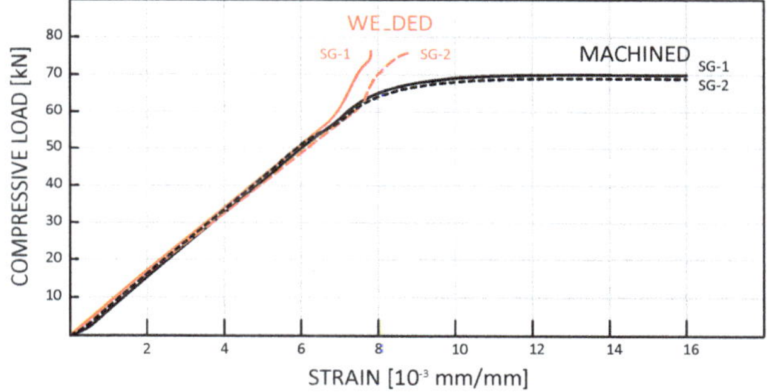

Figure 17. Load-strain diagram, a comparison among machined and welded test-articles.

3.4. Improving the Buy-to-Fly Ratio

The net weight of the test-article as required for the actual application is approximately 1.05 kg. Three methods to produce the component are compared in the following in terms of buy-to-fly: machining from bulk, LBW of plates of increased thickness (i.e., 8 and 5 mm) with eventual machining to nominal size, LBW of plates of nominal thickness (i.e., 5 and 3 mm). Both the theoretical and actual buy-to-fly ratio have been evaluated. Namely, the theoretical buy-to-fly ratio has been measured based on the net weight and the expected allowance at the contour to cut each single part to be welded; the actual buy-to-fly ratio has been measured with additional metal allowance taken into account instead, as a consequence of mutual arranging of the parts on a larger metal sheet to be cut. Both of them are hence larger than 1, the latter being higher. Both the measures have been made for each processing method and are given in Table 5.

Table 5. Theoretical and actual buy-to-fly ratio as resulting for each processing method.

Technology	Theoretical Buy-to-Fly	Actual Buy-to-Fly
Machining from bulk	11.0	15.1
LBW and machining	2.5	3.4
LBW	1.5	2.1

Saving of metal in a measure of at least 80% is benefited when considering LBW as an alternative to machining; moreover, a 50% reduction on the overall cost of the component results in the case of LBW with eventual machining to nominal size. Although additional cost saving would be expected from LBW of plates of nominal thickness to prevent machining, wire feeding or assisting gun for hybrid welding should be required to address the issue of geometrical imperfections such as undercut and shrinkage groove: costs have been found to remain the same, with limited actual gain, but proper managing of the welding paths would be prevented due to additional devices. Given this, the solution of LBW and machining to nominal size is deemed to offer a valid method to be shifted to any complex real part where L- and T-joints are usually required.

4. Conclusions

Grounds have been given for the application of LBW for the purpose of joining titanium parts for aerospace. Comparisons among laser beam welding and machining from bulk of a given test-article have been conducted: the same behavior in static testing under compressive loading has been found in the operating nominal loading window; moreover, improved strength resulted from welding. This must be ascribed to a non-diffusional transformation into a harder martensitic α' microstructure upon processing, as proven by optical microscopy in combination with Vickers microhardness testing. Thermal treating has been required and performed to the purpose of stress relieving: the resulting microhardness in the fusion zone is not deemed to be affected, whereas an improvement in terms of fracture toughness is expected.

Given the thickness of the test-article, the ideal scenario to reduce the buy-to-fly ratio would have been joining 3 mm to 5 mm thick plates; wire feeding or hybrid welding would be required, resulting in critical managing of the laser head along the welding path due to additional devices. This issue is crucial and should be specifically addressed depending on the geometry of the test-article. Therefore, autogenous joining of parts of increased size would be desirable, with final machining upon welding. This approach has been considered with the test-article, and an actual buy-to-fly of 3.5, resulted on average a 50% reduction on the overall cost of the component as a further benefit. These findings are deemed to be reasonable for any complex component where a number of multiple welding passes are required, both on L- and T-joints.

Acknowledgments: The authors gratefully acknowledge the Italian Ministry for University and Research (MIUR) for funding this research activity through PON01_01269 ELIOS, as well as all the people, including students and referee, who contributed to the successful outcome of the research.

Author Contributions: Fabrizia Caiazzo, Giuseppe Barbieri and Francesco Acerra conceived and designed the experiments; Vittorio Alfieri, Gaetano Corrado and Paolo Argenio performed welding; Vincenzo Innaro performed and analyzed the tests on the test-article; all of the authors discussed the data and gave contributions to design the device for clamping and shielding; Vittorio Alfieri wrote the paper.

Conflicts of Interest: The authors declare no conflict of interest.

References

1. Saha, P.K. *Aerospace Manufacturing Processes*, 2nd ed.; CRC Press: Boca Raton, FL, USA, 2016.
2. Hong, K.; Shin, Y.C. Prospects of laser welding technology in the automotive industry: A review. *J. Mater. Process. Technol.* **2017**, *245*, 46–69. [CrossRef]
3. Liu, S.; Mi, G.; Yan, F.; Wang, C.; Jiang, P. Correlation of high power laser welding parameters with real weld geometry and microstructure. *Opt. Laser Technol.* **2017**, *94*, 59–67. [CrossRef]
4. Caiazzo, F.; Alfieri, V.; Cardaropoli, F.; Sergi, V. Investigation on edge joints of Inconel 625 sheets processed with laser welding. *Opt. Laser Technol.* **2017**, *93*, 180–186. [CrossRef]
5. Duley, W.W. *Laser Welding*, 1st ed.; Wiley: New York, NY, USA, 1998.
6. Steen, W.M.; Mazumder, J. *Laser Material Processing*; Springer: London, UK, 2010.
7. Cardaropoli, F.; Alfieri, V.; Caiazzo, F.; Sergi, V. Manufacturing of porous biomaterials for dental implant applications through Selective Laser Melting. *Adv. Mater. Res.* **2012**, *535–537*, 1222–1229. [CrossRef]

8. Cao, X.; Jahazi, M. Effect of welding speed on butt joint quality of Ti–6Al–4V alloy welded using a high-power Nd:YAG laser. *Opt. Laser Eng.* **2009**, *47*, 1231–1241. [CrossRef]

9. Donachie, M.J. *Titanium: A Technical Guide*, 2nd ed.; ASM International: Materials Park, OH, USA, 2000.

10. Schneider, A.; Gumenyuk, A.; Lammers, M.; Maletschek, A.; Rethmeier, M. Laser beam welding of thick titanium sheets in the field of marine technology. *Phys. Procedia* **2014**, *56*, 582–590. [CrossRef]

11. Gao, X.; Zhang, L.; Zhang, J. A comparative study of pulsed Nd:YAG laser welding and TIG welding of thin Ti6Al4V titanium alloy plate. *Mater. Sci. Eng. A* **2013**, *559*, 14–21. [CrossRef]

12. Akman, E.; Demir, A.; Canel, T.; Sınmazc, T. Laser welding of Ti6Al4V titanium alloys. *J. Mater. Process. Technol.* **2009**, *209*, 3705–3713. [CrossRef]

13. Caiazzo, F.; Alfieri, V.; Corrado, G.; Cardaropoli, F. Sergi, V. Investigation and Optimization of Laser Welding of Ti–6Al–4V titanium alloy plates. *J. Manuf. Sci. Eng.* **2013**, *135*, 061012. [CrossRef]

14. Panwisawas, C.; Perumal, B.; Ward, R.M.; Turner, N.; Turner, R.P.; Brooks, J.W.; Basoalto, H.C. Keyhole formation and thermal fluid flow-induced porosity during laser fusion welding in titanium alloys: Experimental and modeling. *Acta Mater.* **2017**, *125*, 251–263. [CrossRef]

15. Li, C.; Li, B.; Wu, Z.; Qi, X.; Ye, B.; Wang, A. Stitch welding of Ti–6Al–4V titanium alloy by fiber laser. *Trans. Nonferr. Met. Soc.* **2017**, *27*, 91–101. [CrossRef]

16. EN ISO 6520-1. *Welding and Allied Processes—Classification of Geometric Imperfections in Metallic Materials—Part 1: Fusion Welding*; ISO: Geneva, Switzerland, 2005.

17. Alfieri, V.; Cardaropoli, F.; Caiazzo, F.; Sergi, V. Investigation on porosity content in 2024 aluminum alloy welding by Yb:YAG disk laser. *Adv. Mater. Res.* **2012**, *383–390*, 6265–6269. [CrossRef]

18. Unt, A.; Poutiainen, I.; Salminen, A. Influence of filler wire feed rate in laser-arc hybrid welding of T-butt joint in shipbuilding steel with different optical setups. *Phys. Procedia* **2015**, *78*, 45–52. [CrossRef]

19. Li, C.; Muneharua, K.; Takao, S.; Kouji, H. Fiber laser-GMA hybrid welding of commercially pure titanium. *Mater. Des.* **2009**, *30*, 109–114. [CrossRef]

20. Ahmed, T.; Rack, H.J. Phase transformations during cooling in $\alpha + \beta$ titanium alloys. *Mater. Sci. Eng. A* **1998**, *243*, 206–211. [CrossRef]

21. Balasubramanian, T.S.; Balakrishnan, M.; Balasubramanian, V.; Muthu Manickam, M.A. Influence of welding processes on microstructure, tensile and impact properties of Ti–6al–4V alloy joints. *Trans. Nonferr. Met. Soc.* **2011**, *21*, 1253–1262. [CrossRef]

22. Sieniawski, J.; Ziaja, W.; Kubiak, K.; Motyka, M. Microstructure and Mechanical Properties of High Strength Two-Phase Titanium Alloys. In *Titanium Alloys—Advances in Properties Control*; Sieniawski, J., Ziaja, W., Eds.; InTechOpen: Rijeka, Croatia, 2013.

23. Kabir, A.S.H.; Cao, X.; Gholipour, J.; Wanjara, P.; Cuddy, J.; Birur, A.; Medraj, M. Effect of postweld heat treatment on microstructure, hardness and tensile properties of laser-welded Ti–6Al–4V. *Met. Mater. Trans. A* **2012**, *43*, 4171–4184. [CrossRef]

24. Lu, M.Y.; Tsay, L.W.; Chen, C. Notched Tensile Fracture of Ti6Al4V laser welds at elevated temperature. *Mater. Trans.* **2012**, *6*, 1042–1047. [CrossRef]

25. Caiazzo, F.; Alfieri, V.; Astarita, A.; Squillace, A.; Barbieri, G. Investigation on laser welding of Ti–6Al–4V in corner joint. *Adv. Mech. Eng.* **2017**, *9*, 1–9. [CrossRef]

26. Caiazzo, F.; Alfieri, V.; Fierro, I.; Sergi, V. Investigation and Optimization of Disk-Laser Welding of 1 mm thick Ti–6Al–4V titanium alloy sheets. *Adv. Mech. Eng.* **2015**, *7*, 1–8. [CrossRef]

27. Caiazzo, F.; Sergi, V.; Corrado, G.; Alfieri, V.; Cardaropoli, F. Automated Apparatus of Laser Beam Welding. Patent EP2,931,468(A1), 21 October 2015.

28. EN ISO 6507-1:2005. *Metallic Materials—Vickers Hardness Test—Part 1: Test Method*; ISO: Geneva, Switzerland, 2005.

29. American Welding Society (AWS). *D17.1—Specification for Fusion Welding for Aerospace Applications*; AWS: Miami, FL, USA, 2001.

30. Caiazzo, F.; Cardaropoli, F.; Alfieri, V.; Sergi, V.; Argenio, P.; Barbieri, G. Disk-laser welding of Ti–6Al–4V titanium alloy plates in T-joint configuration. *Procedia Eng.* **2017**, *183*, 219–226. [CrossRef]

MDPI

St. Alban-Anlage 66

4052 Basel

Switzerland

Tel. +41 61 683 77 34

Fax +41 61 302 89 18

www.mdpi.com

Metals Editorial Office

E-mail: metals@mdpi.com

www.mdpi.com/journal/metals